Uday K. Chakraborty (Ed.)

Advances in Differential Evolution

Studies in Computational Intelligence, Volume 143

Editor-in-Chief
Prof. Janusz Kacprzyk
Systems Research Institute
Polish Academy of Sciences
ul. Newelska 6
01-447 Warsaw
Poland
E-mail: kacprzyk@ibspan.waw.pl

Further volumes of this series can be found on our homepage: springer.com

Vol. 122. Tomasz G. Smolinski, Mariofanna G. Milanova and Aboul-Ella Hassanien (Eds.)
Applications of Computational Intelligence in Biology, 2008
ISBN 978-3-540-78533-0

Vol. 123. Shuichi Iwata, Yukio Ohsawa, Shusaku Tsumoto, Ning Zhong, Yong Shi and Lorenzo Magnani (Eds.)
Communications and Discoveries from Multidisciplinary Data, 2008
ISBN 978-3-540-78732-7

Vol. 124. Ricardo Zavala Yoe
Modelling and Control of Dynamical Systems: Numerical Implementation in a Behavioral Framework, 2008
ISBN 978-3-540-78734-1

Vol. 125. Larry Bull, Bernadó-Mansilla Ester and John Holmes (Eds.)
Learning Classifier Systems in Data Mining, 2008
ISBN 978-3-540-78978-9

Vol. 126. Oleg Okun and Giorgio Valentini (Eds.)
Supervised and Unsupervised Ensemble Methods and their Applications, 2008
ISBN 978-3-540-78980-2

Vol. 127. Régie Gras, Einoshin Suzuki, Fabrice Guillet and Filippo Spagnolo (Eds.)
Statistical Implicative Analysis, 2008
ISBN 978-3-540-78982-6

Vol. 128. Fatos Xhafa and Ajith Abraham (Eds.)
Metaheuristics for Scheduling in Industrial and Manufacturing Applications, 2008
ISBN 978-3-540-78984-0

Vol. 129. Natalio Krasnogor, Giuseppe Nicosia, Mario Pavone and David Pelta (Eds.)
Nature Inspired Cooperative Strategies for Optimization (NICSO 2007), 2008
ISBN 978-3-540-78986-4

Vol. 130. Richi Nayak, Nikhil Ichalkaranje and Lakhmi C. Jain (Eds.)
Evolution of the Web in Artificial Intelligence Environments, 2008
ISBN 978-3-540-79139-3

Vol. 131. Roger Lee and Haeng-Kon Kim (Eds.)
Computer and Information Science, 2008
ISBN 978-3-540-79186-7

Vol. 132. Danil Prokhorov (Ed.)
Computational Intelligence in Automotive Applications, 2008
ISBN 978-3-540-79256-7

Vol. 133. Manuel Graña and Richard J. Duro (Eds.)
Computational Intelligence for Remote Sensing, 2008
ISBN 978-3-540-79352-6

Vol. 134. Ngoc Thanh Nguyen and Radoslaw Katarzyniak (Eds.)
New Challenges in Applied Intelligence Technologies, 2008
ISBN 978-3-540-79354-0

Vol. 135. Hsinchun Chen and Christopher C. Yang (Eds.)
Intelligence and Security Informatics, 2008
ISBN 978-3-540-69207-2

Vol. 136. Carlos Cotta, Marc Sevaux and Kenneth Sörensen (Eds.)
Adaptive and Multilevel Metaheuristics, 2008
ISBN 978-3-540-79437-0

Vol. 137. Lakhmi C. Jain, Mika Sato-Ilic, Maria Virvou, George A. Tsihrintzis, Valentina Emilia Balas and Canicious Abeynayake (Eds.)
Computational Intelligence Paradigms, 2008
ISBN 978-3-540-79473-8

Vol. 138. Bruno Apolloni, Witold Pedrycz, Simone Bassis and Dario Malchiodi
The Puzzle of Granular Computing, 2008
ISBN 978-3-540-79863-7

Vol. 139. Jan Drugowitsch
Design and Analysis of Learning Classifier Systems, 2008
ISBN 978-3-540-79865-1

Vol. 140. Nadia Magnenat-Thalmann, Lakhmi C. Jain and N. Ichalkaranje (Eds.)
New Advances in Virtual Humans, 2008
ISBN 978-3-540-79867-5

Vol. 141. Christa Sommerer, Lakhmi C. Jain and Laurent Mignonneau (Eds.)
The Art and Science of Interface and Interaction Design, 2008
ISBN 978-3-540-79869-9

Vol. 142. George A. Tsihrintzis, Maria Virvou, Robert J. Howlett and Lakhmi C. Jain (Eds.)
New Directions in Intelligent Interactive Multimedia, 2008
ISBN 978-3-540-68126-7

Vol. 143. Uday K. Chakraborty (Ed.)
Advances in Differential Evolution, 2008
ISBN 978-3-540-68827-3

Uday K. Chakraborty
(Ed.)

Advances in Differential Evolution

Uday K. Chakraborty
Mathematics & Computer Science Department
University of Missouri
St. Louis, MO 63121
USA
Email: chakrabortyu@umsl.edu

ISBN 978-3-642-08839-1 e-ISBN 978-3-540-68830-3

DOI 10.1007/978-3-540-68830-3

Studies in Computational Intelligence ISSN 1860949X

Softcover reprint of the hardcover 1st edition 2008

Printed in acid-free paper
9 8 7 6 5 4 3 2 1
springer.com

Foreword

It is a great pleasure to see Uday Chakraborty, a foremost researcher in evolutionary computation, bring out an edited volume on differential evolution.

Differential evolution, one of the newest members of the evolutionary algorithm family, has over the past few years been shown to be a simple yet versatile heuristic for global optimization, particularly real-parameter optimization. The spurt in interest in differential evolution is evident from the breathtakingly wide array of application areas, particularly in engineering, that have benefited from these algorithms, and also from the great number of differential evolution publications in scientific journals and conference proceedings.

Despite its demonstrated success in many areas, differential evolution has many open problems. Optimization with differential evolution will continue to remain a challenging research area for years to come.

This book brings together an outstanding collection of recent research contributions, all by top-notch researchers, to this emerging field. This is an excellent comprehensive resource that captures the state of the art. This volume should be of interest to both theoreticians and practitioners and is a must-have resource for researchers interested in stochastic optimization.

Zbigniew Michalewicz

Preface

Differential evolution is arguably one of the hottest topics in today's computational intelligence research. This book seeks to present a comprehensive study of the state of the art in this technology and also directions for future research.

The fourteen chapters of this book have been written by leading experts in the area. The first seven chapters focus on algorithm design, while the last seven describe real-world applications.

Chapter 1 introduces the basic differential evolution (DE) algorithm and presents a broad overview of the field. Chapter 2 presents a new, rotationally invariant DE algorithm. The role of self-adaptive control parameters in DE is investigated in Chapter 3. Chapters 4 and 5 address constrained optimization; the former develops suitable stopping conditions for the DE run, and the latter presents an improved DE algorithm for problems with very small feasible regions. A novel DE algorithm, based on the concept of "opposite" points, is the topic of Chapter 6. Chapter 7 provides a survey of multi-objective differential evolution algorithms. A review of the major application areas of differential evolution is presented in Chapter 8. Chapter 9 discusses the application of differential evolution in two important areas of applied electromagnetics. Chapters 10 and 11 focus on applications of hybrid DE algorithms to problems in power system optimization. Chapter 12 applies the DE algorithm to computer chess. The use of DE to solve a problem in bioprocess engineering is discussed in Chapter 13. Chapter 14 describes the application of hybrid differential evolution to a problem in control engineering.

I am truly grateful for the unstinting support and inspiration I received from Janusz Kacprzyk, Series Editor, Springer, who did me a great honor by inviting me to edit this book. Among the other people whose help, advice, encouragement and support I gratefully acknowledge, I must specially mention Thomas Ditzinger, Senior Editor, Engineering and Applied Sciences, Springer; Heather King, Editorial Assistant at Springer's Heidelberg office, for her extraordinary help during the production process (including cheerfully accommodating changes in schedule necessitated by my two lapses with regard to deadlines); Zbigniew Michalewicz, of the University of Adelaide; Xin Yao, of the University of Birmingham; the Office of Research Administration, University of Missouri St. Louis; my departmental colleagues, particularly A. P. Rao; Nasser Arshadi, Vice-Provost for Research, UMSL; Mark Burkholder, Dean of the College of Arts and Sciences, UMSL; all the contributing authors for being so cooperative and patient through the unduly long review process; and particularly Rainer Storn and Kenneth Price for their interest and help with reviews.

St. Louis, March 2008 Uday K. Chakraborty

Contents

Differential Evolution Research – Trends and Open Questions
Rainer Storn 1

Eliminating Drift Bias from the Differential Evolution Algorithm
Kenneth V. Price 33

An Analysis of the Control Parameters' Adaptation in DE
Janez Brest, Aleš Zamuda, Borko Bošković, Sašo Greiner, Viljem Žumer 89

Stopping Criteria for Differential Evolution in Constrained Single-Objective Optimization
Karin Zielinski, Rainer Laur 111

Constrained Optimization by ε Constrained Differential Evolution with Dynamic ε-Level Control
Tetsuyuki Takahama, Setsuko Sakai 139

Opposition-Based Differential Evolution
Shahryar Rahnamayan, Hamid R. Tizhoosh, Magdy M.A. Salama 155

Multi-objective Optimization Using Differential Evolution: A Survey of the State-of-the-Art
Efrén Mezura-Montes, Margarita Reyes-Sierra, Carlos A. Coello Coello 173

A Review of Major Application Areas of Differential Evolution
V.P. Plagianakos, D.K. Tasoulis, M.N. Vrahatis 197

The Differential Evolution Algorithm as Applied to Array Antennas and Imaging
A. Massa, M. Pastorino, A. Randazzo 239

Applications of Differential Evolution in Power System Optimization
L. Lakshminarasimman, S. Subramanian 257

Self-adaptive Differential Evolution Using Chaotic Local Search for Solving Power Economic Dispatch with Nonsmooth Fuel Cost Function
Leandro dos Santos Coelho, Viviana Cocco Mariani 275

An Adaptive Differential Evolution Algorithm with Opposition-Based Mechanisms, Applied to the Tuning of a Chess Program
Borko Bošković, Sašo Greiner, Janez Brest, Aleš Zamuda, Viljem Žumer 287

Differential Evolution for the Offline and Online Optimization of Fed-Batch Fermentation Processes
Rui Mendes, Isabel Rocha, José P. Pinto, Eugénio C. Ferreira, Miguel Rocha 299

Worst Case Analysis of Control Law for Re-entry Vehicles Using Hybrid Differential Evolution
P.P. Menon, D.G. Bates, I. Postlethwaite, A. Marcos, V. Fernandez, S. Bennani 319

Index 335

Author Index 339

Differential Evolution Research – Trends and Open Questions

Rainer Storn

Rohde & Schwarz GmbH & Co KG
P.O. Box 80 14 69
81614 München, Germany
rainer.storn@freenet.de

Summary. Differential Evolution (DE), a vector population based stochastic optimization method has been introduced to the public in 1995. During the last 10 years research on and with DE has reached an impressive state, yet there are still many open questions, and new application areas are emerging. This chapter introduces some of the current trends in DE-research and touches upon the problems that are still waiting to be solved.

1 Introduction

It has been more than ten years since Differential Evolution (DE) was introduced by Ken Price and Rainer Storn in a series of papers that followed in quick succession [1, 2, 3, 4, 5] and by means of an Internet page [6]. DE is a population-based stochastic method for global optimization. Throughout this chapter the term optimization shall always be equated with minimization without loss of generality. The original version of DE can be defined by the following constituents.

1) The population

$$\begin{aligned} \mathrm{P}_{\mathbf{x},g} &= \left(\mathbf{x}_{i,g}\right), \quad i = 0,1,\ldots,Np-1, \quad g = 0,1,\ldots,g_{\max}, \\ \mathbf{x}_{i,g} &= \left(x_{j,i,g}\right), \; j = 0,1,\ldots,D-1. \end{aligned} \tag{1}$$

where Np denotes the number of population vectors, g defines the generation counter, and D the dimensionality, i.e. the number of parameters.

2) The initialization of the population via

$$x_{j,i,0} = \mathrm{rand}_j[0,1) \cdot \left(b_{j,\mathrm{U}} - b_{j,\mathrm{L}}\right) + b_{j,\mathrm{L}}. \tag{2}$$

The D-dimensional initialization vectors, $\mathbf{b}_{\mathrm{L}}$ and $\mathbf{b}_{\mathrm{U}}$ indicate the lower and upper bounds of the parameter vectors $\mathbf{x}_{i,j}$. The random number generator, $\mathrm{rand}_j[0,1)$, returns a uniformly distributed random number from within the range [0,1), i.e., $0 \le \mathrm{rand}_j[0,1) < 1$. The subscript, j, indicates that a new random value is generated for each parameter.

U.K. Chakraborty (Ed.): Advances in Differential Evolution, SCI 143, pp. 1–31, 2008.
springerlink.com

3) The perturbation of a base vector $\mathbf{y}_{i,g}$ by using a difference vector based mutation

$$\mathbf{v}_{i,g} = \mathbf{y}_{i,g} + F \cdot \left(\mathbf{x}_{r1,g} - \mathbf{x}_{r2,g}\right). \tag{3}$$

to generate a mutation vector vi,g. The difference vector indices, r1 and r2, are randomly selected once per base vector. Setting yi,g = xr0,g defines what is often called classic DE where the base vector is also a randomly chosen population vector. The random indexes r0, r1, and r2 should be mutually exclusive. There are also variants of perturbations which are different to Eq. (3) and some of them will be described later. For example, setting the base vector to the current best vector or a linear combination of various vectors is also popular. Employing more than one difference vector for mutation has also been tried but has never gained a lot of popularity so far.

4) Diversity enhancement

The classic variant of diversity enhancement is crossover [1, 2, 3, 4, 5, 6, 7] which mixes parameters of the mutation vector $\mathbf{v}_{i,g}$ and the so-called *target vector* $\mathbf{x}_{i,g}$ in order to generate the *trial vector* $\mathbf{u}_{i,g}$. The most common form of crossover is uniform and is defined as

$$\mathbf{u}_{i,g} = u_{j,i,g} = \begin{cases} v_{j,i,g} & \text{if } \left(\text{rand}_j[0,1) \le Cr\right) \\ x_{j,i,g} & \text{otherwise.} \end{cases} \tag{4}$$

In order to prevent the case $\mathbf{u}_{i,g} = \mathbf{x}_{i,g}$ at least one component is taken from the mutation vector $\mathbf{v}_{i,g}$, a detail that is not expressed in Eq. (4). Other variants of crossover are described by Price, Storn and Lampinen [7].

5) Selection

DE uses simple one-to-one survivor selection where the trial vector ui,g competes against the target vector xi,g. The vector with the lowest objective function value survives into the next generation g+1.

$$\mathbf{x}_{i,g+1} = \begin{cases} \mathbf{u}_{i,g} & \text{if } f\left(\mathbf{u}_{i,g}\right) \le f\left(\mathbf{x}_{i,g}\right) \\ \mathbf{x}_{i,g} & \text{otherwise.} \end{cases} \tag{5}$$

Please note that the presentation as well as notation has been chosen slightly different from the original papers[1, 2, 3, 4, 5]. Along with the DE algorithm came a notation [5] to classify the various DE-variants. The notation is defined by DE/x/y/z where x denotes the base vector, y denotes the number of difference vectors used, and z representing the crossover method. For example, DE/rand/1/bin is the shorthand notation for Eq. (1) through Eq. (5) with yi,g = xr0,g. DE/best/1/bin is the same except for yi,g = xbest,g. In this case xbest,g represents the vector with the lowest objective function value evaluated so far. With today's extensions of DE the shorthand notation DE/x/y/z is not sufficient any more, but a more appropriate notation has not been defined yet.

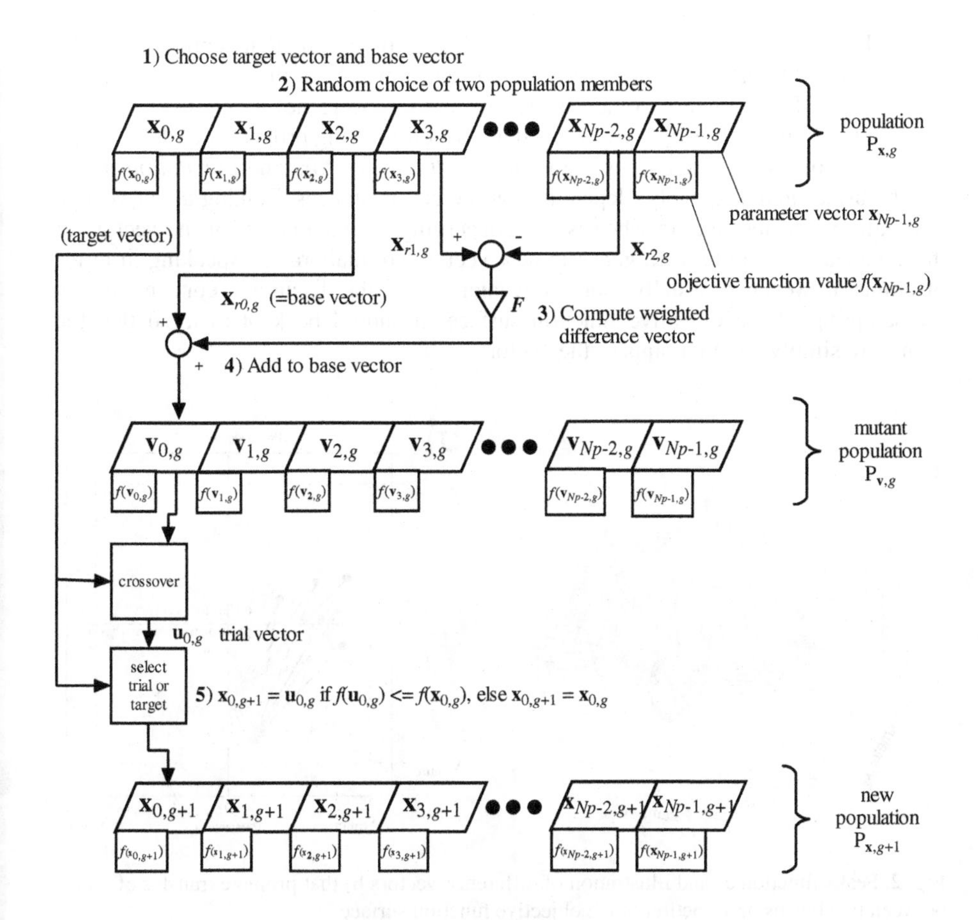

Fig. 1. Flowchart of classical DE [7]

In order to be able to represent DE pictorially the flowgraph representation in Figure 1 was contrived because flowgraphs are very common in the engineering world. The DE-flowgraph representation seemed to be ideal to convey DE's simplicity and first appeared in the DE-article published in the Dr. Dobb's journal at 1997 [4], a magazine for computer programmers. The article spawned a lot of interest for DE among practitioners which Kenneth Price and Rainer Storn concluded from the large number of e-mails they received in which DE was attributed very good convergence along with simplicity. Simplicity is an asset which is very important to anyone who considers optimization to be a necessary but not the primary task. DE's simplicity allowed many practicing engineers and researchers from very diverse disciplines to use global optimization without the need to be an optimization expert.

Interestingly enough, DE received attention only very slowly from fellow researchers in the evolutionary computation community, even though it performed very well on the first international contest on evolutionary computation in Nagoya as early as 1996 [2]. DE's lack of attention might have been due to a lack of understanding concerning its inner workings. More light was shed upon these in 2002 when Daniela Zaharie published a beautiful article [8] that enlightened the convergence of DE from theoretical point of view. Also the contour matching properties of DE, a phrase coined by Kenneth Price, were not explicitly advocated until 2005 [7] and only vaguely described as the self-steering property of DE.

For the following discussions the terms *population vectors* and *population points* will be used interchangeably, depending on the circumstances. Talking about vectors is usually more appropriate when issues concerning the parameters of the vectors or the vector arithmetic used to generate new vectors are elaborated. Speaking in terms of points, however, is usually more convenient when the discussion concentrates on the sampling of the objective function surface. It should be kept in mind that the points are simply the endpoints of the vectors.

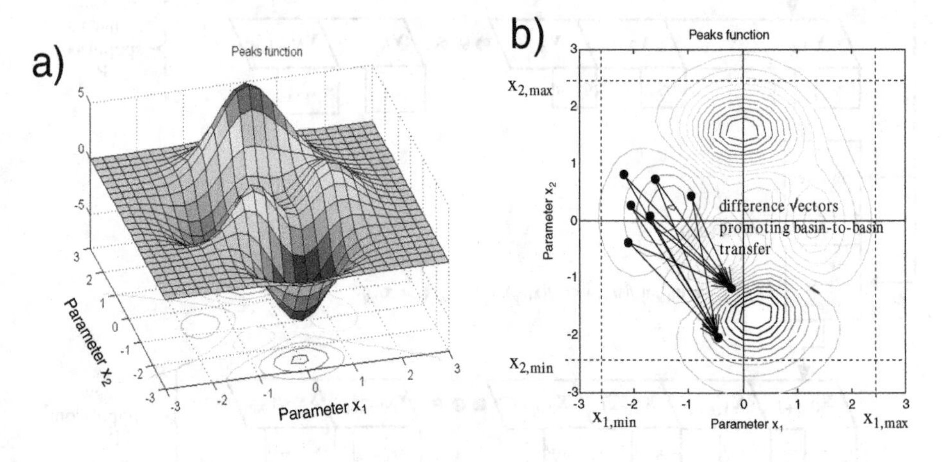

Fig. 2. Peaks function a) and illustration of difference vectors b) that promote transfer of points between two basins of attraction of the objective function surface

Some explanations concerning contour matching are in order. Contour matching means that the vector population adapts such that promising regions of the objective function surface are investigated automatically once they are detected. To this end an important ingredient besides selection is the promotion of *basin-to-basin* transfer where search points may move from one basin of attraction, i.e. a local minimum, to another one. Figure 2 illustrates that DE in fact supports basin-to-basin transfer by yielding a certain amount of difference vectors that are able to generate new trial points in the lower basin of attraction when the base points stem from the upper left basin of attraction.

Professor Jouni Lampinen from Lappeenranta University of Technology, Finland, [7] was one of the first scientists who was intrigued by DE's potential and not only did a lot of seminal DE-research but also started to maintain a bibliography of DE-related papers [9]. A look into this bibliography on the Internet reveals that its maintenance has been halted after 2002. The reason for this stop was that the number of papers began to increase at such a large rate after the year 2002 that it was impossible to keep the bibliography up-to-date and complete.

A simple search for "Differential Evolution" using any kind of search engine on the Internet supports the above statement and shows that DE-research is in full swing.

There are basically four main directions of DE-research that can be identified:

- Basic DE research
  Here the inner workings and theoretical aspects of DE are investigated. Objective functions involved are usually unconstrained. The goal is to better understand DE, identify its weaknesses and improve it in an overall fashion. It is worth mentioning that the majority of the research in this category is empirical. Purely theoretical treatments are rare as it is the case for EAs in general. This is probably due to the situation that those scenarios which lend themselves to feasible theoretical investigations rarely represent the complexity of real-world problems.
- Problem Domain Specific research
  In this area the problem formulation and how DE can be adapted to it is under scrutiny. For example, constraints, time variations, and number of objectives of an objective function are of importance but also effects of dimensionality and parameter granularity, i.e. discreteness are considered.
- Application Specific research
  This research domain is similar to the problem domain specific case, however, certain applications can be much more specific than the general problem domain they belong to. For example, the traveling salesman problem belongs to the problem domain of combinatorial optimization, but its specifics narrow down the heuristics one may use in order to solve the problem.
- Computing Environment related research
  In the real world computational efficiency of DE is often crucial to make design problems tractable. Some problems call for parallel computations while others have to deal with limited memory or processing power.

In the following selected areas from the research domains mentioned above will be discussed. I have tried to highlight those that still exhibit many open questions and hence constitute rewarding research topics, but I am aware that I did not provide a complete picture. For example, the vast domain of multi-objective optimization using DE has not been covered at all, but it will be treated in another part of the book.

2 Basic DE Research

In the early days DE was only marginally understood concerning its strengths and weaknesses. By twisting and tuning the various constituents of DE, i.e. initialization, mutation, diversity enhancement, and selection of DE as well as the choice of the control variables it was tried to make it a foolproof and fast optimization method for any kind of objective function, even though the *No Free Lunch Theorem* (NFL) by Wolpert and Macready (1997) [10] suggested already that such a panacea could not exist. Nevertheless, many real-world problems seem to be of the kind that they are very well amenable to be treated by DE. And even though there will be no cure-all-optimization for every problem, DE can nevertheless be improved also in a general sense.

2.1 The Control Variables Np, F, and Cr

Trying to tune the three main control variables Np, F, and Cr and finding bounds for their values has been a topic of intensive research [7, 8, 11, 12, 13, 14, 15, 16]. An important result was presented by Daniela Zaharie [8] where she proved that the mutation scale factor F should never be smaller than F_{crit} where

$$F_{crit} = \sqrt{\frac{\left(1 - \frac{Cr}{2}\right)}{Np}}. \tag{6}$$

Another important result from Price, Storn, Lampinen [7] was that only high values of Cr guarantee the contour matching properties of DE. In addition, only when Cr=1 is the mean number of function evaluations for an objective function and its rotated counterpart the same, i.e. in this case DE is called *rotationally invariant.* This does not mean, however, that low values of Cr should always be avoided. Low values of Cr are advantageous for separable functions, since the search concentrates on the axes of the coordinate system as outlined in [7]. The rule of thumb values for the control variables given by Storn and Price [5]:

1. $F \in [0.5, 1.0]$
2. $Cr \in [0.8, 1.0]$
3. $Np = 10 \cdot D$

are valid for many practical purposes but still lack generality. Gämperle [11] reported that the control variable settings for F, Cr, and Np can be quite difficult to find, and some objective functions are sensitive to the proper setting. This finding was also stated by Liu and Lampinen [12]. Therefore research trends go towards finding the best settings of F, Cr, and Np automatically [13, 14, 15, 16, 17]. One recent approach by Brest *et al.* [17] uses F and Cr as additional parameters to evolve for each population vector, an idea pioneered by Schwefel [18]. Hence each parameter vector has D + 2 parameters with the last two parameters containing an individual F and Cr for the particular vector. If the trial vector wins in the selection process either both F and Cr from the base vector are transported into the winner vector or the individual F and Cr are randomly determined. It is claimed that the most appropriate values for F and Cr will survive in the long run. The results on a reasonably-sized testbed show that the scheme yields improved objective function values after a fixed set of function evaluations, compared to classical DE with F=0.5 and Cr=0.9, and compared to some other DE-variants. However, it is unclear whether the scheme would maintain it superiority if not a fixed number of evaluations but a fixed value-to-reach (VTR) would have been chosen as a goal. It may also be that the encouraging results are due to the occasionally occurring random selection of F and Cr. This kind of randomness known as dither [7] has been found to be advantageous as will be elaborated later. Furthermore the question remains whether the surviving F and Cr gear the optimization towards fast and therefore possibly premature convergence. Hence the area of automatic control parameter determination remains very interesting and a fruitful area of research.

2.2 Perturbation

Perturbation of the base vector by mutation has been treated very early and has lead to various variants of DE such as the one belonging to classical DE/rand/1/bin

$$\mathbf{v}_{i,g} = \mathbf{x}_{r0,g} + F \cdot \left(\mathbf{x}_{r1,g} - \mathbf{x}_{r2,g}\right), \tag{7}$$

the mutation being used in DE/best/1/bin

$$\mathbf{v}_{i,g} = \mathbf{x}_{best,g} + F \cdot \left(\mathbf{x}_{r1,g} - \mathbf{x}_{r2,g}\right), \tag{8}$$

the mutation for DE/current-to-best/1/bin

$$\mathbf{v}_{i,g} = \mathbf{x}_{i,g} + F \cdot \left(\mathbf{x}_{best,g} - \mathbf{x}_{i,g}\right) + F \cdot \left(\mathbf{x}_{r1,g} - \mathbf{x}_{r2,g}\right). \tag{9}$$

and the variant for DE/best/2/bin

$$\mathbf{v}_{i,g} = \mathbf{x}_{best,g} + F \cdot \left(\mathbf{x}_{r1,g} - \mathbf{x}_{r2,g} + \mathbf{x}_{r3,g} - \mathbf{x}_{r4,g}\right). \tag{10}$$

In fact many more linear combinations of vectors may be used for mutation, a generalization of which can be written as

$$\mathbf{v}_{i,g} = \mathbf{y}_{i,g} + F \cdot \frac{1}{N} \sum_{n=0}^{N-1} \left(\mathbf{x}_{r(2n+1),g} - \mathbf{x}_{r(2n+2),g}\right). \tag{11}$$

with $\mathbf{y}_{i,g}$ being the base vector. The base vector should be distinct from the other vectors in Eq. (11). Most commonly used are the mutation schemes represented by Eq. (7) and Eq. (8) with the latter being more greedy. Recently Price and Rönkkönen [19] investigated Eq. (11) for the case $\mathbf{y}_{i,g} = \mathbf{x}_{i,g}$ and N=0. In [7] the effect of recombination as a perturbation method

$$\mathbf{v}_{i,g} = \mathbf{y}_{i,g} + F_{rec} \cdot \left(\frac{1}{N} \sum_{n=0}^{N-1} \left(\mathbf{x}_{r(2n+1),g} + \mathbf{x}_{r(2n+2),g}\right) - \mathbf{y}_{i,g} \right). \tag{12}$$

for the case $\mathbf{y}_{i,g} = \mathbf{x}_{i,g}$ and N=0 has been elaborated. Eq. (12) is a generalization of the Nelder and Mead reflection operation [20] and defines a point between $\mathbf{y}_{i,g}$ and the centroid of the vectors used in the recombination sum. So far no single perturbation method has turned out to be best for all problems which, of course, doesn't come as a surprise with regard to the NFL [10]. Nevertheless all the various methods need further investigation under which circumstances they perform well. In practice this information can be very important because it may save many computations or may even be crucial for the solution of a certain problem.

2.3 Diversity Enhancement

One of the most fundamental aspects of mutation-based DE is the fact that vector perturbations are generated from the Np·(Np-1) nonzero difference vectors of the population rather than employing a predetermined probability density function. This leads to one of the main assets of DE: *contour matching* [7]. The contour matching property can be observed in Figure 3 through Figure 5 which show the DE-population and the difference vector distribution for Np=8 on the peaks function the latter of which is defined by

$$f(x_1,x_2)=3(1-x_1)^2\cdot\exp\left(x_1^2+(x_2+1)^2\right)-10\left(\frac{x_1}{5}-x_1^3-x_2^5\right)\cdot \\ \cdot\exp\left(x_1^2+x_2^2\right)-\frac{1}{3}\cdot\exp\left((x_1+1)^2+x_2^2\right) \tag{13}$$

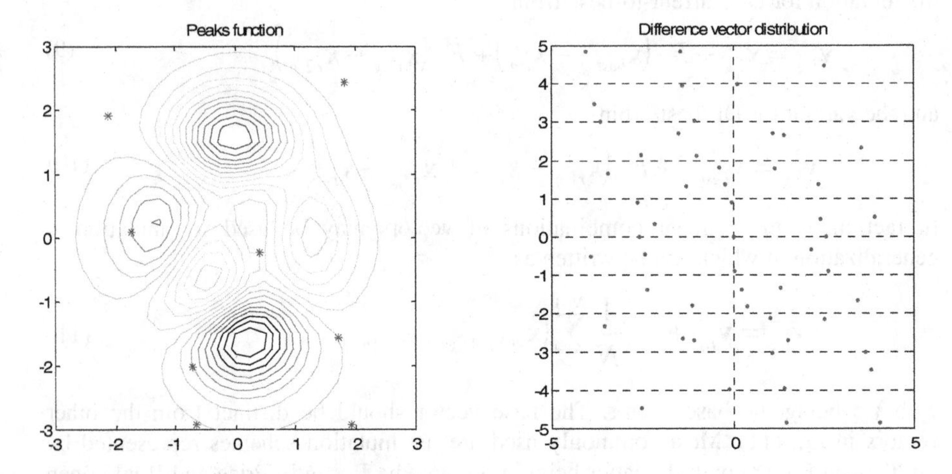

Fig. 3. Generation g=1 using Np = 8

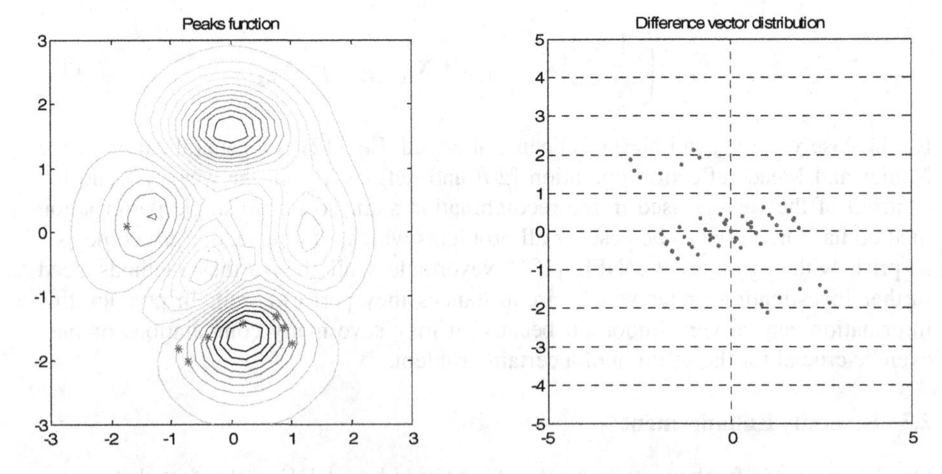

Fig. 4. Generation g=10 using Np = 8. The difference vector distribution (only endpoints shown) exhibits three main clouds where the outer ones promote the transfer between two basins of attraction.

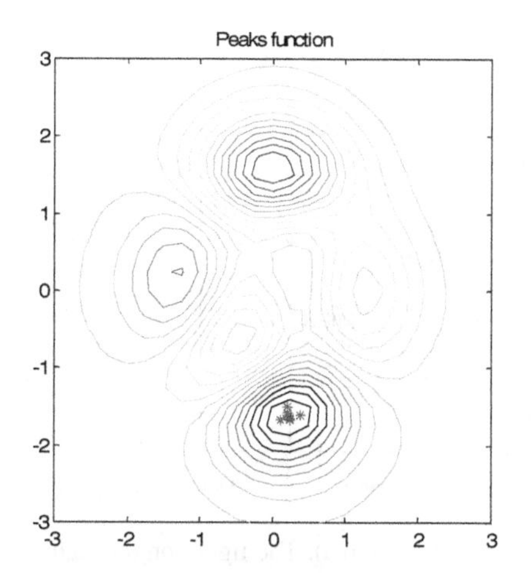

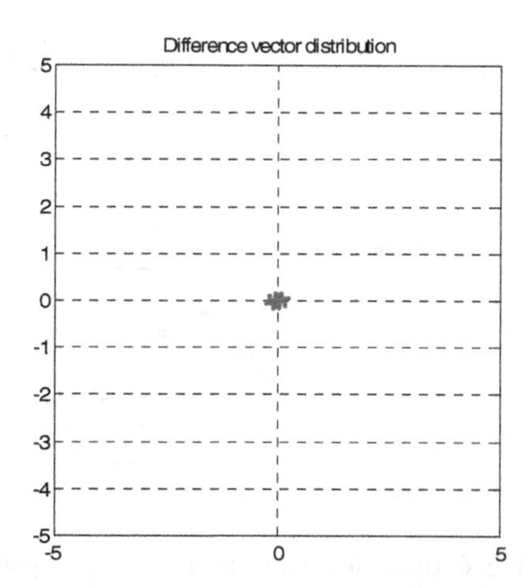

Fig. 5. Generation g=20 using Np = 8. Now the difference vector distribution fosters the local search of the minimum the vector population is enclosing.

It is intriguing to see the difference vector distribution adapt to the landscape of the objective function. This self-adaptivity renders DE's mutation based perturbation superior to mere Gaussian- or Cauchy-based types in most cases. However, Figure 3 through Figure 5 also reveals a weakness:

In the endeavour to obtain fast convergence the population size Np is usually kept low. Due to the limitation of Np·(Np-1) potential perturbation possibilities for a base vector there is a limited possibility to find regions of improvement and hence stagnation [21] can be the price to pay for the low number of Np. The blessing of contour matching may then turn out to be a curse when the contour of the objective function is deceiving and the "matching" leads away from the global optimum.

In order to increase the number of potential points to be searched while still maintaining a low number of Np gives rise to the various strategies for diversity enhancement, certainly one of the most interesting and rewarding areas of DE research today. The basic idea is simply to find some hopefully contour-matching and rotationally invariant way to generate more potential points without increasing the number Np of population members. As has been mentioned above one method for diversity enhancement has always been a part of DE, crossover.

Crossover

Crossover, i.e. mixing parameters of the target and the mutant vector in order to get the trial vector (see Eq. (4)) has been introduced to DE from the beginning [1, 2, 3, 4, 5, 6]. It was felt that mutation alone is too restrictive as a perturbation method, just as genetic algorithms [22] require some random mutation in addition to the dominant recombination mechanism to make the optimization work properly. So both DE and genetic algorithms have some dominant method of change plus an additional ingredient which slightly breaks up the mechanics of the dominant perturbation. Figure 6, however, forebodes that crossover has the potential to destroy the directional information provided by the difference vectors for the sake of increasing diversity. In fact it has been shown [7] that DE's contour matching property is lost

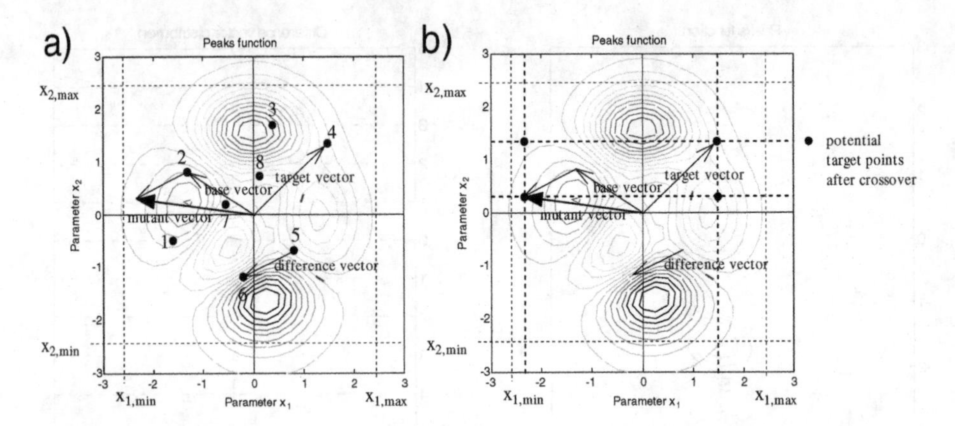

Fig. 6. Example for a population of Np=8 points and a mutation step a). The figure on the right b) shows the potential points when using crossover.

when strong crossover is used (e.g. Cr=0.1) and that in this case DE has a strong tendency to search along the main parameter axes, a property that is in fact beneficial for separable objective functions. Yet for real-world applications separability is rarely present. Parameter dependency seems to be the rule rather than the exception instead.

Another deficiency of crossover is that it is not rotationally invariant, i.e. optimization results obtained for a certain objective function do not directly translate to the rotated counterpart of this function. The differences in potential crossover points for two coordinate systems with the same origin are depicted in Figure 7. Despite its deficiencies the DE-literature reveals that crossover is almost always used. The diversity enhancing features of crossover seem to outweigh its disadvantages, at

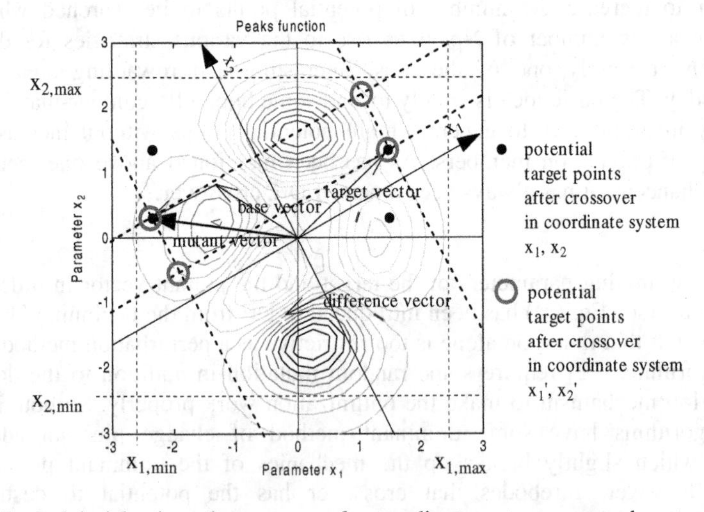

Fig. 7. Potential trial points after crossover for coordinate system x_1, x_2 and system x_1', x_2'

least if used lightly, i.e. if Cr stays close to 1 for the case of non-separable objective functions.

Dither

The term *dither* has been defined in [7] and presumably has been used by early practitioners of DE. The first reported publication advocating its use, however, seems to have been launched by Karaboga and Ökdem [23] in a Turkish Journal even though the term "dither" had not been employed. In [23] the scale factor F was randomized according to

$$F_{dither} = F_l + rand_g(0,1) \cdot (F_h - F_l) \tag{14}$$

for every generation g. Independently of [23] Das, Konar, and Chakraborty [24] have reported improvements in DE's convergence when using dither. In [24] dither had been applied to every difference vector i=0, 1, …, Np-1 rather than on a generational basis.

$$F_{dither} = F_l + rand_i(0,1) \cdot (F_h - F_l) \tag{15}$$

More variants are conceivable, for example changing F using some randomization different from uniform. A pictorial representation of dither is provided in Figure 8.

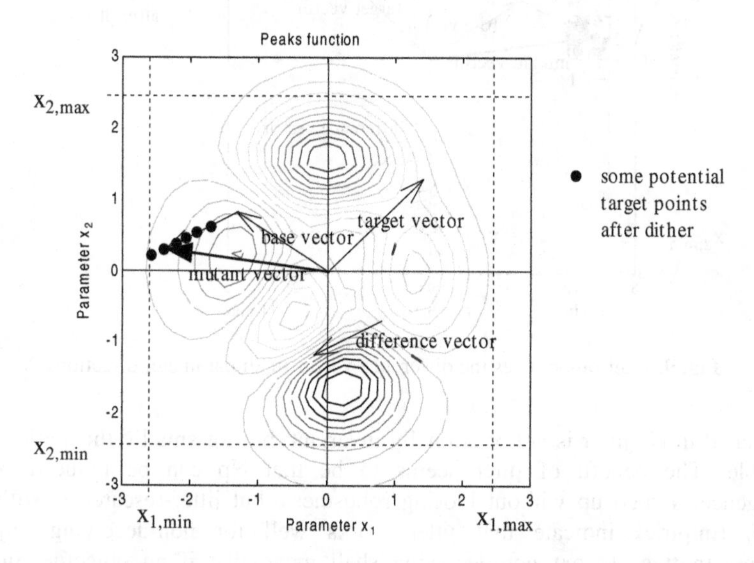

Fig. 8. Pictorial representation of dither which simply randomizes the mutation scale factor F and hence does not compromise DE's contour matching

Besides from improving DE's convergence behavior on time-independent objective functions dither also improves DE's handling of noisy objective functions [25]. Since dither is rotationally invariant and preserves the contour matching property this diversity enhancing method should always be used.

Jitter

Jitter as defined in [7] is somewhat similar to dither in that the scaling factor F is randomized. However, F is randomized for each single parameter j=0, 1, …, D-1 and for every new mutant vector i according to

$$F_{jitter,i} = F \cdot \left(1 + \delta \cdot \left(rand_j[0,1) - 0.5\right)\right) \tag{16}$$

Jitter has not been treated a lot in the literature. Zaharie [8] has used a Gaussian randomized form of jitter for the theoretical convergence proof of DE. Storn [26] has implemented it in a commercial program for digital filter design, and also Lampinen [27] reportedly has used but never published it. For jitter it seems to be very important that δ be small, e.g., δ=0.001. In fact δ may even be randomized itself. Figure 9 visualizes jitter and shows the effect of randomizing all parameter directions. The effect is a square cloud of potential points centered at the tip of the mutant vector.

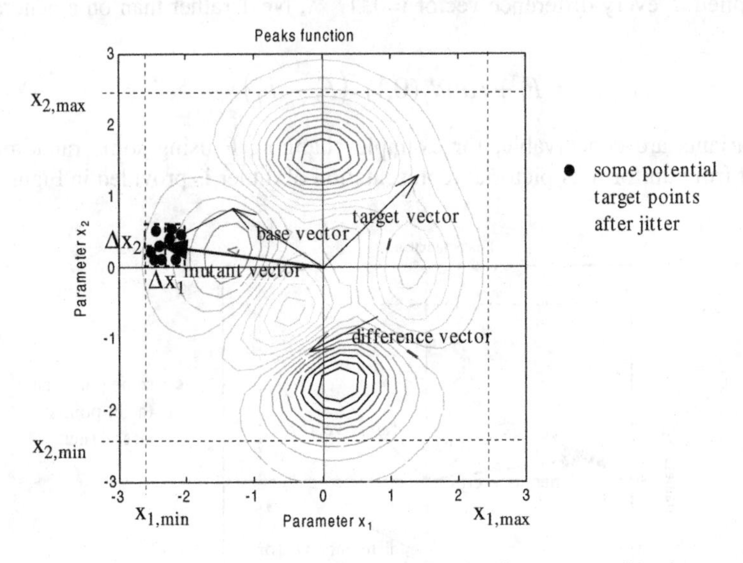

Fig. 9. Jitter randomizes the difference vector in all parameter directions

As stated in [7] jitter is not rotationally invariant, but for small δ this deficiency is negligible. The benefit of jitter seems to be that Np can be reduced so that convergence is sped up without loosing robustness, but jitter-research is still in its infancy. Empirics indicate that jitter works well for non-deceiving objective functions. In this context non-deceiving shall mean that if an objective function posesses a strong global gradient information then it also leads towards the vicinity of the global minimum. For example, Corana's paraboloid [28] which is riddled with small local minima would be non-deceiving.

In addition it also seems to be beneficial to combine jitter with dither [26] as in

$$F_{jitter\&dither,i,g} = \left(F_l + rand_g(0,1) \cdot (F_h - F_l)\right) \cdot \left(1 + \delta \cdot \left(rand_j[0,1) - 0.5\right)\right), \tag{17}$$

but more research is required with regard to jitter to get conclusive results.

Mixing perturbation techniques
Mixing perturbation techniques is another diversity generating technique that has received some attention in the past. One example is the *Either-Or-Algorithm* proposed by Kenneth Price and described in [7]. This technique counteracts stagnation by choosing at random which perturbation method to use, mutation like in Eq. (11) or recombination like in Eq. (12). Figure 10 provides an example for the differences in potential target points.

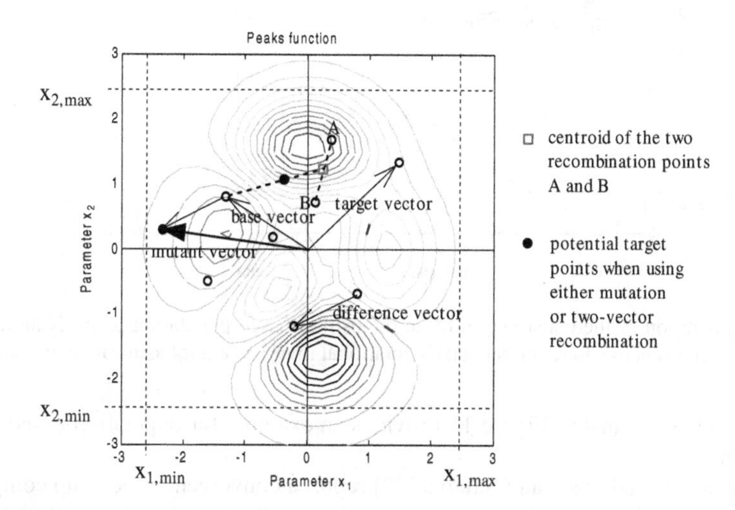

Fig. 10. Potential target points after applying either Eq. (11) or Eq. (12) for N=0

The advantage of this technique is that both rotational invariance and the basin-to-basin transfer property (i.e. contour matching) are preserved. Extensive empirical tests have shown that this approach can be more robust than classical DE while exhibiting a slow-down in convergence. Yet not all varieties of perturbation have been explored to a sufficient extent, e.g. multi-vector mutation or recombination, as well as usage of dither and jitter in addition to mutation and recombination. So there are quite a number of loose ends which need to be investigated further.

Opposition-based points
Another interesting concept for diversity enhancement has been introduced by Rahnamayan, Tizhoosh, and Salama [29] which uses either the mutant vector obtained in the usual way or its opposing point, depending on some probabilistic descision. The scheme is dubbed ODE for *opposition-based DE*. The opposing vector is defined as

$$\mathbf{v}_{i,g,opposed} = \mathbf{x}_{g,\min} + \mathbf{x}_{g,\max} - \mathbf{v}_{i,g}. \tag{18}$$

where $\mathbf{x}_{g,\min}$ and $\mathbf{x}_{g,\max}$ define the momentary extremes for each parameter taken over the entire population at a certain generation g. For the initial generation g=0 the absolute bounds are taken for these extremes. It is interesting to see that the opposing points generation scheme neither fulfills rotational invariance nor has the capability

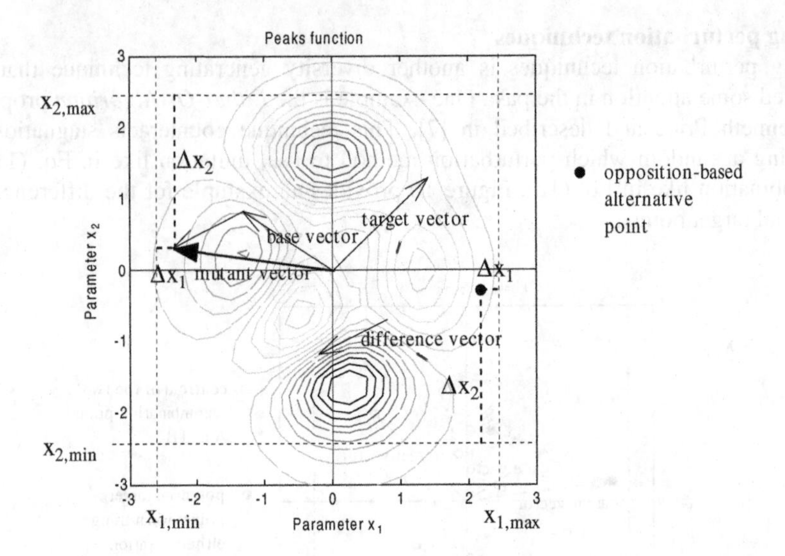

Fig. 11. Illustration of the construction of an opposition-based population point. Note that this point generating scheme does neither satisfy rotational invariance nor basin-to-basin transfer.

for basin to basin transfer. Figure 11 provides an example for opposition based vector generation.

Rahnamayan, Tizhoosh, and Salama [29] report a convergence speed up compared to classical DE. The reason for this may stem from two properties of the scheme:

1. The opposing points are not chosen on an individual basis but an entire population is generated with some probability JR (in [29] with JR=0.3) on a generational basis. Once this population of opposing points is generated there are 2·Np points available, Np from the current population and another Np from the population of points which is opposite to the current population. Out of these 2·Np points the Np best ones are chosen to form the next generation of points. In the evolutionary programming community this selection scheme is called *elitist*, or (μ+λ)-selection [7]. As indicated above this elitist selection is not used for every generation but only with a probability JR. The authors of [29] refer to this scheme as generation jumping. Elitist schemes usually speed up convergence because only the best points are retained. On the downside the chances for premature convergence are increased. Generation jumping might offer a good balance between elitist and one-to-one survivor selection.
2. The occasional generation jumping breaks up the vector generation scheme of DE just like the probabilistically occurring mutation breaks up the crossover scheme for GAs. This may hinder contour matching once in a while, but on the other hand it increases diversity, and the (μ+λ)-selection counteracts the loss of focus towards the optimum by being more greedy than the selection scheme of DE.

Again, the reasons for ODE's success and potential deficiencies need to be investigated further. Also a combination with other diversity enhancing strategies is worth investigating.

The strategies above are not the only diversity enhancing strategies that can be found in the literature. For example, in Ali [30] an extra distribution, the so-called β-distribution is applied to enhance the diversity of DE. Enhancing diversity is certainly a very interesting area of DE-research and hopefully more fruitful ideas in that domain will appear.

2.4 Controlling the Vector Population

One-array vs. Two-array

Classic DE uses two populations in order to allow computation on parallel computers or processors, something which is becoming increasingly important especially since the advent of multicore processors [80]. But in fact the very first algorithm that Kenneth Price came up with used just one single vector population array. This simplified version has been described in [7] in the light of saving memory on limited resource devices. In Feoktistov [31] this scheme is investigated further and extended to transversal DE where an individual may undergo several mutation/evaluation steps before it is compared to the target vector. Unfortunately in [31] the consequences, benefits or drawbacks are not regarded for a sufficiently large test set, so more research is needed to evaluate this idea. Transversal DE bears some similarity with hybrid DE versions that employ gradient algorithms or other greedy techniques for local search (see chapter 4.1.) in that the trial vector undergoes several improvement steps before it is compared to the target vector.

Selection Methods

Selection methods have been extensively discussed in [7]. The main methods of interest are:

1. Elitist $(\mu+\lambda)$-selection where the best μ individuals out of $\mu+\lambda$ individuals are selected. For DE usually $\mu=\lambda=Np$ is used.
2. Tournament selection with one-to-one survivor selection as in classic DE.

There have not been too many investigations on alternatives to DE's one-to-one selection, but it can be said that elitist selection is accelerating convergence while making the optimization more prone to premature convergence. In addition, elitist selection makes parallel computation more difficult. An in-depth numerical comparison of selection methods using an extensive testbed, however, is still lacking.

3 Problem Domain Specific Research

So far we have regarded DE as an optimization method to minimize objective functions without specifying the makeup of these objective functions. The tacit assumption when looking at "basic research" of evolutionary optimization is generally that objective functions to be minimized have a single global minimum, are potentially multimodal and nonlinear, and have a moderate number of parameters. Real-life problems, however, are often more complicated than that. Optimizations may include bounds constraints, inequality constraints, equality constraints, or they may even consist of constraints only without any objective to be optimized. The latter problem is well known as constraint satisfaction problem. If only equality constraints prevail and no objective is present we are encountering a system of equations. So far we have concentrated on parameters stemming from the continuous space, i.e. the floating point domain. Yet problems may also include discrete parameters or consist of discrete parameters only. If there are only discrete parameters and these parameters follow no metric, which means that there is no smaller, equal, or greater relationship, then we are looking at a combinatorial problem in the strict sense [7]. There are even more dimensions to optimization problems which the spider diagram in Figure 12 attempts to visualize. Many of these problem domains have been treated in [7] but still a lot of open questions remain. An entire chapter can be written easily for each case depicted in Figure 12 so only a few problem domain types will be sketched in the following in order to illustrate the research potential.

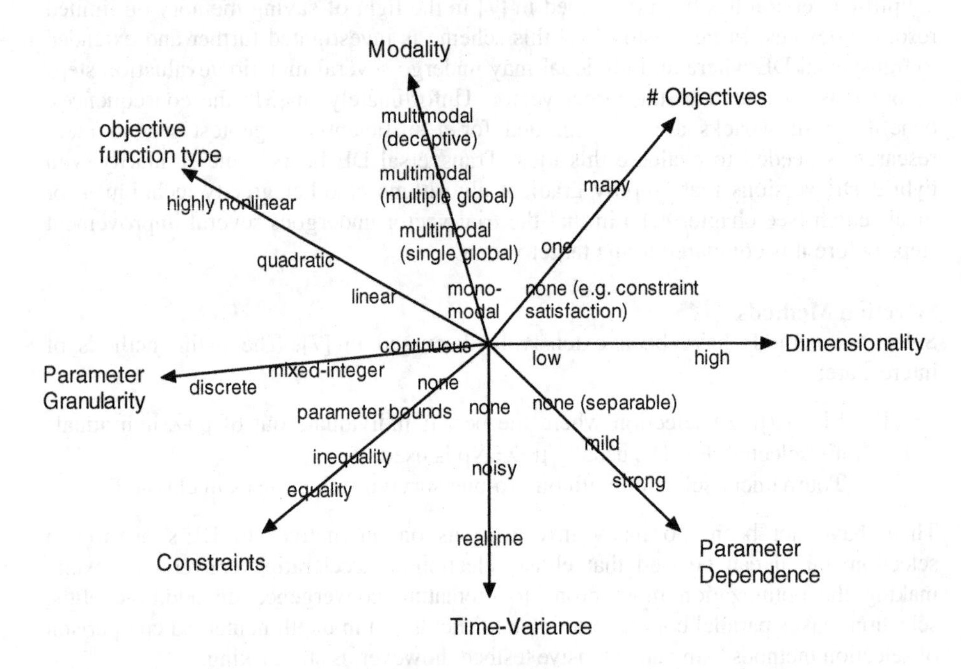

Fig. 12. Spider diagram sketching the various problem domain characteristics

3.1 Objective Functions with Single Objective

Figure 13 shows a classification of objective functions with a single objective regarding three problem domain dimensions, time variance, constraints, and parameter granularity. When available in the literature real-world example applications are provided. Figure 13 illustrates that at present DE is mostly used for time-invariant problems.

Objective Function → Time Variance ↓ ↓ Constraints Granularity		Time-invariant	Noisy	Time-variant (realtime)
Unconstrained	Continuous	Function approximation [38, 39], Channel Capacity Maximization [48]	Parameter identification [40], Object detection [45], Multisensor Fusion [7]	Acoustic Echo Cancellation [41] (quadratic problem, not solved with DE)
	Mixed integer	Circuit optimization with partially discretized components [49]		
	Discrete			
Constrained	Continuous	Electronic Circuit Design [33, 34, 35, 36], Optimization of Si-H Clusters [43]	Design Centering of Manufacturing Processes [7, 37]	
	Mixed integer	Mechanical Design Optimization [32], Compressor Supply Optimization [44]		
	Discrete	Design of Digital Filters in Signal Processing [7, 26]		
	Combinatorial	Traveling Salesman Problem [7], Cryptography applications [46, 47]		Routing Algorithms in Network Communication, not solved with DE [42]

Fig. 13. Classification of objective function types and some example applications. Problem types where DE-research is either in its infancy or not applicable are indicated with solid-frame table entries.

3.2 Combinatorial Problems

Combinatorial problems in the strict sense are problems where the parameters are discrete, the number of discrete states is finite, and the parameters are not associated with a metric, i.e. one cannot tell which of two different values for the same parameter is greater than the other unless an artificial metric is applied. For example the letters A and F of the alphabet may have an order in the alphabet, but one cannot truly say that A is greater than F or the other way round. Many of the well-known combinatorial problems are also highly constrained like the *traveling salesman problem* (TSP) which we will look at for illustration purposes later on. Even though DE has a good reputation of solving discrete or mixed-integer problems [7, 26, 32, 44, 50, 51, 52] there is no good evidence so far that DE is applicable to strict-sense combinatorial problems, at least if they are heavily constrained. In [7] the topic of combinatorial problems has been discussed, and the success of DE-based solutions to combinatorial problems was attributed to well-chosen repair mechanisms in the algorithm rather than DE-mutation. However, the applicability of DE to strict-sense combinatorial problems is neither proven nor disproven and depends also on finding a discrete operator that corresponds to the difference vector in the continuous domain. In addition it is required that the combination of a base vector and a difference vector (or recombination vector) yields a new valid vector. The validity of the newly generated vector is a big problem for most of the classical combinatorial problems like the TSP.

The traveling salesman problem (TSP)

Let us regard the traveling salesman problem (TSP) as an example to see how DE may be used to solve it and what the difficulties are. The TSP is a universal strict-sense combinatorial problem, and many other strict-sense combinatorial problems can be transformed into a TSP formulation [53]. Hence many findings about DE's performance on the TSP can be extrapolated to other strict-sense combinatorial problems.

Let there be M cities c_m, m=1,2, ..., M. Each city c_m has a distance $d_{m,n} = d_{n,m}$ to some other city c_n, n not equal to m, associated with it. The task in the TSP is to find a graph where all cities are visited and where the total distance

$$D = \sum_{m=1}^{M} d_{m,n} \quad with \quad n \neq m \quad and \quad mutually \quad different \tag{19}$$

is minimized. Figure 14 shows an example of a 5-city tour.

An approach using distances as parameters

In order to apply DE we first have to find an appropriate problem formulation. A natural approach would be to set up the problem vector x which contains all M distances $d_{m,n}$ as parameters, because the arithmetic difference of distances has a meaning and hence is suited for DE. For each city c_i there are M-1 distances $d_{m,n}$ to the other cities. For a finite set of parameters it is helpful to have these parameters in ascending order. So for our five city TSP example the list of distances would be as shown in Table 1. If we look at the first row as well as Figure 14, we see that city c_4 is closest to city c_1. The next closest city is c_2, then c_3, and then c_5. The other rows can be checked in a similar way. Since we know that DE prefers to have continuous parameters we take a table-based approach for non-uniform quantization. In this case

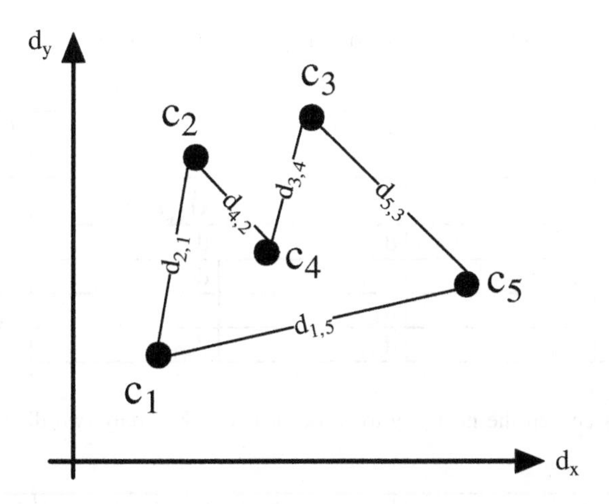

Fig. 14. Example for a city tour in a TSP for five cities [7]

DE's parameters are not the distances themselves but the appropriate array indices. In order to be able to use a continuous parameter vector x for DE we choose

$$\mathbf{x} = (x_1, x_2, x_3, x_4, x_5)^T \tag{20}$$

with x_m representing a reference to the distance between city c_m and its successor along the travel path. The values x_m are continuous and from the range [-0.5,3.5], so the index of the distances is computed via

$$index_m = floor(x_m + 0.5) \tag{21}$$

Table 1. Distance table containing the distances from each city c_m to each other city in ascending order

index $_m$ / city c_m	0	1	2	3
c_1	$d_{1,4}$	$d_{1,2}$	$d_{1,3}$	$d_{1,5}$
c_2	$d_{2,3}$	$d_{2,4}$	$d_{2,1}$	$d_{2,5}$
c_3	$d_{3,2}$	$d_{3,4}$	$d_{3,5}$	$d_{3,1}$
c_4	$d_{4,2}$	$d_{4,3}$	$d_{4,1}$	$d_{4,5}$
c_5	$d_{5,4}$	$d_{5,3}$	$d_{5,2}$	$d_{5,1}$

According to Eq. (21) abd the range information for x_m the $index_m$ can assume the values 0, 1, 2, or 3.

While trying to construct a valid path we easily see the constraints. For the first parameter x_1, which represents the distance associated with city c_1, we may choose a value from all the available table entries in Table 1, e.g. $d_{1,3}$. Since each city may be visited only once we must go to city c_3 in the next step because in the first step a path from city 1 to city 3 was selected. In order to prevent traveling back to city 1 the distance

Table 2. If $d_{1,3}$ is selected for city c_1 then $d_{3,1}$ may not be used for city c_3. This is indicated by greyshading.

index n / city c_m	0	1	2	3
c_1	$d_{1,4}$	$d_{1,2}$	**$d_{1,3}$**	$d_{1,5}$
c_2	$d_{2,3}$	$d_{2,4}$	$d_{2,1}$	$d_{2,5}$
c_3	$d_{3,2}$	$d_{3,4}$	$d_{3,5}$	$d_{3,1}$
c_4	$d_{4,2}$	$d_{4,3}$	$d_{4,1}$	$d_{4,5}$
c_5	$d_{5,4}$	$d_{5,3}$	$d_{5,2}$	$d_{5,1}$

Table 3. If $d_{3,4}$ is chosen the next city to work on is c_4. Now only two distances are free to choose from.

index n / city c_m	0	1	2	3
c_1	$d_{1,4}$	$d_{1,2}$	**$d_{1,3}$**	$d_{1,5}$
c_2	$d_{2,3}$	$d_{2,4}$	$d_{2,1}$	$d_{2,5}$
c_3	$d_{3,2}$	**$d_{3,4}$**	$d_{3,5}$	$d_{3,1}$
c_4	$d_{4,2}$	$d_{4,3}$	$d_{4,1}$	$d_{4,5}$
c_5	$d_{5,4}$	$d_{5,3}$	$d_{5,2}$	$d_{5,1}$

$d_{3,1}$ in the row for c_3 is excluded from the allowed list as indicated in Table 2 by greyshading.

So our search range has been restricted. Let's assume that we will choose $d_{3,4}$ in the next step, then our next city to consider will be c_4 and the associated parameter is x_4.

For x_4 the choice of available distances is even more restricted, as indicated in Table 3. So all the cities which have already been considered are excluded from the further search. The constraints get tighter and tighter until just one city is left. The last city must be connected to the first one in order to complete the tour.

Now the problem DE faces here becomes evident: Not only is the choice of allowable indexes for each city restricted, but also is this restriction dependent on which city we start the tour with and in which direction we go first. There is another fundamental problem with this approach: DE in general relies on the fact that a small difference vector means that the two parameter vectors are close together. This means that two identical solutions should yield the vector difference zero. However, we can immediately verify that the vectors $\mathbf{x}_a = (d_{1,2}\ d_{2,4}\ d_{4,3}\ d_{3,5}\ d_{5,1})$ and $\mathbf{x}_b = (d_{1,5}\ d_{5,3}\ d_{3,4}\ d_{4,2}\ d_{2,1})$ describe exactly the same tour but do not yield the vector difference zero. Hence one of DE's biggest assets, the self-adaptivity of the vector difference distribution is severely disturbed because a converged population still might exhibit large difference vectors.

Additional problems arise due to the heavy constraints inherent in the TSP. For example, even if we have a population of valid vectors, the weighted difference of two vectors added to a third one rarely yields a valid tour. If we want to repair this vector we can do this in many ways so that finally not much of DE's working principles are left. In the above case the constraints are dependent on the selection of the city, so the problem is not invariant. There have been attempts to treat such kinds of problems by not caring about valid solutions at first and simply applying validity as a hard constraint on the vectors [50]. The problem with this approach, however, is that most of the generated vectors are invalid and hence the optimization is very prone to stagnation. There are a number other approaches to optimize the TSP with DE [7], but none of them is really convincing. So successful optimization of heavily constrained strict-sense combinatorial problems using DE still remains to be fairly uncharted territory and leaves substantial room for improvement.

3.3 Design Centering

The problem of *design centering* using DE [37] has been largely left untouched even though this problem is of great importance to manufacturing. The idea is very simple: the design of any technical system usually has to meet certain specifications. Due to imprecisions in the manufacturing process the actual properties of the system often deviate from the nominal ones. The goal of design centering is to estimate the manufacturing imprecisions and to consider them at design time so that the probability that the eventual design violates any of the specifications is minimized.

In mathematical terms the design centering problem can be described as

$$\int_{\mathrm{ROA}} \mathrm{PDF}(\mathbf{x}_0)d\mathbf{x}_0 = \text{maximum}. \tag{22}$$

which means that the D-dimensional parameter vector $\mathbf{x}_0$ should be located such in the so-called *region of acceptability* (ROA) that the deviations of the parameters from their nominal values, which are described by a D-dimensional probability density function, mostly fall into the ROA rather than outside. The ROA is the permitted region in the parameter space within which any actual parameter vector may lie in order to fulfill the design specifications.

As an example Figure 15 shows a nominal magnitude function of a switched-capacitor lowpass filter and also its real-world counterpart after manufacturing. The manufacturing process introduces so-called parasitic capacitors which make the magnitude function violate the tolerance scheme. Figure 16 on the other hand shows the resulting magnitude function after manufacturing if the nominal design has undergone a design centering optimization, i.e. the parasitics are included into the design and the nominal parameter vector is placed within the design center of the ROA. Because the magnitude function lies well between the boundaries of the tolerance scheme it is intuitive that there is some headroom for the parameter values.

In [7] the problem of design centering and its solution via DE has been treated to some extent, and it was suggested that, provided that all population vectors are equally distributed within the ROA, a rough estimate of the design center is the point that maximizes the *center index*:

$$c_j = \sum_{i=1}^{Np} \exp(-d_{ij}), \quad j = 1,2,\ldots,Np, \quad d_{ij} = \left|\mathbf{x}_i - \mathbf{x}_j\right|. \tag{23}$$

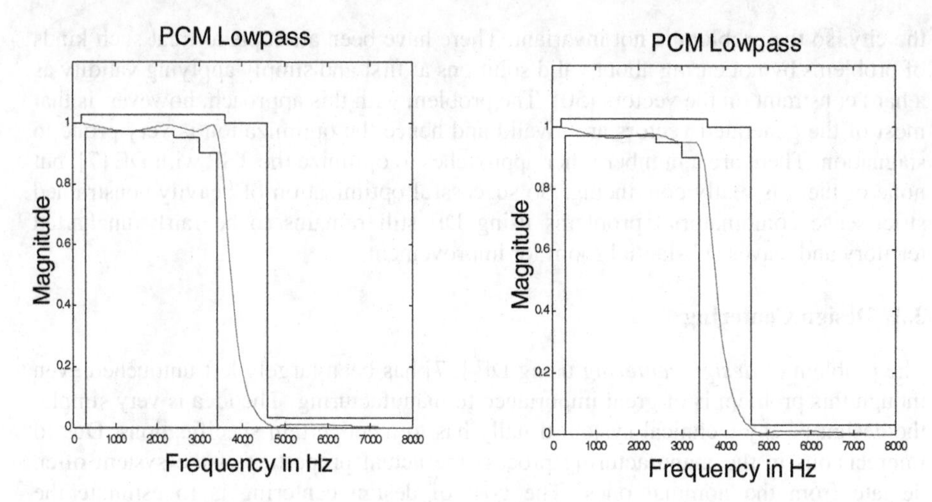

Fig. 15. Lowpass transfer function by standard filter design (left side) and after inclusion of parasitics (right side) [37]

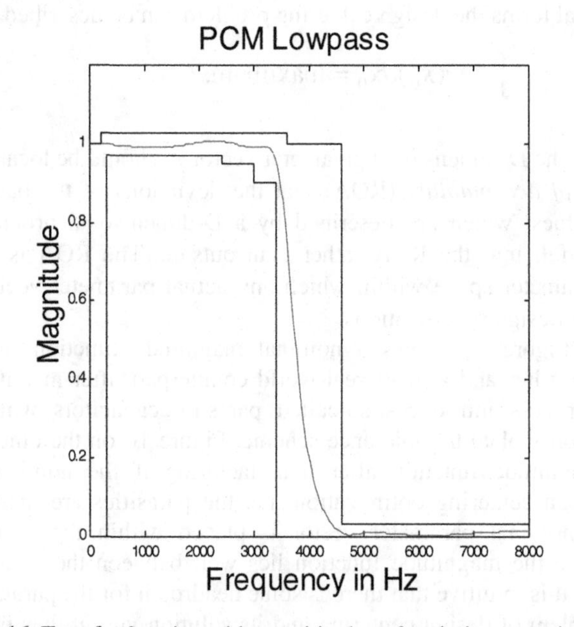

Fig. 16. Transfer function with parasitics but after design centering [37]

The center c_j index increases with the number of vectors that are close to vector $\mathbf{x}_j$. In other words, if a vector has a lot of neighbors then it is probably located fairly close to the design center. For vectors close to the rim of the ROA there is simply not as much space to accommodate many neighbors. Although this claim may appear intuitive, it has not been verified for a large enough testbed, and in fact the usage of DE for design centering problems still offers a lot of research opportunities.

3.4 Time-Variant Objective Functions

All population based optimization routines are mainly geared towards "offline optimization" or, in other words, optimization of time-independent objective functions. Especially when it comes to realtime applications where the minimum-finding process has to occur in a matter of seconds or even milliseconds, like in echo cancellation [41] or routing [42], minimization approaches have to be used that are more greedy and faster converging. Often the minimization problem is simplified to be a quadratic one, so fast gradient methods can be used [41]. Other applications may not necessarily require the global optimum because it will be gone anyway in a matter of seconds [42]. Nevertheless there are problem domains where population based optimizers may be applied. These domains comprise noisy and/or slowly varying multimodal objective functions.

Noisy objective functions

Noisy objective functions frequently occur in practice and are most often due to measurement imprecisions. As an example we may use the parameter identification problem for an induction motor [40]. First certain characteristic voltages of the motor are measured over time and/or over frequency. Then the parameters of a mathematical model of the motor are to be determined such that the absolute difference between the voltages generated by the model and the voltages from the measurements is minimized. The objective function to be minimized is noisy because the voltage measurements are noisy.

It has been reported by Krink, Filipic, Fogel, and Thomsen [54], that classic DE exhibits convergence problems on some noisy objective functions, at least when resampling is used as a noise-mitigating strategy. The idea of resampling is simply to evaluate the same candidate solution m times and to estimate the 'true' fitness value by the mean of the samples [54]. In [54] the question was raised whether thresholding rather than resampling may be a solution. The idea in thresholding [55] is to use a new selection operator for ES, such that a new candidate solution can only replace an existing one if the fitness difference is larger than a threshold τ. A disadvantage, however, is that with τ a new control variable enters the optimization scheme. A hint to the potential solution was already given in [54] when it was noticed that a specific Evolutionary Algorithm, which was compared to DE and showed better performance, used a Gaussian mutation operator. Eventually Chakraborty [24] showed that indeed DE is superior to this particular Evolutionary Algorithm if dither is added to classical DE. Rahnamayan, Tizhoosh, and Salama, [56] provided additional results showing the DE's performance on noisy objective functions can be improved if the evaluation of opposition-based points is added to classic DE resulting in what the authors call ODE. Both methods, dither and ODE, are diversity enhancement methods, so the question arises if there are more diversity enhancement methods which are beneficial for noisy objective function minimization.

Slowly time-variant objective functions

The application of DE to slowly varying objective functions is a very young area of research that has been touched briefly in [7] and which has been more intensely investigated by Mendes and Mohais [57]. The investigation in [57] revolves around the *moving peaks benchmark* (MPB) and suggests a DE-variant called *DynDE* (Dynamic DE) to approach this. The main ingredients in DynDE are:

1. Usage of several populations in parallel
2. Usage of uniform dither for $F \in [0,1]$ as well as $Cr \in [0,1]$
3. To maintain diversity of the populations two approaches may be chosen:
   a. Reinitialization of a population if the best individual of a population gets too close to the best individual of another population. The population with the absolute best individual is kept while the other one is reinitialized. This way the various populations are kept from merging.
   b. Randomization of one or more population vectors by adding a random deviation to the vector components. Various schemes of randomization are suggested.

The authors conclude that DynDE yields reasonable results, but admit that more research is required to improve this particular DE-variant.

The added dimension of time to the optimization problem requires to deal with many objective function instances at different points in time which makes research very expensive in terms of computational effort. This may be the reason why this problem domain has not been covered to a greater extent so far.

Since there is only little available literature there remains a lot of room to further explore the very interesting problem area of slowly varying objective functions.

4 Application Specific Research and Consequences for DE

It has been mentioned before that specific applications may bear some properties that make it worthwhile revisiting or extending DE so that the optimization matches the problem in the best possible way. Generally should any knowledge about the problem be incorporated into the optimization method and/or the objective function in order to make it more efficient. In the following we will look at the problem of digital filter design to illuminate this.

4.1 An Example: Digital Filter Design

Digital Filter design is a field of signal processing where specialized numerical methods govern the field [61]. To perform a filter design in practice is generally a matter of seconds, once the correct specifications are available. In some cases, though, there may be applications that require unconventional designs which cannot be performed using the standard methods [7, 26]. This is where DE can be of help, but it must be kept in mind that the filter designer is used to short design times which is why DE's convergence should be fast. It appears that the objective functions involved in digital filter design are non-deceiving, albeit multimodal. To illustrate this claim we look at the design task where a magnitude function $A(\Omega)$ has to fit into a tolerance scheme. An example for such a scheme is provided in Figure 17 which represents a so-called bandpass filter because only signal-portions with the normalized frequencies

Ω from the passband remain more or less unattenuated while the other portions get suppressed to a large degree. The equations needed to define A(Ω) are:

$$H(z)=\frac{U(z)}{D(z)}=\frac{\sum_{n=0}^{N_z} a(n)\cdot z^{-n}}{1+\sum_{m=1}^{M_p} b(m)\cdot z^{-m}}=A_0\frac{\prod_{n=0}^{N_z-1}\left(z-z_0(n)\right)}{\prod_{m=0}^{M_p-1}\left(z-z_p(m)\right)} \tag{24}$$

with

$$z=e^{j2\pi\Omega}=\cos(2\pi\Omega)+j\cdot\sin(2\pi\Omega),\quad j=\sqrt{-1}\,. \tag{25}$$

From Eq. (25) it is evident that for Ω=0 there must be z=1 and for Ω=0.5 there must be z=ı. Finally we define

$$A(\Omega)=\left|H\left(e^{j2\pi\cdot\Omega}\right)\right|=\sqrt{\mathrm{Re}\left(H\left(e^{j2\pi\cdot\Omega}\right)\right)^2+\mathrm{Im}\left(H\left(e^{j2\pi\cdot\Omega}\right)\right)^2} \tag{26}$$

From Eq. (25) and Figure 17 it becomes clear that a range of $\Omega \in [0, 0.5]$ transforms itself into the upper semi-circle of the z-plane. It is typical for a digital filter that the poles zp(m) are located in the passband while the zeros z0(m) are located in the stopband. So already by defining the tolerance scheme the approximate locations of poles and zeros are known. In [26] the fact is utilized that if the parameters of the objective function are not the coefficients a(n) and b(m) but the zeros z0(n) and the poles zp(m) then applying jitter together with dither in the DE-variant DE/best/1/bin works extremely well. Reasonable values for the control variables are Cr=0.95 and Np = 2·D, …, 5·D. The mutation method used in [26] is described by Eq. (27)

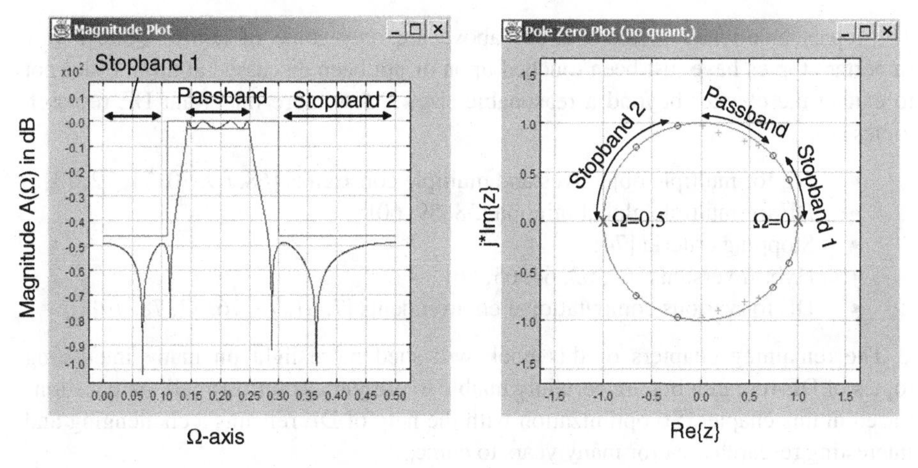

Fig. 17. Tolerance scheme in red and magnitude A(Ω) in blue (left figure) and the corresponding poles (+) and zeros (o) in the z-plane (right figure)

$$u_{j,i,g} = \begin{cases} v_{j,i,g} = x_{j,best,g} + (F_{dither,g} - 0.001 \cdot rand_j^{(1)}[0,1)) \cdot (x_{j,r1,g} - x_{j,r2,g}) & \text{if } rand_j^{(2)}[0,1) \le Cr \\ x_{j,i,g} & \text{otherwise} \end{cases}$$

$$\text{with} \quad F_{dither,g} = 0.5 + rand_g[0,1) \cdot 0.5 \tag{27}$$

The superscripts k on the random number randj(k)[0,1), k=1,2, in Eq. (27) shall denote that the numbers for k=1 and for k=2 are generated independently. Empirical evidence has shown that the positive effect of dither is small while jitter together with low values of Np is considerably speeding up convergence. These results come somewhat unexpected compared to the findings reported in [7] that jitter combined with DE/best/1/bin does not always perform well. The success of the described DE variant for this specific application is probably attributable to the benign nature of the objective functions, but more evidence needs to be gathered to corroborate this assumption.

What can be learned from this example is the advice to not consider the findings about DE-variants obtained from a large testbed as being universally applicable. This is also in line with the NFL [10]. For a specific application it may be worthwhile to revisit certain variants of DE depending on the properties of the objective functions at hand.

There are other potential possibilities to accelerate the convergence of DE-based digital filter design, one of which is using hybrid methods. In [62, 63, 64, 65] hybrid methods have already been used successfully. The basic idea usually is to refine one or more points from the DE-population by applying a fast-converging local search method like the Nelder and Mead optimization [20], dynamic hill climbing [66], or gradient type of algorithms [60, 67]. This refinement may take place for every DE-generation or after a certain amount of DE-generations. Gradient algorithms, however, are probably not appropriate since A(Ω) is not always differentiable.

5 More Topics and Outlook

The topics mentioned in the sections above are only some of many. Quite a few important topics have just been touched upon or not been discussed at all in order not to extend the chapter beyond a reasonable size. A few more important DE research topics are:

- DE for multiple objectives and multiple constraints [68, 69, 70, 71, 72, 73]
- DE for multiple global minima [58, 59, 60]
- Stopping criteria [74]
- Hybrid versions [62, 63, 64, 65, 66]
- DE for various computational environments [7, 31, 75, 76, 77, 78, 79]

The remaining chapters of this book will shed more light on many interesting topics of DE-research but are certainly unable to present a solution to all the questions raised in this chapter. So optimization with the help of DE remains a challenging and interesting research area for many years to come.

References

[1] Storn, R., Price, K.V.: Differential evolution - a simple and efficient adaptive scheme for global optimization over continuous spaces (1995) Technical Report TR-95-012, ICSI (March 1995), ftp://ftp.icsi.berkeley.edu/pub/techreports/1995/tr-95-012.ps.Z

[2] Storn, R., Price, K.V.: Minimizing the real functions of the ICEC 1996 contest by differential evolution. In: Proceedings of the 1996 IEEE international conference on evolutionary computation, Nagoya, Japan, pp. 842–844. IEEE Press, New York (1996)

[3] Storn, R.: On the usage of differential evolution for function optimization. In: Smith, M.H., Lee, M.A., Keller, J., Yen, J. (eds.) Proceedings of the 1996 biennial conference of the North American fuzzy information processing society – NAFIPS, Berkeley, CA, USA, June 19–22, pp. 519–523. IEEE Press, New York (1996)

[4] Price, K., Storn, R.: Differential evolution: a simple evolution strategy for fast optimization. Dr. Dobb's Journal 22, 18–24 (1997)

[5] Storn, R., Price, K.V.: Differential Evolution – a Simple and Efficient Heuristic for Global Optimization over Continuous Spaces. Journal of Global Optimization 11(4), 341–359 (1997)

[6] Storn, R.: Homepage of DE (2002), http://www.icsi.berkeley.edu/~storn/code.html

[7] Price, K., Storn, R., Lampinen, J.: Differential Evolution – A Practical Approach to Global Optimization. Springer, Heidelberg (2005)

[8] Zaharie, D.: Critical values for the control parameters of differential evolution algorithms. In: Matoušek, R., Ošmera, P. (eds.) Proceedings of MENDEL 2002, 8th international conference on soft computing, Brno, Czech Republic. Brno University of Technology, Faculty of Mechanical Engineering, June 5–7, pp. 62–67. Institute of Automation and Computer Science, Brno (2002)

[9] Lampinen, J.: A bibliography of differential evolution algorithms. Technical report, Lappeenranta University of Technology, Department of Information Technology, Laboratory of Information Processing(October 16, 1999), http://www.lut.fi/~jlampine/debiblio.htm

[10] Wolpert, D.H., Macready, W.G.: No free lunch theorems for optimization. IEEE transactions on evolutionary computation 1(1), 67–82 (1997)

[11] Gämperle, R., Müller, S.D., Koumoutsakos, P.: A Parameter Study for Differential Evolution. In: Grmela, A., Mastorakis, N.E. (eds.) Advances in Intelligent Systems, Fuzzy Systems, Evolutionary Computation, pp. 293–298. WSEAS Press (2002)

[12] Liu, J., Lampinen, J.: On setting the control parameters of the differential evolution method. In: Matoušek, R., Ošmera, P. (eds.) Proc. of Mendel 2002, 8th International Conference on Soft Computing, pp. 11–18 (2002)

[13] Liu, J., Lampinen, J.: Adaptive Parameter Control of Differential Evolution. In: Matoušek, R., Ošmera, P. (eds.) Proc. of Mendel 2002, 8th International Conference on Soft Computing, pp. 19–26 (2002)

[14] Liu, J., Lampinen, J.: A fuzzy adaptive differential evolution algorithm Soft Computing – A Fusion of Foundations. Methodologies and Applications 9(6), 448–462 (2005)

[15] Rönkkönen, J., Lampinen, J.: On using normally distributed mutation step length for the differential evolution algorithm. In: 9th Int. Conf. Soft Computing (MENDEL 2002), Brno, Czech Republic, June 5-7, 2002, pp. 11–18 (2003)

[16] Qin, A.K., Suganthan, P.N.: Self-adaptive differential evolution algorithm for numerical optimization. In: 2005 IEEE Congress Evolutionary Computation, Edinburgh, UK, September 2-5, vol. 2, pp. 1785–1791 (2005)

[17] Brest, J., Greiner, S., Bošković, B., Mernik, M., Žumer, V.: Self-Adapting Control Parameters in Differential Evolution, A Comparative Study on Numerical Benchmark Problems. IEEE Trans. on Evol. Comp. 10(6), 646–657 (2006)
[18] Schwefel, H.-P.: Numerical optimization of computer models. Wiley, New York (1981)
[19] Price, K.V., Rönkkönen, J.I.: Comparing the Uni-Modal Scaling Performance of Global and Local Selection in a Mutation-Only Differential Evolution Algorithm. In: IEEE Congress on Evolutionary Computation, 2006. CEC 2006, July 16-21, pp. 2034–2041 (2006)
[20] Nelder, J.A., Mead, R.: A simplex method for function minimization. Computer Journal 7, 308–313 (1965)
[21] Lampinen, J., Zelinka, I.: On Stagnation of the Differential Evolution Algorithm. In: Ošmera, P. (ed.) Proceedings of MENDEL 2000, 6th International Mendel Conference on Soft Computing, Brno, Czech Republic. Brno University of Technology, Faculty of Mechanical Engineering, June 7–9, pp. 76–83. Institute of Automation and Computer Science, Brno (Czech Republic) (2000)
[22] Goldberg, D.E.: Genetic algorithms in search optimization and machine learning. Addison-Wesley, Reading (1989)
[23] Karaboga, D., Ökdem, S.: A simple and global optimization algorithm for engineering problems: differential evolution algorithm. Turkish Journal of Electrical Engineering & Computer Sciences 12(1), 53–60 (2004)
[24] Das, S., Konar, A., Chakraborty, U.K.: Two improved differential evolution schemes for faster global search. In: Proceedings of the 2005 conference on Genetic and evolutionary computation (GECCO 2005), pp. 991–998 (2005)
[25] Das, S., Konar, A., Chakraborty, U.K.: Improved differential evolution algorithms for handling noisy optimization problems. In: Proc. IEEE Congress on Evolutionary Computation, Edinburgh (September 2005)
[26] Storn, R.: Digital Filter Design Program FIWIZ (2000), `http://www.icsi.berkeley.edu/~storn/fiwiz.html`
[27] Storn, R., Lampinen, J.: New DE Strategy, private Email communication (2000)
[28] Corana, A., Marchesi, M., Martini, C., Ridella, S.: Minimizing multimodal functions for continuous variables with the simulated annealing algorithm. ACM Transactions on Mathematical Software, 272–280 (March 1987)
[29] Rahnamayan, S., Tizhoosh, H.R., Salama, M.M.: Opposition-Based Differential Evolution Algorithms. In: 2006 IEEE Congress on Evolutionary Computation, Vancouver, July 16-21, pp. 2010–2017 (2006)
[30] Ali, M.M.: Synthesis of the β-distribution as an aid to stochastic global optimization. Computational Statistics and Data Analysis (accepted for publication, 2006)
[31] Feoktistov, V.: Differential Evolution - In Search of Solutions. Springer, Heidelberg (2006)
[32] Onwubolu, G.C., Babu, B.V.: New Optimization Techniques in Engineering. In: Lampinen, J., Storn, R. (eds.) Differential Evolution, ch. 6, pp. 123–166. Springer, Heidelberg (2004)
[33] Martinek, P., Maršík, J.: Optimized Design of Analogue Circuits Using DE Algorithms. In: EDS 2005 IMAPS CS International Conference Proceedings, pp. 385–389. Vysoké učení technické v Brně, Brno (2005)
[34] Martinek, P., Tichá, D.: Analog Filter Design Based on Evolutionary Algorithms. In: AEE 2005 - Proceedings of the 4th WSEAS International Conference on: Applications of Electrical Engineering, vol. 1, pp. 111–115. WSEAS, Athens (2005)
[35] Vancorenland, P.J., De Ranter, C., Steyaert, M., Gielen, G.G.E.: Optimal RF design using smart evolutionary algorithms. In: Proceedings of 37th Design Automation Conference, Los Angeles, June 5-9, pp. 7–10 (2000)

[36] Francken, K., Vancorenland, P., Gielen, G.: DAISY: a simulation-based high-level synthesis tool for Delta Sigma modulators. In: Proceedings of IEEE/ACM International Conference on Computer Aided Design. ICCAD 2000, San Jose, CA, USA, November 5-9, pp. 188–192 (2000)

[37] Storn, R.M.: System design by constraint adaptation and differential evolution. IEEE Transactions on Evolutionary Computation 3(1), 22–34 (1999)

[38] Storn, R.: On the usage of differential evolution for function optimization. In: Smith, M.H., Lee, M.A., Keller, J., Yen, J. (eds.) Proceedings of the North American Fuzzy Information Processing Society, pp. 519–523. IEEE Press, New York (1996)

[39] Report NDT3-04-2006: Differential Evolution for a Better Approximation to the Arctangent Function (April 26, 2006), http://www.nanodottek.com/Documents.htm

[40] Ursem, R.K., Vadstrup, P.: Parameter identification of induction motors using differential evolution. In: 2003 Congress on Evolutionary Computation, CEC 2003, vol. 2, pp. 790–796 (2003)

[41] Madisetti, V.K., Williams, D.B.: The Digital Signal Processing Handbook. Section VI, Adaptive Filtering. CRC Press, IEEE Press (1998)

[42] Murthy, C.S.R., Manoj, B.S.: Ad Hoc Wireless Networks: Architectures and Protocols. Prentice Hall, Englewood Cliffs (2004)

[43] Price, K., Storn, R., Lampinen, J.: Differential Evolution – A Practical Approach to Global Optimization. In: Chakraborty, N. (ed.) Genetic Algorithms and Related Techniques for Optimizing Si–H Clusters: A Merit Analysis for Differential Evolution, ch. 7.1. Springer, Berlin (2005)

[44] Price, K., Storn, R., Lampinen, J.: Differential Evolution – A Practical Approach to Global Optimization. In: Hancox, E.P., Derksen, R.W. (eds.) Optimization of an Industrial Compressor Supply System, ch. 7.3. Springer, Berlin (2005)

[45] Kasemir, K.U., Betzler, K.: Detecting ellipses of limited eccentricity in images with high noise levels. Image and Vision Computing 21(2), 221–227 (2003)

[46] Laskari, E.C., Meletiou, G.C., Vrahatis, M.N.: The Discrete Logarithm Problem as an Optimization Task: A First Study. In: Proceedings of the IASTED International Conference on Artificial Intelligence and Applications (AIA 2004) (IASTED 2004), Innsbruck, Austria. ACTA Press (2004) ISBN: 0-88986-375-X, ISSN: 1027-2666

[47] Laskari, E.C., Meletiou, G.C., Vrahatis, M.N.: Utilizing Evolutionary Computation Methods for the Design of S-boxes. In: Wang, Y., Cheung, Y.-m., Liu, H. (eds.) CIS 2006. LNCS (LNAI), vol. 4456. Springer, Heidelberg (2007)

[48] Henkel, W., Kessler, T.: Maximizing the Channel Capacity of Multicarrier Transmission by Suitable Adaptation of the Time-Domain Equalizer. IEEE Transactions on Communications 48(12), 2000–2004 (2000)

[49] Storn, R.: Differential Evolution – Ein praktischer Ansatz zur globalen Parameteroptimierung, Vortrag an der TU München, Seminar Elektronische Bauelemente (May 17, 2004)

[50] Onwubolu, G.C., Babu, B.V.: New Optimization Techniques in Engineering. Springer, Heidelberg (2004)

[51] Ruettgers, M.: Differential evolution: a method for optimization of real scheduling problems. Technical report at the International Computer Science Institute, TR-97-013, pp. 1–8 (1997)

[52] Babu, B.V., Rakesh, A.: A Differential Evolution Approach for Global Optimization of MINLP Problems. In: Proceedings of 4th Asia-Pacific Conference on Simulated Evolution And Learning (SEAL 2002), Singapore, November 18-22, Paper No. 1033, vol. 2, pp. 880–884 (2002)

[53] Syslo, M.M., Deo, N., Kowalik, J.S.: Discrete optimization algorithms with Pascal programs. Prentice Hall, New Jersey (1983)
[54] Krink, T., Filipic, B., Fogel, G.B., Thomsen, R.: Noisy Optimization Problems - A Particular Challenge for Differential Evolution? In: Proceedings of 2004 Congress on Evolutionary Computation, pp. 332–339. IEEE Press, Piscataway (2004)
[55] Markon, S., Arnold, D.V., Baeck, T., Beielstein, T., Beyer, H.-G.: Thresholding - a selection operator for noisy ES. In: Proceedings of the 2001 Congress on Evolutionary Computation CEC 2001, May 27-30, pp. 465–472. IEEE Press, Piscataway (2001)
[56] Rahnamayan, S., Tizhoosh, H.R., Salama, M.M.: Opposition-Based Differential Evolution for Optimization of Noisy Problems. In: 2006 IEEE Congress on Evolutionary Computation, IEEE World Congress on Computational Intelligence, Vancouver, July 16-21, pp. 1865–1872 (2006)
[57] Mendes, R., Mohais, A.S.: DynDE: a differential evolution for dynamic optimization problems. In: IEEE Congress on Evolutionary Computation, vol. 3, pp. 2808–2815 (September 2005)
[58] Crutchley, D.A., Zwolinski, M.: Using Evolutionary and Hybrid Algorithms for DC Operating Point Analysis of Nonlinear Circuits. In: Proceedings of the 2002 Congress on Evolutionary Computation, CEC 2002, Honolulu, Hawaii, May 12-17, vol. 1, pp. 753–758 (2002) ISBN 0-7803-7282-4
[59] Crutchley, D., Zwolinski, M.: Globally convergent algorithms for dc operating point analysis of nonlinear circuits. IEEE Transactions on Evolutionary Computing 7(1), 2–10 (2003)
[60] Crutchley, D.: Globally Convergent Algorithms for DC Operating Point Analysis of Nonlinear Electronic Circuits, PhD Dissertation, University of Southampton (2003)
[61] Antoniou, A.: Digital Filters – Analysis, Design, and Applications. McGraw-Hill, New York (1993)
[62] Dos Santos Coelhol, L., Mariani, V.C.: Combining of Differential Evolution and Implicit Filtering Algorithm Applied to Electromagnetic Design Optimization, Pontifical Catholic University of Parana, Technical Report
[63] Rogalsky, T., Derksen, R.W.: Hybridization of Differential Evolution for Aerodynamic Design. In: Proceedings of the 8th Annual Conference of the Computational Fluid Dynamics Society of Canada, June 11–13, pp. 729–736 (2000)
[64] Mydur, R.: Application of Evolutionary Algorithms & Neural Networks to Electromagnetic Inverse Problems. M.Sc. thesis, Texas A&M University, Texas, USA (2000)
[65] Nasimul Noman, N., Iba, H.: Enhancing Differential Evolution Performance with Local Search for High Dimensional Function Optimization. In: Proceedings of the 2005 conference on Genetic and evolutionary computation (GECCO 2005), pp. 967–974 (2005)
[66] Yuret, D., de la Maza, M.: Dynamic hill climbing: Overcoming the limitations of optimization techniques. In: The Second Turkish Symposium on Artifcial Intelligence and Neural Networks, pp. 208–212 (1993)
[67] Scales, L.E.: Introduction to non-linear optimization. Macmillan, London (1985)
[68] Chang, C.S., Xu, D.Y., Quek, H.B.: Pareto-optimal set based multi-objective tuning of fuzzy automatic train operation for mass transit system. IEEE Proceedings on Electric Power Applications 146(5), 577–583 (1999)
[69] Wang, F.-S., Sheu, J.-W.: Multi-objective parameter estimation problems of fermentation processes using a high ethanol tolerance yeast. Chemical Engineering Science 55(18), 3685–3695 (2000)

[70] Abbass, H.A., Sarker, R., Newton, C.: PDE: a Pareto-frontier differential evolution approach for multi-objective optimization problems. In: Proceedings of the 2001 congress on evolutionary computation, vol. 2, pp. 971–978. IEEE Press, Piscataway (2001)
[71] Abbass, H.A.: The self-adaptive Pareto differential evolution algorithm. In: Proceedings of the 2002 Congress on Evolutionary Computation (CEC 2002), Honolulu, Hawaii, May 2002, pp. 831–836 (2002b)
[72] Madavan, N.K.: Multiobjective optimization using a Pareto Differential Evolution approach. In: Proceedings of the 2002 Congress on Evolutionary Computation (CEC 2002), Honolulu, Hawaii, May 2002, pp. 1145–1150 (2002)
[73] Kukkonen, S., Lampinen, J.: GDE3: The third evolution step of generalized differential evolution. In: CEC 2005, Edinburgh, Scotland, pp. 443–450. IEEE Service Center (2005)
[74] Zielinski, K., Weitkemper, P., Laur, R., Kammeyer, K.D.: Examination of Stopping Criteria for Differential Evolution based on a Power Allocation Problem. In: 10th International Conference on Optimization of Electrical and Electronic Equipment, Brasov, Romania, May 18-19 (2006)
[75] Zaharie, D., Petcu, D.: Adaptive Pareto Differential Evolution and its Parallelization. In: Proc. of 5th International Conference on Parallel Processing and Applied Mathematics, Czestochowa, Poland (September 2003)
[76] Zaharie, D., Petcu, D.: Parallel implementation of multi-population differential evolution. In: Sinaia, R., Grigoras, D., et al. (eds.) Proc. of 2nd Workshop on Concurrent Information Processing and Computing (CIPC 2003) (2003)
[77] Tasoulis, D.K., Pavlidis, N.G., Plagianakos, V.P., Vrahatis, M.N.: Parallel Differential Evolution. In: Proceedings of the 2004 congress on evolutionary computation (CEC 2004), Portland OR, June 19-23, pp. 2023–2029 (2004)
[78] Kwedlo, W., Bandurski, K.: A Parallel Differential Evolution Algorithm. In: International Symposium on Parallel Computing in Electrical Engineering, 2006. PAR ELEC 2006, pp. 319–324 (2006)
[79] Angira, R., Babu, B.V.: Performance of modified differential evolution for optimal design of complex and non-linear chemical processes. Journal of Experimental & Theoretical Artificial Intelligence 18(4), 501–512 (2006)
[80] Goldstein, H.: Cure For The Multicore Blues. IEEE Spectrum, 36–39 (January 2007)

Eliminating Drift Bias from the Differential Evolution Algorithm

Kenneth V. Price

Summary. Differential evolution (DE) is an evolutionary algorithm designed for global numerical optimization. This chapter presents a new, rotationally invariant DE algorithm that eliminates drift bias from its trial vector generating function by projecting randomly chosen vector differences along lines of recombination. In this way, the natural distribution of vector differences drives both mutation and recombination. The new method also eliminates drift bias from survivor selection, leaving recombination as the only migration pathway. A suite of scalable test functions benchmarks the performance of drift-free DE against that of the algorithm from which it was derived.

1 Introduction

It has been ten years since the first differential evolution (DE) algorithm was published (Price and Storn 1997). Since then, DE has been applied to a multitude of optimization tasks, often with great success (Chap. 7, Price et al. 2005). Despite these successes, both theory and extended testing have exposed inadequacies in the "classic DE" algorithm (Storn and Price 1997; Price 1999). One of these deficiencies – the algorithm's rotation-dependent performance – was addressed in (Chap. 2, Price et al. 2005), but other problems remain. In particular, DE's generating and selection operators both exhibit drift bias. The goal of this chapter is to define drift bias, identify its sources and to empirically test the effectiveness of the modifications that eliminate it.

For background, Sect. 2 describes both classic DE and a second, more recent algorithm (DE/ran/1/either-or) that is comparatively unproven on real-world problems, but which the benchmark comparison in (Price et al. 2005) showed to be much more effective than classic DE on hard (rotated) benchmark problems. After discussing the drawbacks inherent in these previous methods, Sect. 3 identifies recombination as the source of drift bias in DE's trial vector generating scheme, then Sect. 4 proposes a simple solution for eliminating it. Section 5 both computes the drift bias attributable to DE's selection operator and shows how to eradicate it. Section 6 presents a drift-free DE algorithm that incorporates the bias-free solutions described in Sects. 4 and 5. Section 7 not only describes the test bed and performance measures that benchmark both the new algorithm and its predecessor, but also presents and discusses test results. Section 8 offers guidelines for selecting the new algorithm's control parameters, while Sect. 9 summarizes this study, presents conclusions and proposes topics for further research.

2 Background

Given an objective function $f(\mathbf{x})$ of D, real-valued parameters $x_j = \mathbf{x}$, $j = 1,2,\ldots,D$, DE attempts to find the vector $\mathbf{x}^*$ at which $f(\mathbf{x})$ attains its minimum (or maximum) value

U.K. Chakraborty (Ed.): Advances in Differential Evolution, SCI 143, pp. 33–88, 2008.
springerlink.com

over the D-dimensional space of real numbers $\Re^D$. More specifically, the global minimization problem is

$$\text{Find}: \mathbf{x}^* \Big| f(\mathbf{x}^*) \leq f(\mathbf{x}) \forall \mathbf{x} \in \Re^D . \qquad (1)$$

Ordinarily, the search is restricted to $S \subset \Re^D$ and a solution is often considered satisfactory if it lies some small distance from the (known) optimal vector $\mathbf{x}^*$, or if $f(\mathbf{x})$ is greater than the (known) optimal value $f(\mathbf{x}^*)$ by some small amount.

Like many evolutionary algorithms (EAs), DE simultaneously explores $\Re^D$ at multiple points with a population of Np, real-valued vectors $\mathbf{x}_{i,g}$ whose parameters are encoded as floating-point numbers. In addition to its population index $i = 1,2,\ldots,Np$, each vector also carries a generation index $g = 0,1,\ldots,g_{\max}$. Ordinarily, the initial population ($g = 0$) is distributed over S with random uniformity.

Although DE has relied on discrete crossover and arithmetic recombination, it has typically eschewed mutating vectors with deviations sampled from predefined probability distributions, choosing instead to alter them by differential mutation. Differential mutation creates a mutant vector $\mathbf{v}_{i,g}$ by adding the scaled difference of two different, but otherwise randomly chosen population vectors $\mathbf{x}_{r1,g}$ and $\mathbf{x}_{r2,g}$ to a third, distinct vector known as the base vector $\mathbf{x}_{b,g}$ (Eq. 2 and Fig 1).

$$\mathbf{v}_{i,g} = \mathbf{x}_{b,g} + F \cdot \left(\mathbf{x}_{r1,g} - \mathbf{x}_{xr2,g}\right), \quad r1, r2, b \in [1,2,\ldots, Np], \quad r1 \neq r2 \neq b \qquad (2)$$

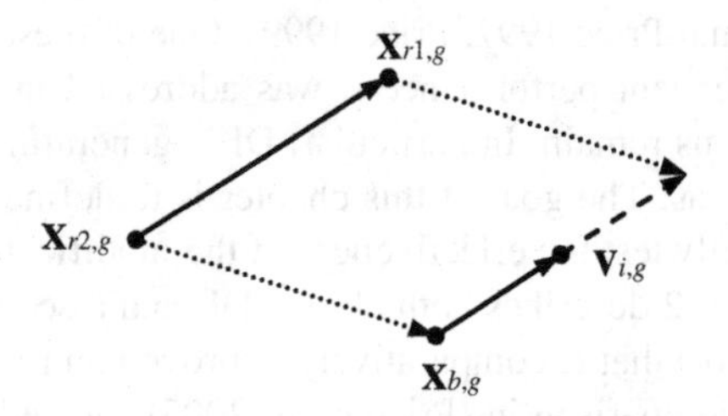

Fig. 1. Differential mutation. The randomly chosen vector difference $\mathbf{x}_{r1,g} - \mathbf{x}_{r2,g}$ is scaled and added to the base vector $\mathbf{x}_{b,g}$ to create a mutant $\mathbf{v}_{i,g}$ that competes with $\mathbf{x}_{i,g}$ (not shown).

The base vector index b can be chosen in a variety of ways. In the "classic DE" algorithm described below, it is randomly selected.

2.1 Classic DE

As its name suggests, the DE/ran/1/bin algorithm ("classic DE") pits each vector $\mathbf{x}_{i,g}$ in the current population against a trial vector $\mathbf{u}_{i,g}$ to whose composition it contributes through uniform crossover with a randomly ("/ran/") chosen base vector $\mathbf{x}_{r1,g}$ that has been mutated by the addition of a single ("/1/") scaled and randomly chosen difference vector $F \cdot (\mathbf{x}_{r2,g} - \mathbf{x}_{r3,g})$. The appellation "bin" refers to the fact that the number of parameters inherited by the trial vector $\mathbf{u}_{i,g}$ from the mutant vector $\mathbf{v}_{i,g}$ approximates a binomial distribution. During survivor selection, $\mathbf{u}_{i,g}$ replaces $\mathbf{x}_{i,g}$ if $f(\mathbf{u}_{i,g}) \leq f(\mathbf{x}_{i,g})$; otherwise, $\mathbf{x}_{i,g}$ retains its place in the population. Figure 2 outlines the classic DE algorithm.

```
for (i = 1; i ≤ Np; i = i + 1) // Initialize population
   for (j = 1; j ≤ D; j = j + 1) x_{j,i,g} = x_j^(lower) + U_j(0,1)·(x_j^(upper) – x_j^(lower));
   end for
   f(i) = f(x_{i,g}); // Evaluate and store f(x_{i,g})
end for
for (g = 1; g ≤ g_max; g = g +1) // Generation loop
   for (i = 1; i ≤ Np; i = i + 1) // Generate a trial population
      jrand_i = floor[U_i(0,1)·D] + 1; // Randomly select a parameter
      do r1=floor(U(0,1)_i·Np)+1; while (r1 = i); // Select 3 distinct indices
      do r2=floor(U(0,1)_i·Np)+1; while (r2 = i or r2 = r1);
      do r3=floor(U(0,1)_i·Np)+1; while (r3 = i or r3 = r1 or r3 = r2);
      for (j = 1; j ≤ D; j = j + 1) // Generate a trial vector
         if (U_j(0,1) ≤ Cr or j = jrand_i) u_{j,i,g} = v_{j,i,g} = x_{j,r1,g} + F·(x_{j,r2,g} – x_{j,r3,g});
         else u_{j,i,g} = x_{j,i,g};
      end for
   end for
   for (i = 1; i ≤ Np; i = i + 1) // Select new population
      if f(u_{i,g}) ≤ f(x_{i,g}) // Evaluate trial vector and compare with target vector
      {
         for (j = 1; j ≤ Np; j = j + 1) x_{j,i,g} = u_{j,i,g}; // Replace inferior target
         end for
         f(i) = f(u_{i,g});
      }
   end for
end for // End
```

Fig. 2. Pseudo-code for classic DE (DE/ran/1/bin). Input $Np \geq 4$, $F \in (0,1+)$ and $Cr \in [0,1]$. Try $Np \geq 4D$, $F = 0.7$ and $Cr = 0.9$ initially. Also, $F = 1$ works well for many multimodal problems.

The scale factor $F \in (0,1+)$ controls the mutation step size and, as a consequence, the population's convergence speed. The control variable $Cr \in [0,1]$ is a probability that mediates crossover by determining the average number of parameters that the trial vector $\mathbf{u}_{i,g}$ inherits from the mutant vector $\mathbf{v}_{i,g}$. The function $U_j(0,1)$ is a random number generator that returns a uniformly distributed value in the range [0,1). The subscript j indicates that the random number generator is sampled anew for each parameter. The values $x_j^{(upper)}$ and $x_j^{(lower)}$ are the initial upper and lower parameter bounds, respectively, for the jth parameter. The base vector index $r1$ and both difference vector indices, $r2$ and $r3$, are different but otherwise randomly chosen population indices each of which also differs from the *target* vector index i. To prevent $\mathbf{u}_{i,g}$ from duplicating the target vector with which it will be compared, the trial vector always inherits the parameter with the randomly chosen index $jrand_i$ from $\mathbf{v}_{i,g}$. Its subscript i indicates that $jrand_i$ is generated anew for each vector. Figure 3 illustrates the trial vector generating process.

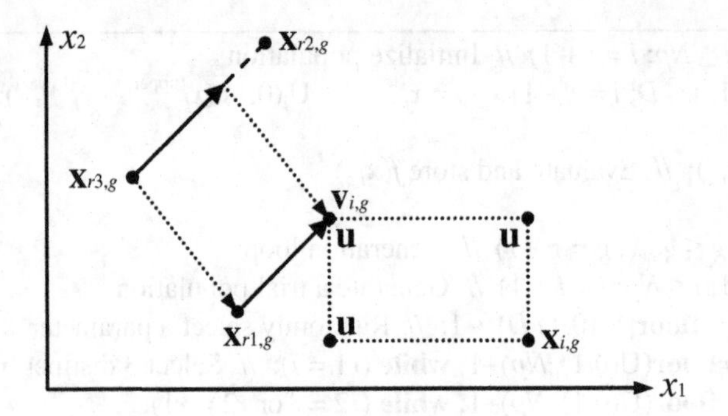

Fig. 3. The scaled difference of two randomly chosen vectors is added to the randomly selected base vector $\mathbf{x}_{r1,g}$ to produce a mutant $\mathbf{v}_{i,g}$ that is then uniformly crossed with the target vector $\mathbf{x}_{i,g}$. In this example, there are three possible trial vectors labeled "**u**". When $Cr = 1$, $\mathbf{u} = \mathbf{v}_{i,g}$.

The algorithm halts when termination criteria are met. In this study, trials were terminated either after reaching a preset number of generations, or once the objective function value of the best vector in the population was greater than the known minimum by less than a preset tolerance (see Sect. 7).

2.2 Decomposability and the Role of *Cr*

As long as $Cr < 1$, classic DE's performance will depend on the orientation of the coordinate system in which vectors are represented. As Fig. 4 illustrates, uniform crossover with $Cr < 1$ generates trial vectors whose positions change as the coordinate system rotates. If the coordinate axes and the function's principal axes are unfavorably aligned, e.g., if the optima of a multimodal function lie on coordinate diagonals, then classic DE may fail when $Cr < 1$ (Salomon 1996; Price 1999).

Only when $Cr = 1$ does classic DE's performance become invariant under a coordinate system rotation. This is because setting $Cr = 1$ transforms classic DE's generating

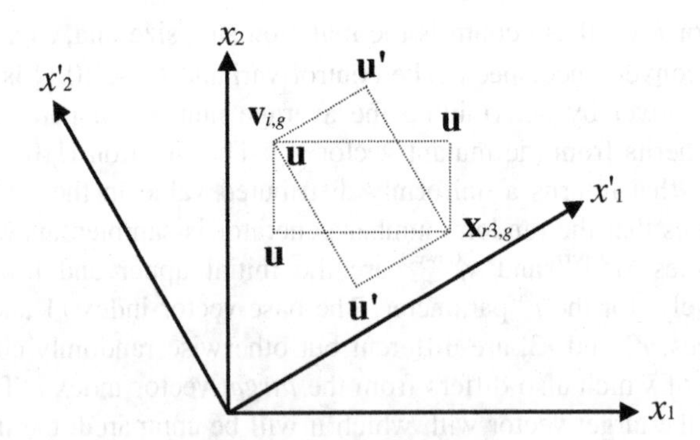

Fig. 4. Rotating the coordinate system relocates some trial vectors, but the position of the mutant trial vector $\mathbf{u} = \mathbf{v}_{i,g}$ is rotationally invariant ($Cr = 1$)

equation into a purely vector relationship that eliminates uniform crossover with the target vector and equates the trial vector with the differentially mutated base vector

$$\mathbf{u}_{i,g} = \mathbf{v}_{i,g} = \mathbf{x}_{r1} + F \cdot (\mathbf{x}_{r2} - \mathbf{x}_{r3}). \tag{3}$$

Since it preserves real vector inner products like the Euclidean distance (Hassani 1999), an orthogonal rotation affects neither the position of one vector with respect to another, nor the vector differences on which differential mutation is based. Rotation also leaves unchanged the position of the vector population with respect to objective function's contours. Consequently, a coordinate system rotation does not affect classic DE's performance when $Cr = 1$ because it does not affect the placement of trial vectors with respect to the objective function's contours.

Without crossover, however, the classic DE algorithm performs poorly on multimodal functions (Chap. 3, Price et al. 2005). The next sub-section presents a rotationally invariant algorithm that replaces uniform crossover with three-vector arithmetic recombination.

2.3 The DE/ran/1/either-or Algorithm

The goal when designing the DE/ran/1/either-or algorithm (Price et al. 2005) was to make DE's performance rotationally invariant while also providing a way to recombine vectors. The DE/ran/1/either-or algorithm assumes that trial vectors are normalized linear combinations of three randomly chosen vectors (two vectors are not enough to create a mutant and four vectors are more than is necessary). Normalizing the linear combination eliminates scale bias from the generating equation. For example, if the trial vector $\mathbf{u}_{i,g}$ is a linear combination of only one randomly chosen vector, then

$$\mathbf{u}_{i,g} = a_1 \cdot \mathbf{x}_{r1,g}, \tag{4}$$

where a_1 is real-valued. Unless $a_1 = 1$, $\mathbf{u}_{i,g}$ will be subject to a scale bias that will either consistently enlarge ($a_1 > 1$) or shrink ($a_1 < 1$) trial vectors. In this case, normalization, i.e., $a_1 = 1$, transforms Eq. 4 into a cloning operation that has no search capability.

Adding a second randomly selected vector to the combination and normalizing leads to the familiar equation for two-vector arithmetic recombination, also known as line recombination.

$$\begin{gathered}
\mathbf{u}_{i,g} = a_1 \cdot \mathbf{x}_{r1,g} + a_2 \cdot \mathbf{x}_{r2,g} \\
a_1 + a_2 = 1, \quad a_1 = 1 - a_2 \\
\mathbf{u}_{i,g} = (1 - a_2) \cdot \mathbf{x}_{r1,g} + a_2 \cdot \mathbf{x}_{r2,g} \\
\mathbf{u}_{i,g} = \mathbf{x}_{r1,g} + a_2 \cdot (\mathbf{x}_{r2,g} - \mathbf{x}_{r1,g})
\end{gathered} \tag{2.5}$$

Two-vector arithmetic recombination places trial vectors along the line joining $\mathbf{x}_{r1,g}$ and $\mathbf{x}_{r2,g}$ at a point determined by the coefficient of combination a_2.

Vector differences like the one in the last line of Eq. 5 are *recombination differentials* because the base vector – in this case $\mathbf{x}_{r1,g}$ – also appears in the difference term. By contrast, *mutation differentials* are vector differences that do not contain the base vector. Since the base vector and both difference vectors must be distinct, differential mutation requires three vectors.

After normalization, a linear combination of three randomly chosen vectors is

$$\mathbf{u}_{i,g} = \mathbf{x}_{r1,g} + a_2 \cdot \left(\mathbf{x}_{r2,g} - \mathbf{x}_{r1,g}\right) + a_3 \cdot \left(\mathbf{x}_{r3,g} - \mathbf{x}_{r1,g}\right) \tag{6}$$

Written this way, a_2 and a_3 each control a two-vector arithmetic recombination operation, i.e., both difference terms contain the base vector (Fig. 5).

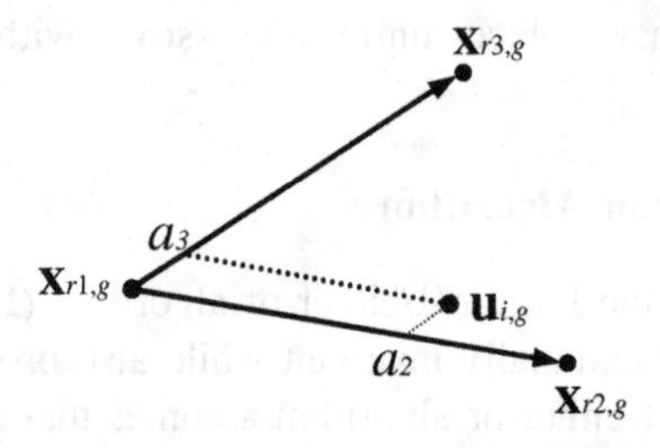

Fig. 5. In this decomposition, the trial vector's "coordinates" are measured from the base vector $\mathbf{x}_{r1,g}$ along the two (two-vector arithmetic) recombination differentials.

With a change of variables, however,

$$F \equiv \frac{(a_2 - a_3)}{2}, \quad K \equiv \frac{(a_2 + a_3)}{2} \tag{7}$$

the same combination also can be written as:

$$\mathbf{u}_{i,g} = \mathbf{x}_{r1,g} + F \cdot \left(\mathbf{x}_{r2,g} - \mathbf{x}_{r3,g}\right) + K \cdot \left(\mathbf{x}_{r2,g} + \mathbf{x}_{r3,g} - 2\mathbf{x}_{r1,g}\right) \tag{8}$$

This change of variables splits the normalized, three-vector linear combination into separate mutation and recombination components that can be independently controlled (Fig. 6). This decomposition also illustrates that three-vector arithmetic recombination is the natural counterpart to (three-vector) differential mutation.

In Fig. 6, $F \neq 0$ and $K \neq 0$, so both recombination and mutation play a role creating the trial vector. When mutation and recombination are simultaneously applied, the value chosen for F usually affects the best value for K and *vice versa*. Since F and K control distinctly different operations, any dependence between them should be minimized. To reduce control parameter dependence between F and K and because of empirical evidence supporting the approach, the DE/ran/1/either-or algorithm creates trial vectors that are *either* mutants *or* three-vector recombinants (Fig. 7).

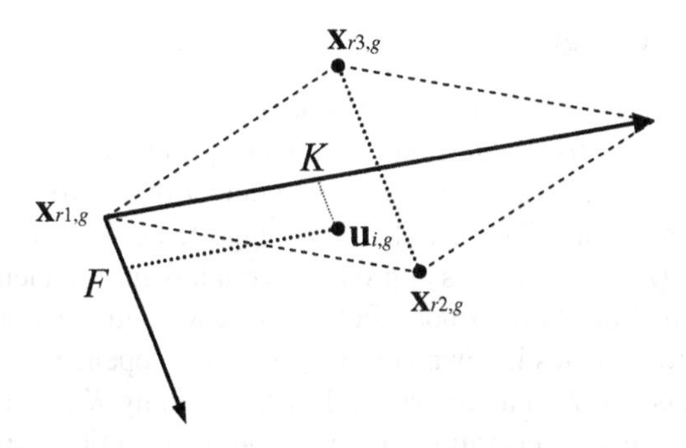

Fig. 6. The trial vector's location can also be decomposed into recombination and mutation components. Positive values for K are measured from the base vector $\mathbf{x}_{r1,g}$ along the medial line that passes midway between $\mathbf{x}_{r2,g}$ and $\mathbf{x}_{r3,g}$. Positive values for F are measured from the base vector in the direction $\mathbf{x}_{r2,g} - \mathbf{x}_{r3,g}$. In this scheme, trial vectors can be located anywhere in the plane defined by the mutation and recombination differentials.

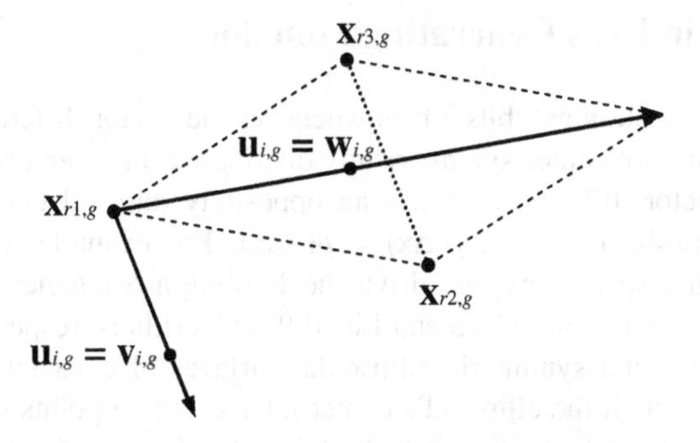

Fig 7. In the DE/ran/1/either-or algorithm, trial vectors are either mutants $\mathbf{v}_{i,g}$ or three-vector arithmetic recombinants $\mathbf{w}_{i,g}$. In contrast to Fig. 6, trial vectors are restricted to the mutation and recombination axes.

Equation 9 shows how the mutation probability p_F determines whether the target vector competes against a mutant $\mathbf{v}_{i,g}$ or a three-vector arithmetic recombinant $\mathbf{w}_{i,g}$.

$$\mathbf{u}_{i,g} = \mathbf{x}_{r1,g} + \begin{cases} \mathbf{v}_{i,g} = F \cdot \left(\mathbf{x}_{r2,g} - \mathbf{x}_{r3,g}\right) \text{ if } \mathrm{U}_i(0,1) \le p_F, \\ \mathbf{w}_{i,g} = K \cdot \left(\mathbf{x}_{r2,g} + \mathbf{x}_{r3,g} - 2\mathbf{x}_{r1,g}\right) \text{ otherwise.} \end{cases}$$

$$\mathbf{x}_{i,g+1} = \begin{cases} \mathbf{u}_{i,g} & \text{if } f\left(\mathbf{u}_{i,g}\right) \le f\left(\mathbf{x}_{i,g}\right) \\ \mathbf{x}_{i,g} & \text{otherwise} \end{cases} \tag{9}$$

$$i = 1,2,\ldots,Np, \quad r1, r2, r3 \in [1,2,\ldots,Np], \quad r1 \ne r2 \ne r3 \ne i,$$

$$F \in (0,1+), \quad p_F \in [0,1].$$

2.4 A Rationale for Redesign

Despite outperforming classic DE on rotated benchmark functions (Chap. 3, Price et al. 2005), the DE/ran/1/either-or algorithm retains old problems and introduces new ones. In particular, the addition of p_F increases the number of control variables to four (Np, F, K, p_F). To compensate, (Price et al. 2005) suggested a default relation linking F and K (i.e., $K = 0.5 \cdot (F + 1)$), but this approach surrenders independent control over mutation and recombination. Furthermore, both p_F and K control recombination and this duplication of effort creates its own control parameter dependence: the choice of p_F affects the best value for K and *vice versa.* Finally, holding K constant introduces drift bias into the trial vector generating process. The next section explores this last problem in more detail before the subsequent section presents a simple solution that not only eradicates drift bias from the DE/ran/1/either-or algorithm's trial vector generating scheme, but also reduces the number of control variables from four to three by eliminating K.

3 Drift Bias in DE's Generating Function

DE's generating function exhibits a bias whenever the vector differences that perturb the base vector are not center-symmetrically distributed. In a *center-symmetric* distribution, every vector difference $\Delta\mathbf{x}$ has an oppositely directed counterpart of equal magnitude and probability, i.e., $p(\Delta\mathbf{x}) = p(-\Delta\mathbf{x})$. For example, the Gaussian and Cauchy mutation distributions that drive the Evolution Strategies (Schwefel 1995) and Fast Evolution Strategies (Yao and Liu 1997) algorithms, respectively, characteristically generate center-symmetric ellipsoidal surfaces of constant probability. Any line that passes through the ellipsoid's center intersects two points on its surface that are equally probable and equidistant from the center, i.e., center-symmetric counterparts are related by a reflection through the ellipsoid's center (Fig. 8).

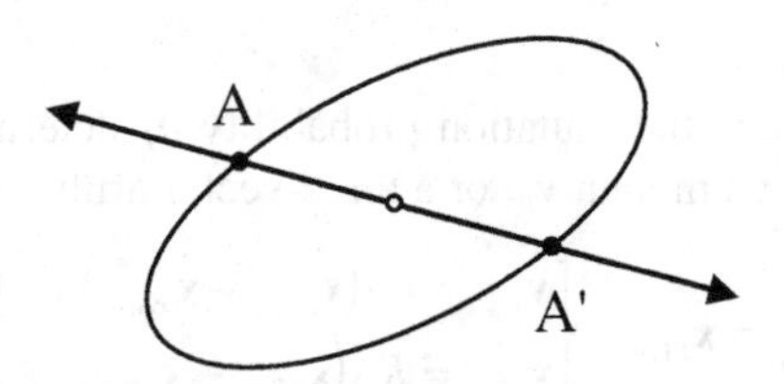

Fig. 8. The points A and A' are center-symmetric

A distribution's center-symmetry more reliably indicates that it is free of bias than does its expected value because there are zero-mean distributions that are not center-symmetric (e.g., Fig. 9) and also center-symmetric distributions (e.g., Cauchy) whose expectations do not converge. When it does converge, however, a distribution's expected (vector) value quantifies both the direction and the magnitude of its long-term *drift bias.*

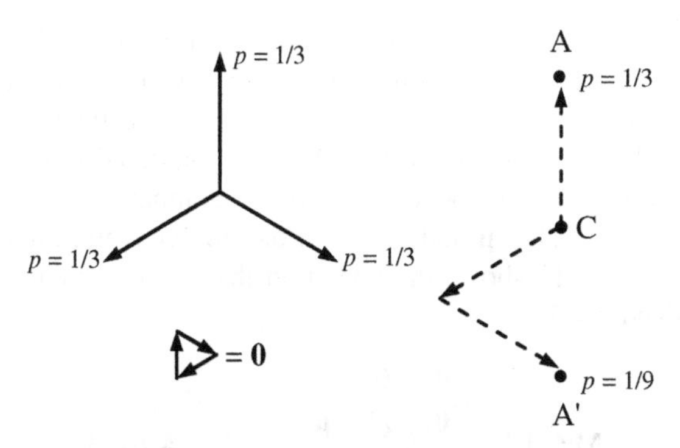

Fig. 9. The distribution depicted in the top left diagram contains three, equally probable vectors of equal magnitude that are oriented 120° apart. Its expected value is the sum of the probability-weighted vectors (bottom left). Even though its expectation is null, the distribution is not center-symmetric. As the diagram on the right shows, reaching point A from point C takes only one move sampled from this distribution, but to reach A's center-reflected point A' takes two moves, each of which has a probability 1/3, so in this case, $p(\Delta\mathbf{x})/3 = p(-\Delta\mathbf{x})$. Thus, this distribution suffers a short-term directional bias.

In lieu of a reason for biasing a search, a "black box" optimizer like DE should be drift-free so that optimization is driven by a response to the objective function and not by artifacts of the search process. The goals, therefore, of this section are to compute the expectations of the DE/ran/1/either-or algorithm's generating distributions and to transform those that exhibit drift bias into unbiased, center-symmetric ones. The next two subsections prove that while the mutation distribution is center-symmetric and has a null expectation, the distribution of three-vector arithmetic recombination differentials is asymmetrical and biased toward the population's centroid.

3.1 The Mutation Distribution M

Mutation differentials are drawn from the current distribution of vector differences, which also includes both null differentials and two-vector recombination differentials. More specifically, let **M** be defined as the matrix whose Np^2 elements, $\mathbf{m}_{k,l}$, comprise the *current* generation of vector differences $\mathbf{x}_{k,g} - \mathbf{x}_{l,g}$, $k,l = 1,2,\ldots,Np$.

$$\mathbf{M} = \begin{bmatrix} \mathbf{m}_{1,1} & \mathbf{m}_{1,2} & \cdots & \mathbf{m}_{1,Np} \\ \mathbf{m}_{2,1} & \ddots & & \vdots \\ \vdots & & & \\ \mathbf{m}_{Np,1} & \cdots & & \mathbf{m}_{Np,Np} \end{bmatrix}, \quad \mathbf{m}_{k,l} = \mathbf{x}_{k,g} - \mathbf{x}_{l,g}, \tag{10}$$

M is a skew-symmetric matrix because $\mathbf{m}_{k,l} = (\mathbf{x}_{k,g} - \mathbf{x}_{l,g}) = -(\mathbf{x}_{l,g} - \mathbf{x}_{k,g}) = -\mathbf{m}_{l,,k}$.

$$\mathbf{M} = \begin{bmatrix} \mathbf{0} & \mathbf{m}_{1,2} & \cdots & \mathbf{m}_{1,Np} \\ -\mathbf{m}_{1,2} & \ddots & & \\ \vdots & & & \\ -\mathbf{m}_{1,Np} & \cdots & & \mathbf{0} \end{bmatrix}, \quad \mathbf{m}_{k,l} = -\mathbf{m}_{l,k}, \tag{11}$$

Since $\mathbf{x}_{k,g} - \mathbf{x}_{k,g} = \mathbf{0} =$ (0,0,…,0), the differentials that comprise the leading diagonal are null vectors, i.e., $\mathbf{m}_{k,k} = \mathbf{0}$. The remaining $Np^2 - Np$ non-zero differentials can be partitioned into a set $\mathbf{M}^{(\chi 2)}$ that contains $2(Np - 1)$ two-vector recombination differentials and a set $\mathbf{M}^{(\mu)}$ that contains $Np^2 - 3Np + 2$ mutation differentials. The two-vector recombination differentials occupy row b and column b in $\mathbf{M}$, where b is the index of the base vector. Using $\boldsymbol{\mu}$ and $\boldsymbol{\chi 2}$ to denote, respectively, mutation and two-vector recombination, Eq. 12 shows the operation that each element of $\mathbf{M}$ performs for the case $Np = 4$ and $b = 3$.

$$\mathbf{M}: \begin{bmatrix} \mathbf{0} & \boldsymbol{\mu} & \boldsymbol{\chi 2} & \boldsymbol{\mu} \\ \boldsymbol{\mu} & \mathbf{0} & \boldsymbol{\chi 2} & \boldsymbol{\mu} \\ \boldsymbol{\chi 2} & \boldsymbol{\chi 2} & \mathbf{0} & \boldsymbol{\chi 2} \\ \boldsymbol{\mu} & \boldsymbol{\mu} & \boldsymbol{\chi 2} & \mathbf{0} \end{bmatrix}, \quad Np = 4, b = 3. \tag{12}$$

When opting for mutation, the combination $k = l$ is forbidden so that trial vectors do not simply clone the base vector by adding a null vector to it. Similarly, forbidding the combinations $k = b$ and $l = b$ prevents mutation from degenerating into two-vector recombination. The otherwise random selection of the remaining vector index combinations ensures that all mutation differentials are chosen with the same probability

$$p_{k,l} = \begin{cases} 0 & \text{if } k = l \vee k = b \vee l = b, \\ \dfrac{1}{Np^2 - 3Np + 2} & \text{otherwise.} \end{cases} \tag{13}$$

Substituting the appropriate values of $p_{k,l}$ into the general expression for a distribution's expectation eliminates the leading diagonal terms as well as the two-vector recombination differentials in column b and row b (Eq. 14). Because the remaining (mutation) differentials are equally probable, the expected vector value $\mathbf{E}(F \cdot \mathbf{M}^{(\mu)})$ of their scaled distribution reduces to a simple average, i.e., the sum of all $F \cdot \mathbf{m}_{k,l}$, $k \neq l$, $k \neq b$, $l \neq b$ divided by $Np^2 - 3Np + 2$.

$$\begin{aligned} \mathbf{E}\left(F \cdot \mathbf{M}^{(\mu)}\right) &= \sum_{l=1}^{Np} \sum_{k=1}^{Np} p_{k,l} \cdot F \cdot \mathbf{m}_{k,l} \\ &= \frac{F \cdot \sum\limits_{\substack{l=1 \\ l \neq k,b}}^{Np} \sum\limits_{\substack{k=1 \\ k \neq l,b}}^{Np} \mathbf{m}_{k,l}}{Np^2 - 3Np + 2} + F \cdot \left(\sum_{\substack{k=1 \\ k \neq b}}^{Np} \left(0 \cdot \mathbf{m}_{k,b} + 0 \cdot \mathbf{m}_{b,k}\right) + \sum_{k=1}^{Np} 0 \cdot \mathbf{m}_{k,k} \right) \\ &= \frac{F \cdot \sum\limits_{\substack{l=1 \\ l \neq k,b}}^{Np} \sum\limits_{\substack{k=1 \\ k \neq l,b}}^{Np} \mathbf{m}_{k,l}}{Np^2 - 3Np + 2}. \end{aligned} \tag{14}$$

By pairing each element in $\mathbf{M}^{(\mu)}$ with its additive inverse, Eq. 15 rewrites the sum of the $Np^2 - 3Np + 2$ mutation differentials as a sum of $(Np^2 - 3Np + 2)/2$ null vectors.

$$\begin{aligned}\sum_{\substack{l=1\\ l\neq k,b}}^{Np} \sum_{\substack{k=1\\ k\neq l,b}}^{Np} \mathbf{m}_{l,k} &= \sum_{\substack{l=1\\ l\neq k,b}}^{Np} \sum_{\substack{k=l+1\\ k\neq b}}^{Np} \left(\mathbf{m}_{k,l} + \mathbf{m}_{l,k}\right) \\ &= \sum_{\substack{l=1\\ l\neq k,b}}^{Np} \sum_{\substack{k=l+1\\ k\neq b}}^{Np} \left(\mathbf{m}_{k,l} - \mathbf{m}_{k,l}\right) \\ &= \mathbf{0}.\end{aligned} \tag{15}$$

Since the sum of the vectors in $\mathbf{M}^{(\mu)}$ is null, the expectation for the distribution of mutation differentials is also null.

$$\mathbf{E}\left(F \cdot \mathbf{M}^{(\mu)}\right) = \frac{F}{Np^2 - 3Np + 2} \cdot \mathbf{0} = \mathbf{0}. \tag{16}$$

Simply put, its skew-symmetry ensures that $\mathbf{M}^{(\mu)}$ is both center-symmetric and free of drift bias when mutation differentials are randomly chosen.

Like the elements of $\mathbf{M}^{(\mu)}$, the elements of $\mathbf{M}^{(\chi 2)}$ also sum to zero because every element in row b can be paired with its additive inverse in column b, i.e., if $\mathbf{m}_{k,b}$ is a two-vector recombination differential, then so is $\mathbf{m}_{b,k} = -\mathbf{m}_{k,b}$. Two-vector recombination differentials also appear as degenerate combinations in the matrix of three-vector recombination differentials, but without their additive inverses.

3.2 The Three-Vector Recombination Distribution R

The matrix of three-vector recombination differentials $\mathbf{R}$ consists of Np^2 elements $\mathbf{r}_{k,l} = \mathbf{x}_{k,g} + \mathbf{x}_{l,g} - 2\cdot\mathbf{x}_{b,g}$, where $b \in [1,2,\ldots,Np]$ is the base vector index, g is the current generation and $k,l = 0,1,\ldots,Np$.

$$\mathbf{R} = \begin{bmatrix} \mathbf{r}_{1,1} & \mathbf{r}_{1,2} & \cdots & \mathbf{r}_{1,Np} \\ \mathbf{r}_{2,1} & \ddots & & \vdots \\ \vdots & & & \\ \mathbf{r}_{Np,1} & \cdots & & \mathbf{r}_{Np,Np} \end{bmatrix}, \quad \mathbf{r}_{k,l} = \mathbf{x}_{k,g} + \mathbf{x}_{l,g} - 2\mathbf{x}_{b,g}. \tag{17}$$

Unlike $\mathbf{M}$, which is skew-symmetric, $\mathbf{R}$ is symmetric because $\mathbf{r}_{k,l} = (\mathbf{x}_{k,g} + \mathbf{x}_{l,g} - 2\cdot\mathbf{x}_{b,g}) = (\mathbf{x}_{l,g} + \mathbf{x}_{k,g} - 2\cdot\mathbf{x}_{b,g}) = \mathbf{r}_{l,k}$.

$$\mathbf{R} = \begin{bmatrix} \mathbf{r}_{1,1} & \mathbf{r}_{1,2} & \cdots & \mathbf{r}_{1,Np} \\ \mathbf{r}_{1,2} & \ddots & & \vdots \\ \vdots & & & \\ \mathbf{r}_{1,Np} & \cdots & & \mathbf{r}_{Np,Np} \end{bmatrix}, \quad \mathbf{r}_{k,l} = \mathbf{r}_{l,k} \tag{18}$$

$\mathbf{R}$ contains a single null vector $\mathbf{r}_{b,b} = \mathbf{0}$ corresponding to the case $k = l = b$. The remaining $Np^2 - 1$ non-zero differentials can be partitioned into a set $\mathbf{R}^{(\chi 2)}$ that contains $3(Np - 1)$ two-vector recombination differentials and a set $\mathbf{R}^{(\chi 3)}$ consisting of

$Np^2 - 3Np + 2$ three-vector recombination differentials. The two-vector recombination differentials occupy column b, row b *and* the leading diagonal of $\mathbf{R}$ and respectively correspond to the degenerate combinations: $k = b$, $l = b$ and $k = l$. Denoting two- and three-vector recombination by $\chi\mathbf{2}$ and $\chi\mathbf{3}$, respectively, Eq. 19 shows which operation each element of $\mathbf{R}$ performs for the case $Np = 4$ and $b = 3$.

$$\mathbf{R}: \begin{bmatrix} \chi 2 & \chi 3 & \chi 2 & \chi 3 \\ \chi 3 & \chi 2 & \chi 2 & \chi 3 \\ \chi 2 & \chi 2 & \mathbf{0} & \chi 2 \\ \chi 3 & \chi 3 & \chi 2 & \chi 2 \end{bmatrix}, \quad Np = 4, b = 3. \tag{19}$$

Transposing $\mathbf{x}_{k,g}$ and $\mathbf{x}_{l,g}$ does not change the sign of the recombination differential because $\mathbf{r}_{k,l}$ forms their sum, not their difference. Consequently, $\mathbf{R}$ does not contain the additive inverse for any (non-zero) element, so randomly sampling the permitted combinations of k and l does not automatically generate an unbiased distribution as it does for $\mathbf{M}^{(\mu)}$.

The expected value of $\mathbf{R}^{(\chi 3)}$, which reveals both the magnitude and the direction of its drift bias, must exclude the combinations $k = l$, $k = b$ and $l = b$ so that null and two-vector recombination differentials are not counted. The otherwise random selection of the remaining vector index combinations ensures that all three-vector recombination differentials are chosen with the same probability

$$p_{k,l} = \begin{cases} 0 \text{ if } & k = l \vee k = b \vee l = b, \\ \dfrac{1}{Np^2 - 3Np + 2} & \text{otherwise.} \end{cases} \tag{20}$$

Substituting the appropriate values of $p_{k,l}$ into the general expression for a distribution's expectation eliminates the null element $\mathbf{r}_{k,k}$ as well as the two-vector recombination differentials in column b, row b and on the leading diagonal. Because the remaining three-vector recombination differentials are equally probable, the expected vector value $\mathbf{E}(K \cdot \mathbf{R}^{(\chi 3)})$ reduces to a simple average, i.e., the sum of all $K \cdot \mathbf{r}_{k,l}$, $k \neq l$, $k \neq b$, $l \neq b$ divided by $Np^2 - 3Np + 2$.

$$\begin{aligned} \mathbf{E}\left(K \cdot \mathbf{R}^{(\chi 3)}\right) &= \sum_{l=1}^{Np} \sum_{k=1}^{Np} p_{k,l} \cdot K \cdot \mathbf{r}_{k,l} \\ &= \frac{K \cdot \sum\limits_{\substack{l=1 \\ l \neq k,b}}^{Np} \sum\limits_{\substack{k=1 \\ k \neq l,b}}^{Np} \mathbf{r}_{k,l}}{Np^2 - 3Np + 2} + K \cdot \left(\sum_{\substack{k=1 \\ k \neq b}}^{Np} \left(0 \cdot \mathbf{r}_{k,b} + 0 \cdot \mathbf{r}_{b,k}\right) + \sum_{k=1}^{Np} 0 \cdot \mathbf{r}_{k,k} \right) \\ &= \frac{K \cdot \sum\limits_{\substack{l=1 \\ l \neq k,b}}^{Np} \sum\limits_{\substack{k=1 \\ k \neq l,b}}^{Np} \mathbf{r}_{k,l}}{Np^2 - 3Np + 2}. \end{aligned} \tag{21}$$

To compute the sum of the elements in $\mathbf{R}^{(\chi 3)}$, Eq. 22 subtracts all two-vector recombination differentials from the sum of the elements in $\mathbf{R}$.

$$\sum_{\substack{l=1 \\ l \neq k,b}}^{Np} \sum_{\substack{k=1 \\ k \neq l,b}}^{Np} \mathbf{r}_{k,l} = \sum_{l=1}^{Np} \sum_{k=1}^{Np} \mathbf{r}_{k,l} - \sum_{k=1}^{Np} \left(\mathbf{r}_{k,b} + \mathbf{r}_{b,k} + \mathbf{r}_{k,k} \right) + 2\mathbf{r}_{b,b}$$

$$= \sum_{l=1}^{Np} \sum_{k=1}^{Np} \mathbf{r}_{k,l} - \sum_{k=1}^{Np} \left(2\mathbf{r}_{k,b} + \mathbf{r}_{k,k} \right) \tag{22}$$

Equation 22 adds the term $2\mathbf{r}_{b,b}$ because otherwise, the element $\mathbf{r}_{b,b}$ is counted three times in the rightmost summation – once as an element of column b, once as an element of row b and once as an element of the leading diagonal. Even so, $\mathbf{r}_{b,b}$ contributes nothing to the sum because it is null. Since $\mathbf{R}$ is symmetric, the rightmost summation can be further simplified by replacing $(\mathbf{r}_{k,b} + \mathbf{r}_{b,k})$ with $2\mathbf{r}_{k,b}$.

To compute the sum of the elements of $\mathbf{R}$, Eq. 23 treats the base vector as a constant that can be extracted from the double sum and multiplied by Np^2. Similarly, $\mathbf{x}_{l,g}$ can be extracted from the innermost summation over k and multiplied by Np. Next, the sum of all vectors in the population is replaced by $Np \cdot \langle \mathbf{x} \rangle_g$, where $\langle \mathbf{x} \rangle_g$ is the mean population vector, i.e., its centroid in generation g.

$$\sum_{l=1}^{Np} \sum_{k=1}^{Np} \mathbf{r}_{k,l} = \sum_{l=1}^{Np} \sum_{k=1}^{Np} \left(\mathbf{x}_{k,g} + \mathbf{x}_{l,g} - 2\mathbf{x}_{b,g} \right)$$

$$= -2Np^2 \cdot \mathbf{x}_{b,g} + \sum_{l=1}^{Np} \sum_{k=1}^{Np} \left(\mathbf{x}_{k,g} + \mathbf{x}_{l,g} \right)$$

$$= -2Np^2 \cdot \mathbf{x}_{b,g} + \sum_{l=1}^{Np} \left(Np \cdot \mathbf{x}_{l,g} + \sum_{k=1}^{Np} \mathbf{x}_{k,g} \right)$$

$$= -2Np^2 \cdot \mathbf{x}_{b,g} + \sum_{l=1}^{Np} \left(Np \cdot \mathbf{x}_{l,g} + Np \cdot \langle \mathbf{x} \rangle_g \right)$$

$$= -2Np^2 \cdot \mathbf{x}_{b,g} + Np^2 \cdot \langle \mathbf{x} \rangle_g + Np \cdot \sum_{l=1}^{Np} \cdot \mathbf{x}_{l,g}$$

$$= -2Np^2 \cdot \mathbf{x}_{b,g} + Np^2 \cdot \langle \mathbf{x} \rangle_g + Np^2 \cdot \langle \mathbf{x} \rangle_g$$

$$= 2Np^2 \cdot \left(\langle \mathbf{x} \rangle_g - \mathbf{x}_{b,g} \right) \tag{23}$$

The second term in Eq. 22 sums the contribution from all two-vector recombination differentials in **R**. Substituting population vectors for recombination differentials and combining terms yields:

$$\sum_{k=1}^{Np}\left(2\mathbf{r}_{k,b}+\mathbf{r}_{k,k}\right)=\sum_{k=1}^{Np}2\left(\mathbf{x}_{k,g}+\mathbf{x}_{b,g}-2\mathbf{x}_{b,g}+\mathbf{x}_{k,g}-\mathbf{x}_{b,g}\right)$$

$$=2\sum_{k=1}^{Np}\left(2\mathbf{x}_{k,g}-2\mathbf{x}_{b,g}\right)$$

$$=4\sum_{k=1}^{Np}\mathbf{x}_{k,g}-4Np\cdot\mathbf{x}_{b,g} \tag{24}$$

$$=4Np\cdot\left(\langle\mathbf{x}\rangle_g-\mathbf{x}_{b,g}\right)$$

Combining the results of Eq. 23 and Eq. 24 gives the sum of the elements in $\mathbf{R}^{(\chi 3)}$.

$$\sum_{\substack{l=1\\ l\neq k,b}}^{Np}\sum_{\substack{k=1\\ k\neq l,b}}^{Np}\mathbf{r}_{k,l}=2Np^2\cdot\left(\langle\mathbf{x}\rangle_g-\mathbf{x}_{b,g}\right)-4Np\cdot\left(\langle\mathbf{x}\rangle_g-\mathbf{x}_{b,g}\right)$$

$$=2Np\cdot(Np-2)\cdot\left(\langle\mathbf{x}\rangle_g-\mathbf{x}_{b,g}\right) \tag{25}$$

Plugging this value into the expression for the expected value of the three-vector recombination distribution shows that there is a residual bias in $\mathbf{R}^{(\chi 3)}$ that points toward or away from the current population centroid depending on the sign of K.

$$\mathbf{E}\left(K\cdot\mathbf{R}^{(\chi 3)}\right)=2K\cdot\frac{Np^2-2Np}{Np^2-3Np+2}\cdot\left(\langle\mathbf{x}\rangle_g-\mathbf{x}_{b,g}\right) \tag{26}$$

Dividing both the numerator and denominator by Np^2 shows that the expected value of the three vector recombination distribution approaches $2K\cdot(\langle\mathbf{x}\rangle_g-\mathbf{x}_{b,g})$ in the limit of large Np.

$$\mathbf{E}\left(K\cdot\mathbf{R}^{(\chi 3)}\right)=2K\cdot\frac{1-\frac{2}{Np}}{1-\frac{3}{Np}+\frac{2}{Np^2}}\cdot\left(\langle\mathbf{x}\rangle_g-\mathbf{x}_{b,g}\right)$$

$$\lim_{Np\to\infty}\mathbf{E}\left(K\cdot\mathbf{R}^{(\chi 3)}\right)=2K\cdot\left(\langle\mathbf{x}\rangle_g-\mathbf{x}_{b,g}\right) \tag{27}$$

Equation 27 indicates that when K ($\neq 0$) is held constant, setting $\mathbf{x}_{b,g} = \langle\mathbf{x}\rangle_g$ generates a distribution of recombinant trial vectors that has a null expectation. Despite having a null expectation, the recombinant distribution generated by the equation $\mathbf{w}_{i,g} = \langle\mathbf{x}\rangle_g + K\cdot(\mathbf{x}_{k,g} + \mathbf{x}_{l,g} - 2\langle\mathbf{x}\rangle_g)$ is not center-symmetric. Like the vectors depicted in Fig. 9, differentials in this centroid-centric distribution radiate outward from the center, but not necessarily in the opposite direction.

To be center-symmetric, the distribution of recombinant trial vectors must contain an equally probable additive inverse for each recombination differential. Since the additive inverse of $K\cdot(\mathbf{x}_{k,g} + \mathbf{x}_{l,g} - 2\cdot\mathbf{x}_{b,g})$ is $-K\cdot(\mathbf{x}_{k,g} + \mathbf{x}_{l,g} - 2\cdot\mathbf{x}_{b,g})$, sampling K from any distribution that is symmetric about zero ensures that each recombination differential can be paired with an equally probable and oppositely directed counterpart of the same magnitude. For example, K could be drawn from the bimodal distribution in which $+K$ and $-K$ are equally likely possibilities, or from a Cauchy distribution centered on zero. The next section shows how to harness the vector differences that drive mutation to distribute K symmetrically and transform $K\cdot\mathbf{R}^{(\chi 3)}$ into a center-symmetric distribution with a null expectation.

4 Naturally Distributed *K*

Just as adding a mutation differential to the base vector locates a mutant, projecting $\mathbf{m}_{k,l}$ onto $\mathbf{r}_{m,n}$ can locate a recombinant (Fig. 10). In this scheme, K is the ratio of the length of the projected mutation differential to the length of $\mathbf{r}_{m,n}$. More specifically, K is the inner (dot) product of $\mathbf{m}_{k,l}$ and $\mathbf{r}_{m,n}$ divided by the squared length of the recombination differential (Eq. 28). Population indices k, l, m, and n are randomly chosen except that $k \neq l$, $k \neq b$, $l \neq b$ and $m \neq n$, $m \neq b$, $n \neq b$.

$$K_{k,l,m,n} = \sqrt{D} \cdot \left(\frac{\mathbf{m}_{k,l} \bullet \mathbf{r}_{m,n}}{\left\| \mathbf{r}_{m,n} \right\|^2} \right). \tag{28}$$

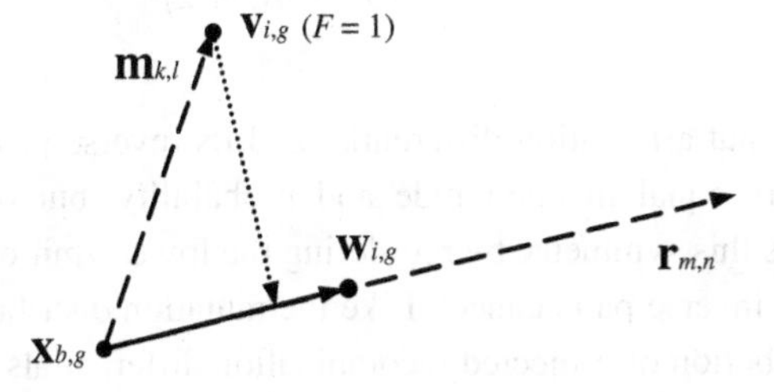

Fig. 10. The recombinant trial vector $\mathbf{w}_{i,g}$ is located along $\mathbf{r}_{m,n}$ at a distance from the base vector equal to $\sqrt{D}$ ($\sqrt{2}$ in this two-dimensional case) times the projection of the mutation differential $\mathbf{m}_{k,l}$ onto $\mathbf{r}_{m,n}$

The factor $\sqrt{D}$ compensates for the dimension-dependent foreshortening that occurs when vectors are projected. For example, the length of the component that a diagonal unit vector projects onto orthogonal coordinate axes is $1/\sqrt{D}$, so multiplying projected differentials by $\sqrt{D}$ ensures that recombination does not degenerate into a local search for large D.

In Eq. 29, $p_{k,l,m,n}$ is the probability that $\mathbf{m}_{k,l}$ will be projected onto $\mathbf{r}_{m,n}$. Since there are $Np^2 - 3Np + 2$ equally probable elements of $\mathbf{M}^{(\mu)}$ and just as many equally probably elements in $\mathbf{R}^{(\chi3)}$, there are $(Np^2 - 3Np + 2)^2$ equally probable ways to combine them, so the probability that any particular combination of mutation and recombination differentials is chosen is $1/(Np^2 - 3Np + 2)^2$.

$$p_{k,l,m,n} = \begin{cases} 0 \text{ if } m = n \vee m = b \vee n = b \vee k = l \vee k = b \vee l = b \\ \dfrac{1}{\left(Np^2 - 3Np + 2\right)^2} \text{ otherwise.} \end{cases} \tag{29}$$

Substituting the appropriate value for $p_{k,l,m,n}$ into the general expression for a distribution's expectation gives the expected value of the projected three-vector recombination distribution $\mathbf{E}(K_{k,l,m,n} \cdot \mathbf{R}^{(\chi3)})$ (Eq.30). To save space, Eq. 30 does not enumerate impossible degenerate combinations ($p_{k,l,m,n} = 0$).

$$\mathbf{E}\left(K_{k,l,m,n} \cdot \mathbf{R}^{(\chi3)}\right) = \sum_{n=1}^{Np}\sum_{m=1}^{Np}\sum_{l=1}^{Np}\sum_{k=1}^{Np} p_{k,l,m,n} \cdot K_{k,l,m,n} \cdot \mathbf{r}_{m,n}$$
$$= \frac{\displaystyle\sum_{\substack{n=1 \\ n \neq m,b}}^{Np} \sum_{\substack{m=1 \\ m \neq n,b}}^{Np} \sum_{\substack{l=1 \\ l \neq k,b}}^{Np} \sum_{\substack{k=1 \\ k \neq l,b}}^{Np} K_{k,l,m,n} \cdot \mathbf{r}_{m,n}}{\left(Np^2 - 3Np + 2\right)^2} \tag{30}$$

The components that a mutation differential and its inverse project along any line of recombination are equal in magnitude and probability, but oppositely directed. Equation 31 exploits this symmetry by reordering the lower limit of the inner summation so that additive inverse pairs cancel. Like the mutation distribution from which it is derived, the distribution of projected recombination differentials has a null expectation and is free of drift bias.

$$\mathbf{E}\left(K_{k,l,m,n}\cdot\mathbf{R}^{(\chi 3)}\right)=\frac{\sum\limits_{\substack{n=1\\ n\neq m,b}}^{Np}\sum\limits_{\substack{m=1\\ m\neq n,b}}^{Np}\sum\limits_{\substack{l=1\\ l\neq k,b}}^{Np}\sum\limits_{\substack{k=1\\ k\neq l,b}}^{Np}K_{k,l,m,n}\cdot\mathbf{r}_{m,n}}{\left(Np^2-3Np+2\right)^2}$$

$$=\frac{\sum\limits_{\substack{n=1\\ n\neq m,b}}^{Np}\sum\limits_{\substack{m=1\\ m\neq n,b}}^{Np}\sum\limits_{\substack{l=1\\ l\neq k,b}}^{Np}\sum\limits_{\substack{k=l+1\\ k\neq b}}^{Np}\left(K_{k,l,m,n}+K_{l,k,m,n}\right)\cdot\mathbf{r}_{m,n}}{\left(Np^2-3Np+2\right)^2}$$

$$=\frac{\sum\limits_{\substack{n=1\\ n\neq m,b}}^{Np}\sum\limits_{\substack{m=1\\ m\neq n,b}}^{Np}\sum\limits_{\substack{l=1\\ l\neq k,b}}^{Np}\sum\limits_{\substack{k=l+1\\ k\neq b}}^{Np}\sqrt{D}\cdot\left(\frac{\mathbf{m}_{k,l}\bullet\mathbf{r}_{m,n}+\mathbf{m}_{l,k}\bullet\mathbf{r}_{m,n}}{\left\|\mathbf{r}_{m,n}\right\|^2}\right)\cdot\mathbf{r}_{m,n}}{\left(Np^2-3Np+2\right)^2}$$

$$=\frac{\sum\limits_{\substack{n=1\\ n\neq m,b}}^{Np}\sum\limits_{\substack{m=1\\ m\neq n,b}}^{Np}\sum\limits_{\substack{l=1\\ l\neq k,b}}^{Np}\sum\limits_{\substack{k=l+1\\ k\neq b}}^{Np}\left(\mathbf{m}_{k,l}\bullet\mathbf{r}_{m,n}-\mathbf{m}_{k,l}\bullet\mathbf{r}_{m,n}\right)\cdot\frac{\sqrt{D}\cdot\mathbf{r}_{m,n}}{\left\|\mathbf{r}_{m,n}\right\|^2}}{\left(Np^2-3Np+2\right)^2}$$

$$=\frac{\sum\limits_{\substack{n=1\\ n\neq m,b}}^{Np}\sum\limits_{\substack{m=1\\ m\neq n,b}}^{Np}\sum\limits_{\substack{l=1\\ l\neq k,b}}^{Np}\sum\limits_{\substack{k=l+1\\ k\neq b}}^{Np}0\cdot\frac{\sqrt{D}\cdot\mathbf{r}_{m,n}}{\left\|\mathbf{r}_{m,n}\right\|^2}}{\left(Np^2-3Np+2\right)^2}=\mathbf{0} \tag{31}$$

In the projection method, $\mathbf{r}_{m,n}$ determines the direction of the recombination differential, while the *un-scaled* mutation differential $\mathbf{m}_{k,l}$ determines its length. Projecting the *scaled* mutation differential $F\cdot\mathbf{m}_{k,l}$ onto the medial line would, like the default equation linking F with K in the DE/ran/1/either-or algorithm, make recombination dependent on the mutation control parameter F.

In summary, the projection method unifies DE's approach to mutation and recombination by driving both operations with randomly sampled vector differences. Projecting un-scaled mutation differentials along medial lines not only frees the DE/ran/1/either-or algorithm's trial vector generating scheme from drift bias, it also eliminates the control variable K so that recombination's effect on the population can be independently controlled with a single variable p_F. Because F plays a secondary role as a control variable in the algorithm being developed here, the text that follows replaces p_F with p_μ to denote the probability of generating a mutant.

5 Drift Bias in Survivor Selection

Naturally distributing K restores a neutral bias to DE's trial vector generating function, but selection introduces its own drift bias because the potential trial vector distribution – although center-symmetric about the base vector – is not center-symmetric about the target vector. The move that occurs when a trial vector replaces a target vector defines a difference vector, $\mathbf{u}_{i,g} - \mathbf{x}_{i,g}$, that points from the target vector to the trial vector. The vector differences that are defined by a target vector and its potential adversaries comprise a set of *selection differentials* $\mathbf{S}$. Computing the expectation of the distribution of selection differentials reveals any drift bias in survivor selection.

Selection differentials can be partitioned into a set $\mathbf{S}^{(\mu)}$ that contains the $Np^3 - 3Np^2 + 2Np$ selection differentials that point from the target vector to all possible mutants and a set $\mathbf{S}^{(\chi3)}$ of $Np \cdot (Np^2 - 3Np + 2)^2$ selection differentials that point from the target vector to all possible three-vector recombinants generated by projection.

$$\mathbf{S}^{(\mu)} = \left[\mathbf{s}_{k,l,b}\right], \quad \mathbf{s}_{k,l,b} = \mathbf{x}_{b,g} + F \cdot \mathbf{m}_{k,l} - \mathbf{x}_{i,g}, \qquad k \neq l, k \neq b, l \neq b \tag{32}$$

$$\mathbf{S}^{(\chi3)} = \left[\mathbf{s}_{k,l,m,n,b}\right], \quad \mathbf{s}_{k,l,m,n,b} = \mathbf{x}_{b,g} + K_{k,l,m,n} \cdot \mathbf{r}_{m,n} - \mathbf{x}_{i,g}, \quad k \neq l, k \neq b, l \neq b \quad m \neq n, m \neq b, n \neq b \tag{33}$$

The probability that a trial vector will come from $\mathbf{S}^{(\mu)}$ is p_μ ($= p_F$) and the probability that it will come from $\mathbf{S}^{(\chi3)}$ is $1 - p_\mu$, so the expectation for the full distribution of selection differentials is

$$\mathbf{E}(\mathbf{S}) = p_\mu \cdot \mathbf{E}\left(\mathbf{S}^{(\mu)}\right) + \left(1 - p_\mu\right) \cdot \mathbf{E}\left(\mathbf{S}^{(\chi3)}\right). \tag{34}$$

5.1 Computing E(S$^{(\mu)}$)

If the target vector is included as a possible base vector, then there will be $Np^3 - 3Np^2 + 2Np$ possible mutants with which it can compete because each of the $Np^2 - 3Np + 2$ mutation differentials in $\mathbf{M}^{(\mu)}$ can be paired with one of Np possible base vectors. If both the mutation differential and the base vector are randomly chosen except $l \neq k$, $l \neq b$ and $k \neq b$, then all elements in $\mathbf{S}^{(\mu)}$ have the same probability

$$p_{k,l,b} = \begin{cases} 0 \ \text{ if } \ k = l \vee k = b \vee l = b, \\ \dfrac{1}{Np^3 - 3Np^2 + 2Np} \ \text{ otherwise.} \end{cases} \tag{35}$$

Substituting the appropriate values of $p_{k,l,b}$ into the general expression for a distribution's expectation eliminates those selection differentials that point from the target vector to null elements and to two-vector recombination differentials. Because the remaining selection differentials in $\mathbf{S}^{(\mu)}$ are equally probable, the expected vector value $\mathbf{E}(\mathbf{S}^{(\mu)})$ of their distribution reduces to a simple average, i.e., the sum of all $\mathbf{s}_{k,l,b}$, $k \neq l$, $k \neq b$, $l \neq b$ divided by $Np^3 - 3Np^2 + 2Np$. To save space, Eq. 36 does not list the null contributions from degenerate combinations.

$$\mathbf{E}\left(\mathbf{S}^{(\mu)}\right)=\sum_{b=1}^{Np}\sum_{l=1}^{Np}\sum_{k=1}^{Np}p_{k,l,b}\cdot\left(\mathbf{x}_{b,g}+F\cdot\mathbf{m}_{k,l}-\mathbf{x}_{i,g}\right)$$

$$=\frac{\sum_{b=1}^{Np}\sum_{\substack{l=1\\ l\neq k,b}}^{Np}\sum_{\substack{k=1\\ k\neq l,b.}}^{Np}\left(\mathbf{x}_{b,g}+F\cdot\mathbf{m}_{k,l}-\mathbf{x}_{i,g}\right)}{Np^3-3Np^2+2Np}. \tag{36}$$

In Eq. 37, the target vector $\mathbf{x}_{i,g}$ is a constant that can be extracted from the summations because its net contribution of $Np^3 - 3Np^2 + 2Np$ terms is ultimately divided by $Np^3 - 3Np^2 + 2Np$. Similarly, the base vector is a constant within the inner two summations and the constant Np can be factored out of the triple sum altogether. The inner double sum is just the expected value of $F\cdot\mathbf{M}^{(\mu)}$ which Eq. 16 showed to be $\mathbf{0}$. In the final step, averaging over all base vectors shows that the expected value of the $\mathbf{S}^{(\mu)}$ is the vector that points from the target vector to the population centroid.

$$\mathbf{E}\left(\mathbf{S}^{(\mu)}\right)=\frac{1}{Np}\cdot\sum_{b=1}^{Np}\left(\mathbf{x}_{b,g}+\frac{\sum_{\substack{l=1\\ l\neq k,b}}^{Np}\sum_{\substack{k=1\\ k\neq l,b}}^{Np}F\cdot\mathbf{m}_{k,l}}{Np^2-3Np+2}\right)-\mathbf{x}_{i,g},$$

$$\frac{1}{Np}\cdot\sum_{b=1}^{Np}\left(\mathbf{x}_{b,g}+\mathbf{E}\left(F\cdot\mathbf{M}^{(\mu)}\right)\right)-\mathbf{x}_{i,g}. \tag{37}$$

$$=\frac{1}{Np}\cdot\sum_{b=1}^{Np}\left(\mathbf{x}_{b,g}+\mathbf{0}\right)-\mathbf{x}_{i,g}=\langle\mathbf{x}\rangle_g-\mathbf{x}_{i,g}.$$

Simply put, each element of $\mathbf{S}^{(\mu)}$ is the sum of two components (Fig. 11). One component points from the target vector to a base vector, while the other points from that base vector to one of its mutants. Because the displacements from the base vector to center-symmetric pairs of its mutants cancel, the average of the selection differentials that point to mutants around a given base vector is just the vector that points from the target to that base vector. Furthermore, the average of the Np differentials that point from the target to all base vectors (including the target) is the vector that points from the target to the population's centroid, i.e., $\langle\mathbf{x}\rangle_g - \mathbf{x}_{i,g}$. As the next subsection shows, averaging all selection differentials to center-symmetrical pairs of recombinants also leaves a residual drift bias equal to $\langle\mathbf{x}\rangle_g - \mathbf{x}_{i,g}$.

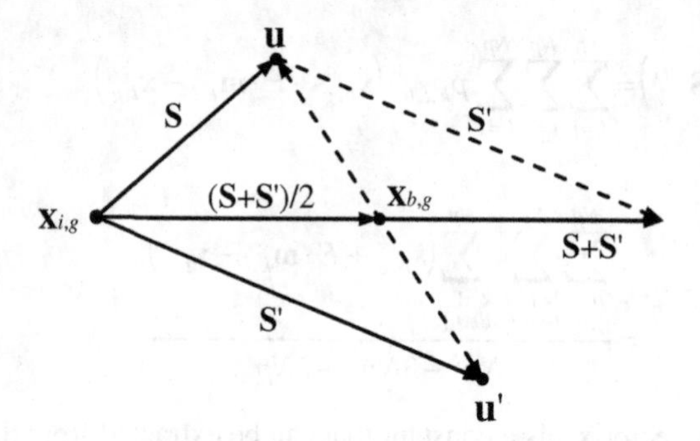

Fig. 11. Because selection differentials **S** and **S'** point to a center-symmetric pair of trial vectors **u** and **u'**, respectively, their components pointing from the base vector cancel, so their average is the vector that points from the target vector to the base vector, i.e., $(\mathbf{S} + \mathbf{S}')/2 = \mathbf{x}_{b,g} - \mathbf{x}_{i,g}$

5.2 Computing $E(S^{(\chi 3)})$

Each of the $Np^2 - 3Np + 2$ mutation differentials $\mathbf{m}_{k,l}$ can be projected onto any one of as many recombination differentials and each of these combined with one of Np possible base vectors. If $\mathbf{x}_{b,g}$, $\mathbf{m}_{k,l}$ and $\mathbf{r}_{m,n}$ are randomly chosen except that $k \neq l,\ k \neq b,\ l \neq b$ and $m \neq n,\ m \neq b,\ n \neq b$, then all elements in $\mathbf{S}^{(\chi 3)}$ have the same probability

$$p_{k,l,m,n,b} = \begin{cases} 0 \text{ if } k = l \vee k = b \vee l = b \vee m = n \vee m = b \vee n = b \\ \dfrac{1}{Np \cdot \left(Np^2 - 3Np + 2\right)^2} \text{ otherwise.} \end{cases} \tag{38}$$

Substituting the appropriate values of $p_{k,l,m,n,b}$ into the general expression for a distribution's expectation eliminates those selection differentials that point from the target vector to both the null element and the two-vector recombination differentials. Because the remaining selection differentials in $\mathbf{S}^{(\chi 3)}$ are equally probable, the expected vector value $\mathbf{E}(\mathbf{S}^{(\chi 3)})$ of their distribution reduces to a simple average, i.e., the sum of all $\mathbf{s}_{k,l,m,n,b}$, $k \neq l,\ k \neq b,\ l \neq b,\ m \neq n,\ m \neq b,\ n \neq b$, divided by $Np \cdot (Np^2 - 3Np + 2)^2$. For simplicity, Eq. 39 does not enumerate the components of $\mathbf{E}(\mathbf{S}^{(\chi 3)})$ corresponding to the forbidden combinations that do not contribute to the expected value.

$$\begin{aligned} \mathbf{E}\left(\mathbf{S}^{(\chi 3)}\right) &= \sum_{b=1}^{Np}\sum_{n=1}^{Np}\sum_{m=1}^{Np}\sum_{l=1}^{Np}\sum_{k=1}^{Np} p_{k,l,m,n,b} \cdot \left(\mathbf{x}_{b,g} + K_{k,l,m,n} \cdot \mathbf{r}_{m,n} - \mathbf{x}_{i,g}\right) \\ &= \frac{\displaystyle\sum_{b=1}^{Np}\sum_{\substack{n=1 \\ n \neq m,b}}^{Np}\sum_{\substack{m=1 \\ m \neq n,b}}^{Np}\sum_{\substack{l=1 \\ l \neq k,b}}^{Np}\sum_{\substack{k=1 \\ k \neq l,b}}^{Np} \left(\mathbf{x}_{b,g} + K_{k,l,m,n} \cdot \mathbf{r}_{m,n} - \mathbf{x}_{i,g}\right)}{Np \cdot \left(Np^2 - 3Np + 2\right)^2}. \end{aligned} \tag{39}$$

Once again, the target vector $\mathbf{x}_{i,g}$ is a constant that can be extracted from the summations because its net contribution of $Np \cdot (Np^2 - 3Np + 2)^2$ terms is divided by the same amount (Eq. 40). Similarly, the base vector is a constant within the inner four summations and the constant Np can be factored out of the quintuple sum altogether. The inner quadruple sum is now just the expected value of $K_{k,l,m,n}\cdot\mathbf{R}^{(\chi3)}$ which Eq. 31 showed to be $\mathbf{0}$. In the final step, averaging over all base vectors shows that, like $\mathbf{E}(\mathbf{S}^{(\mu)})$, the expected value of $\mathbf{S}^{(\chi3)}$ is the vector that points from the target vector to the population's centroid.

$$\mathbf{E}\left(\mathbf{S}^{(\chi3)}\right)=\frac{1}{Np}\cdot\sum_{b=1}^{Np}\left(\mathbf{x}_{b,g}+\frac{\sum\limits_{\substack{n=1\\n\neq m,b}}^{Np}\sum\limits_{\substack{m=1\\m\neq n,b}}^{Np}\sum\limits_{\substack{l=1\\l\neq k,b}}^{Np}\sum\limits_{\substack{k=1\\k\neq l,b}}^{Np}K_{k,l,m,n}\cdot\mathbf{r}_{m,n}}{\left(Np^2-3Np+2\right)^2}\right)-\mathbf{x}_{i,g}$$

$$=\frac{1}{Np}\cdot\sum_{b=1}^{Np}\left[\mathbf{x}_{b,g}+\mathbf{E}\left(K_{k,l,m,n}\cdot\mathbf{R}^{(\chi3)}\right)\right]-\mathbf{x}_{i,g} \tag{40}$$

$$=\frac{1}{Np}\cdot\sum_{b=1}^{Np}\left(\mathbf{x}_{b,g}+\mathbf{0}\right)-\mathbf{x}_{i,g}=\langle\mathbf{x}\rangle_g-\mathbf{x}_{i,g}.$$

5.3 Computing E(S)

The elements of $\mathbf{S}^{(\mu)}$ are chosen with probability p_μ, while those from $\mathbf{S}^{(\chi3)}$ are chosen with probability $1-p_\mu$, so the expected value for the full distribution of selection differentials is

$$\begin{aligned}\mathbf{E}(\mathbf{S})&=p_\mu\cdot\mathbf{E}\left(\mathbf{S}^{(\mu)}\right)+\left(1-p_\mu\right)\cdot\mathbf{E}\left(\mathbf{S}^{(\chi3)}\right)\\&=p_\mu\cdot\left(\langle\mathbf{x}\rangle_g-\mathbf{x}_{i,g}\right)+\left(1-p_\mu\right)\cdot\left(\langle\mathbf{x}\rangle_g-\mathbf{x}_{i,g}\right)\\&=\left(\langle\mathbf{x}\rangle_g-\mathbf{x}_{i,g}\right)\cdot\left(p_\mu+1-p_\mu\right)\\&=\langle\mathbf{x}\rangle_g-\mathbf{x}_{i,g}.\end{aligned}\tag{41}$$

Because $\mathbf{E}(\mathbf{S}^{(\chi3)})$ and $\mathbf{E}(\mathbf{S}^{(\mu)})$ are equal, $\mathbf{E}(\mathbf{S})$ is independent of p_μ.

5.4 Drift-Free Selection

If base vectors could be center-symmetrically distributed about each target vector, then there would be no drift bias due to survivor selection, but if only population vectors can serve as base vectors, then this will not be possible since the same population (of base vectors) would need to be center-symmetrically distributed around *Np* different centers (target vectors). If, however, each target vector serves as its own base vector, drift bias will be eliminated because trial vectors will be center-symmetrically distributed about the target vector. For example, if $\mathbf{x}_{i,g}$ is substituted for $\mathbf{x}_{b,g}$ in the last line of Eq. 41, then the sum divided by *Np* yields $\mathbf{x}_{i,g}$, not $<\mathbf{x}>_g$, so the expected value becomes null. With the target vector as the center of the trial vector distribution, $\mathbf{S}^{(\mu)} = F \cdot \mathbf{M}^{(\mu)}$ and $\mathbf{S}^{(\chi3)} = K_{k,l,m,n} \cdot \mathbf{R}^{(\chi3)}$, i.e., the selection differentials become identical to either mutation or three-vector recombination differentials. Since both generating distributions are center-symmetrical, so too are the corresponding selection distributions.

By centering the trial vector distribution on the target vector, drift-free DE's selection operator improves the population without homogenizing it. For example, when both three-vector arithmetic recombination and differential mutation are turned off (i.e., when $p_\mu = 1$ and $F = 0$), the DE/ran/1/either-or algorithm's selection operator compares the target vector's objective function value to that of a randomly selected base vector $\mathbf{x}_{r1,g}$. After what is known as the takeover time, selection operating alone will have filled the population with *Np* copies of the best vector, i.e., the population will have become homogenous through the action of selection alone (Price and Rönkkönen 2006). Under the same circumstances ($p_\mu = 1$, $F = 0$), drift-free DE pits the target vector against itself, so the population's composition never changes. Without selection's constant background rate of homogenization, only p_μ controls the rate at which the population coalesces.

6 Drift-Free DE

Modifying the DE/ran/1/either-or algorithm to be drift free is straightforward. If $\mathrm{U}_i(0,1) \le p_\mu$, then mutation adds $F \cdot \mathbf{m}_{r1,r2}$ to the *target* vector; otherwise, recombination adds the dimension-compensated component of $\mathbf{m}_{r1,r2}$ that points along $\mathbf{r}_{r3,r4}$.

$$\mathbf{u}_{i,g} = \mathbf{x}_{i,g} + \begin{cases} F \cdot \mathbf{m}_{r1,r2} \text{ if } \mathrm{U}_i(0,1) \le p_\mu \\ \sqrt{D} \cdot \left(\dfrac{\mathbf{m}_{r1,r2} \bullet \mathbf{r}_{r3,r4}}{\left\| \mathbf{r}_{r3,r4} \right\|^2} \right) \cdot \mathbf{r}_{r3,r4} \text{ otherwise.} \end{cases} \tag{42}$$

Population indices *r*1, *r*2, *r*3 and *r*4 are randomly selected except that $r1 \ne r2$, $r1 \ne i$, $r2 \ne i$ and $r3 \ne r4$, $r3 \ne i$, $r4 \ne i$.

If $\mathbf{e}_{r3,r4}$ is defined as the normalized version of the recombination differential $\mathbf{r}_{r3,r4}$,

$$\mathbf{e}_{r3,r4} \equiv \frac{\mathbf{r}_{r3,r4}}{\left\| \mathbf{r}_{r3,r4} \right\|} \tag{43}$$

then

$$\left(\frac{\mathbf{m}_{r1,r2} \bullet \mathbf{r}_{r3,r4}}{\left\|\mathbf{r}_{r3,r4}\right\|^2}\right) \cdot \mathbf{r}_{r3,r4} = \left(\mathbf{m}_{r1,r2} \bullet \frac{\mathbf{r}_{r3,r4}}{\left\|\mathbf{r}_{r3,r4}\right\|}\right) \cdot \frac{\mathbf{r}_{r3,r4}}{\left\|\mathbf{r}_{r3,r4}\right\|}$$

$$= \left(\mathbf{m}_{r1,r2} \bullet \mathbf{e}_{r3,r4}\right) \cdot \mathbf{e}_{r3,r4}, \tag{44}$$

and the drift-free DE generating function can be more compactly written as

$$\mathbf{u}_{i,g} = \mathbf{x}_{i,g} + \begin{cases} F \cdot \mathbf{m}_{r1,r2} \text{ if } \mathrm{U}_i(0,1) \le p_\mu \\ \sqrt{D} \cdot \left(\mathbf{m}_{r1,r2} \bullet \mathbf{e}_{r3,r4}\right) \cdot \mathbf{e}_{r3,r4} \text{ otherwise.} \end{cases} \tag{45}$$

This formulation emphasizes that in the projection method, recombination only determines the direction in which the recombinant is placed. When combined with selection, the complete description for the drift-free DE algorithm becomes:

$$\mathbf{u}_{i,g} = \mathbf{x}_{i,g} + \begin{cases} F \cdot \mathbf{m}_{r1,r2} \text{ if } \mathrm{U}_i(0,1) \le p_\mu \\ \sqrt{D} \cdot \left(\mathbf{m}_{r1,r2} \bullet \mathbf{e}_{r3,r4}\right) \cdot \mathbf{e}_{r3,r4} \text{ otherwise.} \end{cases}$$

$$\mathbf{x}_{i,g+1} = \begin{cases} \mathbf{u}_{i,g} \text{ if } f\left(\mathbf{u}_{i,g}\right) \le f\left(\mathbf{x}_{i,g}\right) \\ \mathbf{x}_{i,g} \text{ otherwise.} \end{cases}$$

$$\mathbf{m}_{r1,r2} = \mathbf{x}_{r1,g} - \mathbf{x}_{r2,g}, \quad \mathbf{e}_{r3,r4} = \frac{\mathbf{x}_{r3,g} + \mathbf{x}_{r4,g} - 2\mathbf{x}_{i,g}}{\left\|\mathbf{x}_{r3,g} + \mathbf{x}_{r4,g} - 2\mathbf{x}_{i,g}\right\|},$$

$$i = 1,2,\ldots,Np, \quad r1, r2, r3, r4 \in [1,2,\ldots,Np], \tag{46}$$

$$r1 \ne r2, r1 \ne i, r2 \ne i, r3 \ne r4, r3 \ne i, r4 \ne i,$$

$$F \in (0,1+), \quad p_\mu \in [0,1].$$

Figure 12 presents pseudo-code for a drift-free DE algorithm that terminates when reaching a preset maximum number of generations, but other halting criteria are possible.

In drift-free DE, each operation has a unique function: mutation explores, recombination homogenizes and selection improves. The drift-free DE algorithm achieves this goal with very few assumptions. In particular, the algorithm assumes that trial vectors are three-vector linear combinations, since three vectors are needed to implement differential mutation and four are more than is necessary. One of the vectors must be the target vector to eliminate selection drift bias and combinations are

```
for (i = 1; i ≤ Np; i = i + 1) // Initialize the population
   for (j = 1; j ≤ D; j = j + 1) x_{i,j,0} = x_j^(lower) + U(0,1)_j· (x_j^(upper) – x_j^(lower));
   end for
   f(i) = f(x_{i,0}); // Evaluate and save f(x_i)
end for

for (g = 1; g ≤ g_max; g = g + 1) // Begin generation loop
   for (i = 1; i ≤ Np; i = i + 1) // Generate Np trial vectors
      do r1 = floor(Np·U(0,1)) + 1; while (r1 = i);
      do r2 = floor(Np·U(0,1)) + 1; while (r2 = i or r2 = r1);
      if (U(0,1)_i ≤ p_μ) // Mutate...
      {
         for (j = 1; j ≤ D; j = j + 1) u_{j,i,g} = x_{j,i,g} + F·(x_{j,r1,g} – x_{j,r2,g});
         end for;
      }
      else // ...or recombine
      {
         do // Compute inner product and K
            do r3 = floor(Np·U(0,1)_i) + 1; while (r3 = i);
            do r4 = floor(Np·U(0,1)_i) + 1; while (r4 = i or r4 = r3);
            sum1 = 0; sum2 = 0;
            for (j = 1; j ≤ D; j = j + 1)
               d1 = x_{j,r1,g} – x_{j,r2,g};
               d2 = x_{j,r3,g} + x_{j,r4,g} – 2x_{j,i,g};
               sum1 = sum1 + d1*d2;
               sum2 = sum2 + d2*d2;
            end for
         while (sum2 = 0); // If ||x_{r3,g} + x_{r4,g} – 2x_{i,g}||^2 = 0, reselect r3 and r4
         K = sum1/sum2;
         for (j = 1; j ≤ D; j = j + 1) u_{j,i,g} = x_{j,i,g} + K· (x_{j,r3,g} + x_{j,r4,g} – 2x_{j,i,g});
         end for
      }
   end for
   for (i = 1; i ≤ Np; i = i +1) // Select new population
      if (f(u_{i,g}) ≤ f(i) // Evaluate the trial vector and compare with target value
      {
         for (j = 1; j ≤ Np; j = j + 1) x_{j,i,g} = u_{j,i,g}; // Replace inferior target
         end for
         f(i) = f(u_{i,g}); // Store trial vector function value
      }
   end for
end for // End
```

Fig. 12. Pseudo-code for the drift-free DE algorithm

normalized to eliminate scale bias. Any linear combination of the target and two randomly chosen vectors can be decomposed into complimentary differential mutation and three-vector recombination operations, controlled by F and K respectively, which are interleaved with probability p_μ to minimize control parameter interaction. Unless K is distributed center-symmetrically about zero, the recombination operator will be biased, but projecting randomly selected, un-scaled mutation differentials onto randomly chosen (medial) lines of recombination eliminates both drift bias from the trial vector generating distribution and K as a control variable. With the aid of a test bed of scalable benchmark functions, the following section compares the algorithmic performance of drift-free DE to that of the DE/ran/1/either-or algorithm.

7 Benchmarking Performance

The comparison in (Chap. 3, Price et al. 2005) found classic DE to be faster and more reliable than DE/ran/1/either-or when optimizing separable functions, but less effective than DE/ran/1/either-or when the objective function was both non-separable and multi-modal. This section compares the algorithmic performance of drift-free DE to that of the DE/ran/1/either-or algorithm. There are only two aspects in which the DE/ran/1/either-or algorithm and the drift-free DE algorithm listed in Fig. 12 differ. In the DE/ran/1/either-or algorithm, K is a constant and the base vector is randomly chosen (except for being distinct from both difference vectors and the target vector), while in drift-free DE, K is a distributed value and the base vector is the target vector. For the DE/ran/1/either-or algorithm, $K = 1/3$ for all experiments in this study – a setting that mimics a contraction operation in a 2-dimensional Nelder–Mead algorithm (Nelder and Mead 1965).

All trials were run with $F = 1$ not only to keep the number of control variable combinations manageable, but also because preliminary experiments indicated that $F = 1$ is effective on multimodal functions. Since K is either preset ($K = 1/3$) or determined by projection, the DE algorithms being tested here are particularly simple since their evolution only depends on two variables: Np and p_μ. The test bed chosen to benchmark performance consists of fourteen scalable objective functions.

7.1 The Test Bed: Scalable Benchmark Functions

Because their dimension is variable, the scalable functions in this test bed can benchmark how an algorithm and its control parameters depend on problem size. Of the fourteen benchmark functions, twelve are multimodal and two are unimodal. Although some functions in this test bed are separable, no attempt has been made to exploit this special knowledge (e.g., by modifying only one target parameter per evaluation as is possible in classic DE when $Cr = 0$).

Included with each function description is a tolerance ε chosen so that $\mathbf{x}$ lies within the basin of attraction containing the optimum $\mathbf{x}^*$ if $f(\mathbf{x}) \leq f(\mathbf{x}^*) + \varepsilon$. As such, ε determines how close a vector's objective function value must be to the minimum before the optimization can be considered to be a success. Each function description also includes lower and upper initial parameter bounds (e.g., $x_j \in [-100,100]$ for $f_1(\mathbf{x})$)

In this study, initial parameter values were distributed with random uniformity over the specified range.

$$x_{j,i,g=0} = \mathrm{U}(0,1)_j \cdot \left(x_j^{(\text{upper})} - x_j^{(\text{lower})}\right) \tag{47}$$

For constrained functions, these bounds remain active during the optimization and any constrained trial parameter $u_{j,i,g}$ that exceeds its bound is reset to a randomly chosen point between the bound that is violated and the base vector's corresponding parameter value. For the DE/ran/1/either-or algorithm, the base vector is a randomly chosen population vector, whereas for drift-free DE, it is the target vector (*base* = *i*).

$$u_{j,i,g} = x_{j,base,g} + \mathrm{U}(0,1)_j \cdot \begin{cases} \left(x_j^{(\text{upper})} - x_{j,base,g}\right) \text{ if } u_{j,i,g} > x_j^{(\text{upper})} \\ \left(x_j^{(\text{lower})} - x_{j,base,g}\right) \text{ if } u_{j,i,g} < x_j^{(\text{lower})} \end{cases} \tag{48}$$

The test bed's first two functions – the multidimensional sphere $f_1(\mathbf{x})$ (Eq. 49) and Schwefel's ridge $f_2(\mathbf{x})$ (Eq. 50) – are unimodal for all D. The sphere provides a baseline for the minimum effort needed to optimize a D-dimensional function, whereas the ridge's diagonally oriented ellipsoidal contours reveal the extent to which conditioning (disparate optimal parameter magnitudes) and coordinate rotation affect a search algorithm's efficiency. The third function $f_3(\mathbf{x})$, (the extended, or modified Rosenbrock function, Eq. 51) is unimodal when $D \leq 3$, but reported to be bimodal for $4 \leq D \leq 30$ (Shang and Qiu 2006). Optimizing its banana-shaped basin of attraction has traditionally proven challenging.

Sphere

$$\begin{gathered} f_1 = \sum_{j=1}^{D} x_j^2, \\ \mathbf{x}^* = (0,0,\ldots,0),\ f_1\left(\mathbf{x}^*\right) = 0,\ \varepsilon = 1.0\times10^{-6},\ x_j \in [-100,100] \end{gathered} \tag{49}$$

Schwefel's Ridge

$$\begin{gathered} f_2 = \sum_{k=1}^{D}\left(\sum_{j=1}^{k} x_j\right)^2, \\ \mathbf{x}^* = (0,0,\ldots,0),\ f_2\left(\mathbf{x}^*\right) = 0,\ \varepsilon = 1.0\times10^{-6},\ x_j \in [-100,100] \end{gathered} \tag{50}$$

Extended Rosenbrock

$$\begin{gathered} f_3 = \sum_{j=1}^{D-1}\left[100\left(x_{j+1} - x_j^2\right)^2 + \left(x_j - 1\right)^2\right], \\ \mathbf{x}^* = (1,1,\ldots,1),\ f_3\left(\mathbf{x}^*\right) = 0,\ \varepsilon = 1.0\times10^{-6},\ x_j \in [-30,30] \end{gathered} \tag{51}$$

Functions $f_4(\mathbf{x})$–$f_{14}(\mathbf{x})$ are multimodal. The function landscapes for $f_4(\mathbf{x})$ (Ackley, Eq. 52), $f_5(\mathbf{x})$ (Griewangk, Eq. 53) and $f_6(\mathbf{x})$ (Rastrigin, Eq. 54) are similar in that the global optimum for each lies at the bottom of a large bowl, or funnel, whose surface is covered with many small dents (local optima). Whitley's function $f_7(\mathbf{x})$ (Eq. 55) is a composite of the two-dimensional Rosenbrock and one-dimensional Griewangk functions with a full matrix expansion that includes diagonal terms (see (Whitley et al. 1996) for details on function composition and matrix expansion). Its function landscape resembles the banana-shaped Rosenbrock function at large scale and the highly multimodal Griewangk function at small scale. The Odd Square $f_8(\mathbf{x})$ (Eq. 56) has a function landscape which, in two dimensions, resembles square ripples on a pond (Equation 56 corrects the erroneous description for the Odd Square given in (Price et al., 2005)). The term $\max(x_j - b_j)^2$ in $f_8(\mathbf{x})$ is the largest squared coordinate difference between $\mathbf{x}$ and $\mathbf{b}$.

Ackley

$$f_4(\mathbf{x}) = -20\exp\left(-0.2\sqrt{\frac{1}{D}\cdot\sum_{j=1}^{D}x_j^2}\right) - \exp\left(\frac{1}{D}\sum_{j=1}^{D}\cos(2\pi\cdot x_j)\right) + 20 + e,$$
$$\mathbf{x}^* = (0,\ldots0),\ f_4(\mathbf{x}^*) = 0,\ \varepsilon = 1.0\times10^{-6},\ x_j \in [-30,30]. \tag{52}$$

Griewangk

$$f_5(\mathbf{x}) = \frac{1}{4000}\sum_{j=1}^{D}x_j^2 - \prod_{j=1}^{D}\cos\left(\frac{x_j}{\sqrt{j}}\right) + 1,$$
$$\mathbf{x}^* = (0,\ldots0),\ f_5(\mathbf{x}^*) = 0,\ \varepsilon = 1.0\times10^{-6},\ x_j \in [-600,600]. \tag{53}$$

Rastrigin

$$f_6(\mathbf{x}) = \sum_{j=1}^{D}\left[x_j^2 - 10\cos(2\pi\cdot x_j) + 10\right],$$
$$\mathbf{x}^* = (0,\ldots0),\ f_6(\mathbf{x}^*) = 0,\ \varepsilon = 1.0\times10^{-6},\ x_j \in [-5.12,5.12]. \tag{54}$$

Whitley

$$f_7(\mathbf{x}) = \sum_{k=1}^{D}\sum_{j=1}^{D}\left[\frac{y_{j,k}^2}{4000} - \cos(y_{j,k}) + 1\right],\quad y_{j,k} = 100\left(x_j - x_k^2\right)^2 + (x_k - 1)^2,$$
$$\mathbf{x}^* = (1,\ldots1),\ f_7(\mathbf{x}^*) = 0,\ \varepsilon = 1.0\times10^{-6},\ x_j \in [-100,100]. \tag{2.55}$$

Odd Square

$$f_8(\mathbf{x}) = -\exp\left(\frac{-d}{2\pi}\right)\cdot\cos(\pi\cdot d)\cdot\left(1 + \frac{0.2h}{d + 0.01}\right),$$
$$d = \sqrt{D\cdot\max(x_j - b_j)^2},\quad h = \sqrt{\sum_{j=1}^{D}(x_j - b_j)^2},\quad D \le 20,$$
$$\mathbf{b} = [1, 1.3, 0.8, -0.4, -1.3, 1.6, -0.2, -0.6, 0.5, 1.4,$$
$$1, 1.3, 0.8, -0.4, -1.3, 1.6, -0.2, -0.6, 0.5, 1.4\,], \tag{56}$$
$$\mathbf{x}^* = \text{many } \mathbf{x}^* \text{ near } \mathbf{b},\ f_8(\mathbf{x}^*) = -1.14383,\ \varepsilon = 0.01,\ x_j \in [-5\pi,5\pi].$$

Functions $f_9(\mathbf{x})$ (Schwefel, Eq. 57), $f_{10}(\mathbf{x})$ (Rana, Eq. 58) and $f_{11}(\mathbf{x})$ (Weierstrass Eq. 59) are bound constrained problems. Both the Schwefel and Rana functions have optima located in the corner of their bounding hypercubes and both functions are highly multimodal.

Schwefel

$$f_9(\mathbf{x}) = \frac{1}{D}\cdot\sum_{j=1}^{D}\left[-x_j\cdot\sin\left(\sqrt{|x_j|}\right)\right],$$
$$\mathbf{x}^* = (s,s...,s),\ s = 420.969,\ f_9(\mathbf{x}^*) = -418.983,\ \varepsilon = .01,\ x_j \in [-500,500]. \quad (57)$$

Rana

$$f_{10}(\mathbf{x}) = \sum_{j=1}^{D}\left[x_j\cdot\sin\left(\sqrt{|a-x_j|}\right)\cdot\cos\left(\sqrt{|a+x_j|}\right) + a\cdot\cos\left(\sqrt{|a-x_j|}\right)\cdot\sin\left(\sqrt{|a+x_j|}\right)\right],\ a = x_{(j+1)\bmod D}+1, \quad (58)$$
$$\mathbf{x}^* = (-512,-512,...,-512),\ f_{10}(\mathbf{x}^*) = -511.708,\ \varepsilon = 0.01,\ x_j \in [-512,511].$$

Weierstrass

$$f_{11}(\mathbf{x}) = \sum_{j=1}^{D}\sum_{k=0}^{20}\left[a^k\cdot\cos\left(b^k\cdot(x_j+0.5)\right)\right] - D\cdot\sum_{k=0}^{20}\left[a^k\cdot\cos\left(\pi\cdot b^k\right)\right],\ a = 0.5, b = 3,$$
$$\mathbf{x}^* = (0,0,...,0),\ f_{11}(\mathbf{x}^*) = 0,\ \varepsilon = 1.0\times10^{-6},\ x_j \in [-0.5,0.5]. \quad (59)$$

Storn's Chebyshev polynomial fitting problem $f_{12}(\mathbf{x})$ (Eq. 60), Hilbert's function $f_{13}(\mathbf{x})$ (Eq. 61) and the Lennard-Jones potential energy function $f_{14}(\mathbf{x})$ (Eq. 62) are only defined for $D = 2n + 1$, $D = n^2$ and $D = 3n$, respectively, where $n = 1,2,\ldots$. Storn's Chebyshev problem asks for the coefficients of a (Chebyshev) polynomial $T(x)$ whose value varies between [0,1] over the range $x = [-1,1]$. The function itself is a sum of squared error terms sampled at $32D + 1$ regularly spaced intervals and also at $x = \pm1.2$.

Storn's Chebyshev

$$f_{12}(\mathbf{x}) = p_1 + p_2 + p_3,$$
$$p_1 = \begin{cases}(u-d)^2 & \text{if } u < d\\ 0 & \text{otherwise}\end{cases},\quad u = \sum_{j=1}^{D} x_j\cdot(1.2)^{D-j},$$
$$p_2 = \begin{cases}(v-d)^2 & \text{if } v < d\\ 0 & \text{otherwise}\end{cases},\quad v = \sum_{j=1}^{D} x_j\cdot(-1.2)^{D-j},$$
$$p_3 = \sum_{k=0}^{m} r_k,\quad r_k = \begin{cases}(w_k-1)^2 & \text{if } w_k > 1\\ (w_k+1)^2 & \text{if } w_k < -1,\\ 0 & \text{otherwise}\end{cases}\quad w_k = \sum_{j=1}^{D} x_j\cdot\left(\frac{2k}{m}-1\right)^{D-j}, \quad (60)$$
$$m = 32D,\ D = 2n+1,\ n = 1,2,...,$$
$$f_{12}(\mathbf{x}^*) = 0,\ \varepsilon = 1.0\times10^{-8},\ x_j \in \left[-2^D, 2^D\right].$$

The constant d and $\mathbf{x}^*$ for Storn's Chebyshev

D	$d = T_{D-1}(1.2)$	$\mathbf{x}^*$
3	1.88000	−1,0,2
5	6.06880	1,0,−8,0,8
7	20.9386	−1,0,18,0,−48,0,32
9	72.6607	1,0,−32,0,160,0,−256,0,128
11	252.265	−1,0,50,0,−400,0,1120,0,−1280,0,512
13	875.857	1,0,−72,0,840,0,−3584,0,6912,0,−6144,0,2048
15	3040.96	−1,0,98,0,−1568,0,9408,0,−26880,0,39424,0,−28672,0,8192

Like Storn's Chebyshev problem, inverting the Hilbert matrix tests an algorithm's ability to efficiently search highly conditioned function landscapes. The matrix $\mathbf{W}$ is the difference between the identity matrix $\mathbf{I}$ and the product of the Hilbert matrix $\mathbf{H}$ and its proposed inverse $\mathbf{Z}$, so summing the absolute value of its elements $w_{j,k}$ measures the error due to $\mathbf{Z}$'s approximate value. The relation $z_{j,k} = x_{j+n\cdot(k-1)}$, $j,k = 1,2,\ldots,n$, $D = n^2$ maps trial vector parameters to matrix elements.

Hilbert

$$f_{13}(\mathbf{x}) = \sum_{k=1}^{n}\sum_{j=1}^{n}\left|w_{j,k}\right|,\quad \left[w_{j,k}\right] = \mathbf{W} = \mathbf{HZ} - \mathbf{I},$$

$$\mathbf{H} = \left[h_{j,k}\right],\ h_{j,k} = \frac{1}{j+k-1},\ \mathbf{Z} = \left[z_{j,k}\right],\ z_{j,k} = x_{j+n\cdot(k-1)},$$

$$\mathbf{I} = \begin{bmatrix} 1 & 0 & \cdots & 0 \\ 0 & 1 & \cdots & 0 \\ \vdots & \vdots & & \vdots \\ 0 & 0 & \cdots & 1 \end{bmatrix},$$

$$f_{13}(\mathbf{x}^*) = 0,\ \varepsilon = 1.0\times10^{-8},\ x_j \in \left[-2^D, 2^D\right],\ D = n^2,\ n = 1,2,\ldots, \tag{61}$$

$$\mathbf{Z}^* = \begin{bmatrix} 4 & -6 \\ -6 & 12 \end{bmatrix} (D=4),\ \mathbf{Z}^* = \begin{bmatrix} 9 & -36 & 30 \\ -36 & 192 & -180 \\ 30 & -180 & 180 \end{bmatrix} (D=9),$$

$$\mathbf{Z}^* = \begin{bmatrix} 16 & -120 & 240 & -140 \\ -120 & 1200 & -2700 & 1680 \\ 240 & -2700 & 6480 & -4200 \\ -140 & 1680 & -4200 & 2800 \end{bmatrix} (D=16).$$

The Lennard-Jones function $f_{14}(\mathbf{x})$ (Eq. 62) simulates the atomic potential arising from a cluster of atoms in three-dimensional space. As such, this function is only

defined for $D = 3n$, $n = 1,2,\ldots$. Minimizing cluster energy is an important problem made difficult by the fact that it is fully parameter dependent (a "many-body" problem). Differing spatial symmetries make some clusters harder to optimize than others. Since the position and orientation of the clusters are not specified, optimal parameter values are not unique. Optima for Lennard-Jones and other clusters are available at the Cambridge cluster database:
http://www-wales.ch.cam.ac.uk/~jon/structures/LJ/ tables.150.html

Lennard-Jones

$$f_{14}(\mathbf{x}) = \sum_{i=1}^{n-1}\sum_{j=i+1}^{n}\left(\frac{1}{d_{i,j}^2} - \frac{2}{d_{i,j}}\right),\ d_{i,j} = \left[\sum_{k=1}^{3}\left(x_{3i+k} - x_{3j+k}\right)^2\right]^3, \tag{62}$$

D	$f_{14}(\mathbf{x}^*)$
3	−1
6	−3
9	−6
12	−12.712062
15	−16.505384

$$D = 3n,\ n = 1,2,\ldots,\ \varepsilon = 0.0001,\ x_j \in [-2,2],$$

7.2 Measuring Performance

This study measured algorithmic performance by the AFES, or average number of function evaluations per success. A trial is counted as a success if the objective function value of the population's best vector falls to within ε of the objective function's known minimum $f(\mathbf{x}^*)$ before reaching the allotted maximum number of generations g_{max}. In this study, $g_{max} = 4000D$. If no trials are successful, then the AFES is undefined.

Measuring algorithmic performance is a multi-objective optimization problem in which speed and reliability are conflicting objectives. For example, reducing *Np* lowers not only the AFES, but also the probability of success. To allow for a fair comparison, the experiments in this section report the AFES at a constant probability of success. In particular, the lowest AFES (AFES*) for which ten-out-of-ten trials were successful and the *Np* at which it occurred (*Np**) were recorded at $p_\mu = 0.25, 0.5, 0.75, 0.875$ for each valid combination of objective function and dimension $2 \le D \le 15$ ($D_{max} = 16$ for Hilbert). Populations ranged from $2D$ (empirically shown to be a minimum for $f_1(\mathbf{x})$and $f_2(\mathbf{x})$ in (Price and Rönkkönen 2006)) up to $16D^2$ in six steps, i.e., $Np = \text{floor}(2D\cdot(8D)^{k/6})$, $k = 0,1,\ldots,6$. The "floor()" operator returns the integer part of its argument.

For example, the AFES for $f_1(\mathbf{x})$ was computed as a ten-trial average for each combination of D, p_μ and Np, with each trial running for no more than $g_{max} = 4000D$

generations. With both D and p_μ held constant, Np was varied over the range $[2D,\ldots,16D^2]$ after which the lowest AFES for which all ten trials were successful (i.e., AFES*) and the Np at which it occurred (i.e., Np^*) were recorded. This experiment (varying Np to find Np^* and AFES* with D and p_μ held constant) was repeated three more times and the results for AFES* and Np^* were averaged. Once the four-trial average for Np^* and AFES* for $f_1(\mathbf{x})$ was determined for all D, p_μ was incremented by 0.25 and the entire process was repeated. Once the four-trial averages of AFES* and Np^* were computed for all values of D and p_μ for $f_1(\mathbf{x})$, $f_2(\mathbf{x})$ was evaluated and so on. This exhaustive search process has the nested loop structure:

```
fk(x): k = 1,2,..14
{
   pμ = 0.25,0.5,0.75,0.875
   {
      D = 2,3,...,15 (16)  // Not all functions are defined for all D in this range
      {
         q = 1,2,3,4       // Compute the four-trial average (AFES*, Np*)
         {
            Np = 2D,...,16D^2   // Compute (AFES_q*, Np_q*)
            {
               t = 1,2,...,10     // Run ten trials at constant k, pμ, D, q and Np
            }
         }
      }
   }
}
```

7.3 Results

Figures 12–15 plot AFES* *vs.* D for p_μ= 0.25, 0.5, 0.75, and 0.875 respectively, for the drift-free DE algorithm. Their non-consecutive dimension means that results for $f_{12}(\mathbf{x})$, $f_{13}(\mathbf{x})$ and $f_{14}(\mathbf{x})$ appear in these plots as isolated points. Figures 16–19 plot the corresponding Np^* as a function dimension for drift-free DE. Similarly, Figs. 20–23 plot AFES* *vs.* D for p_μ= 0.25, 0.5, 0.75, and 0.875 respectively, for the DE/ran/1/either-or algorithm. Figures 24–27 plot the corresponding Np^* as a function dimension for the DE/ran/1/either-or algorithm.

7.4 Discussion of Results

Based on the data in Figs. 12–27, this section categorizes the reliability and speed of both drift-free DE and the DE/ran/1/either-or algorithm. The compiled results suggest guidelines for setting drift-free DE's control parameters.

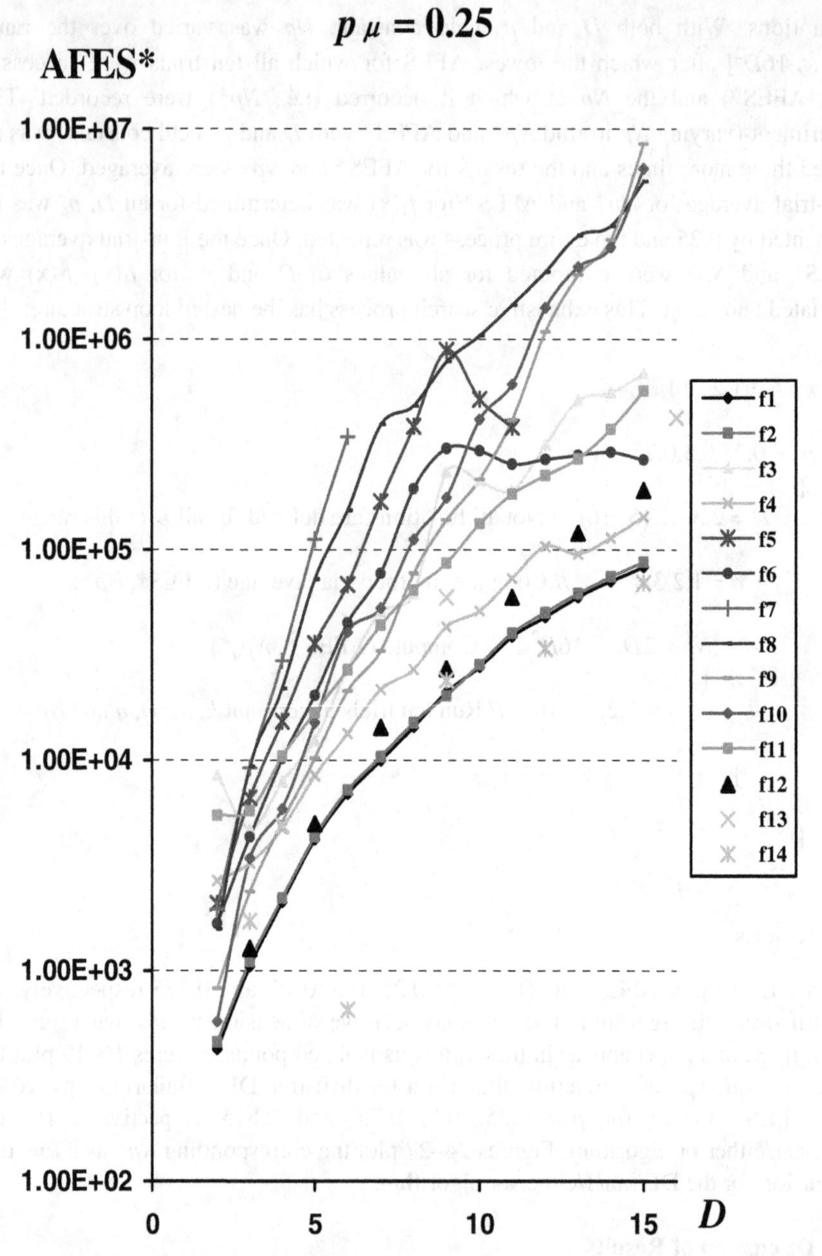

Fig. 12. A semi-log plot of AFES* *vs.* D for drift-free DE with $F = 1$ and $p_\mu = 0.25$

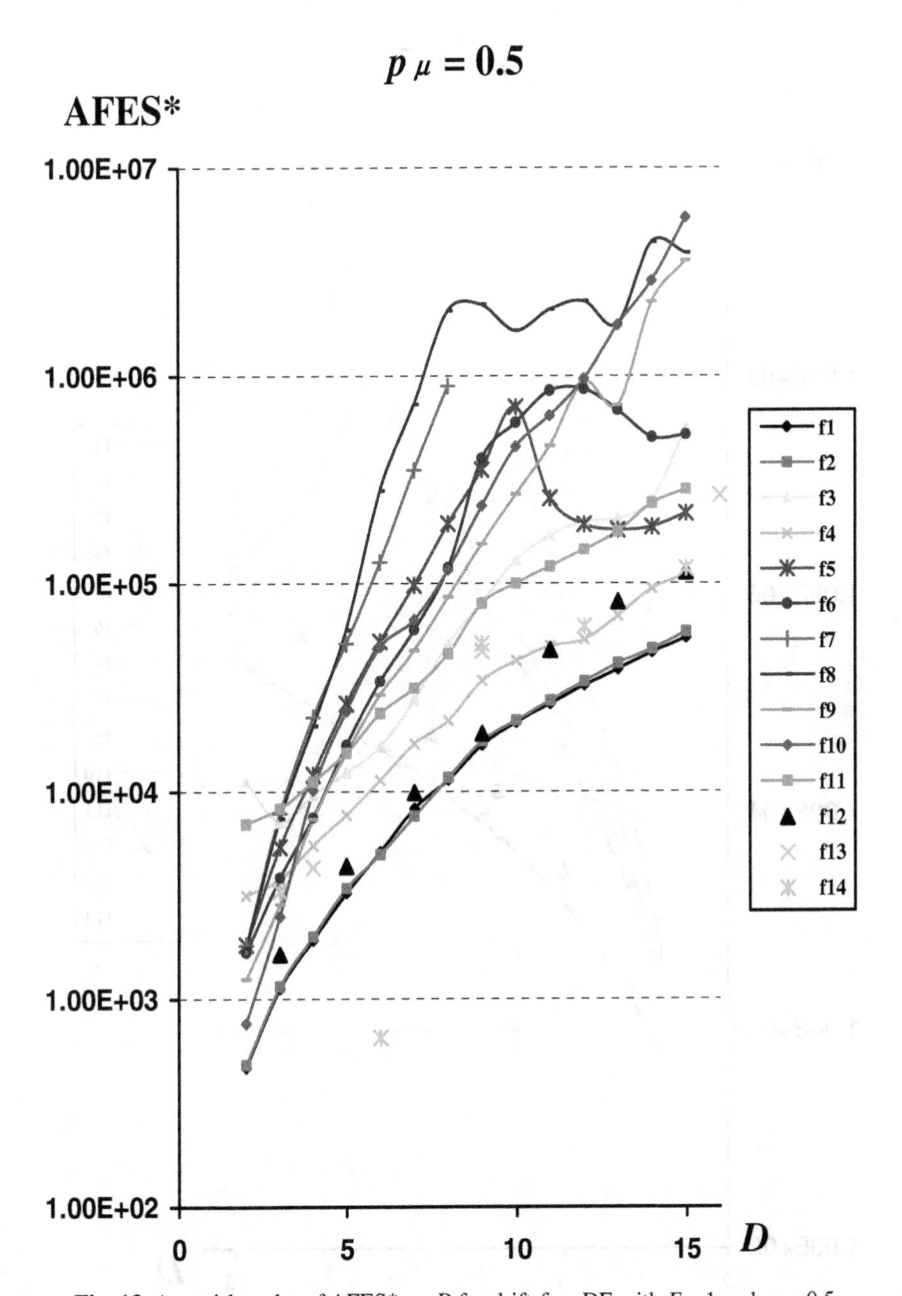

Fig. 13. A semi-log plot of AFES* *vs.* D for drift-free DE with $F = 1$ and $p_\mu = 0.5$

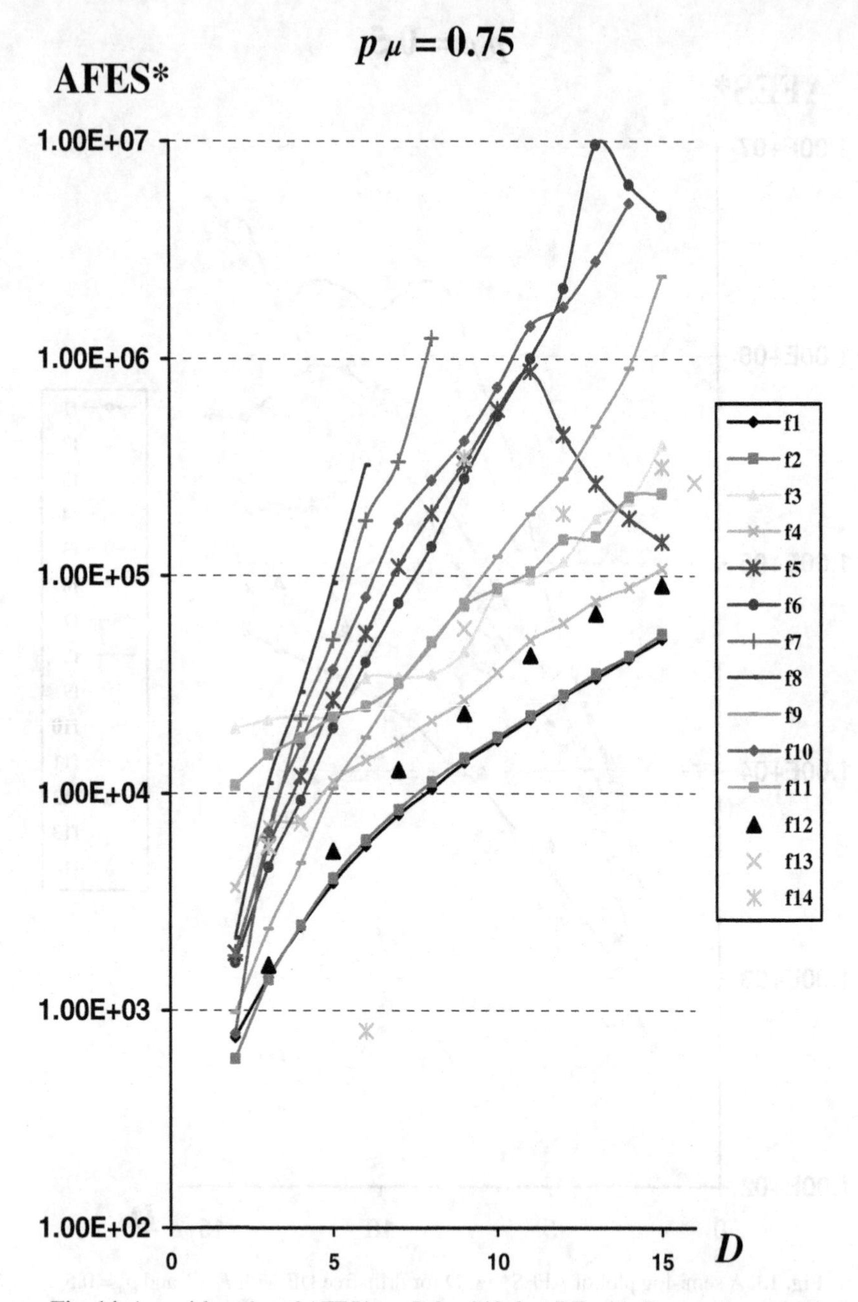

Fig. 14. A semi-log plot of AFES* *vs*. D for drift-free DE with $F = 1$ and $p_\mu = 0.75$

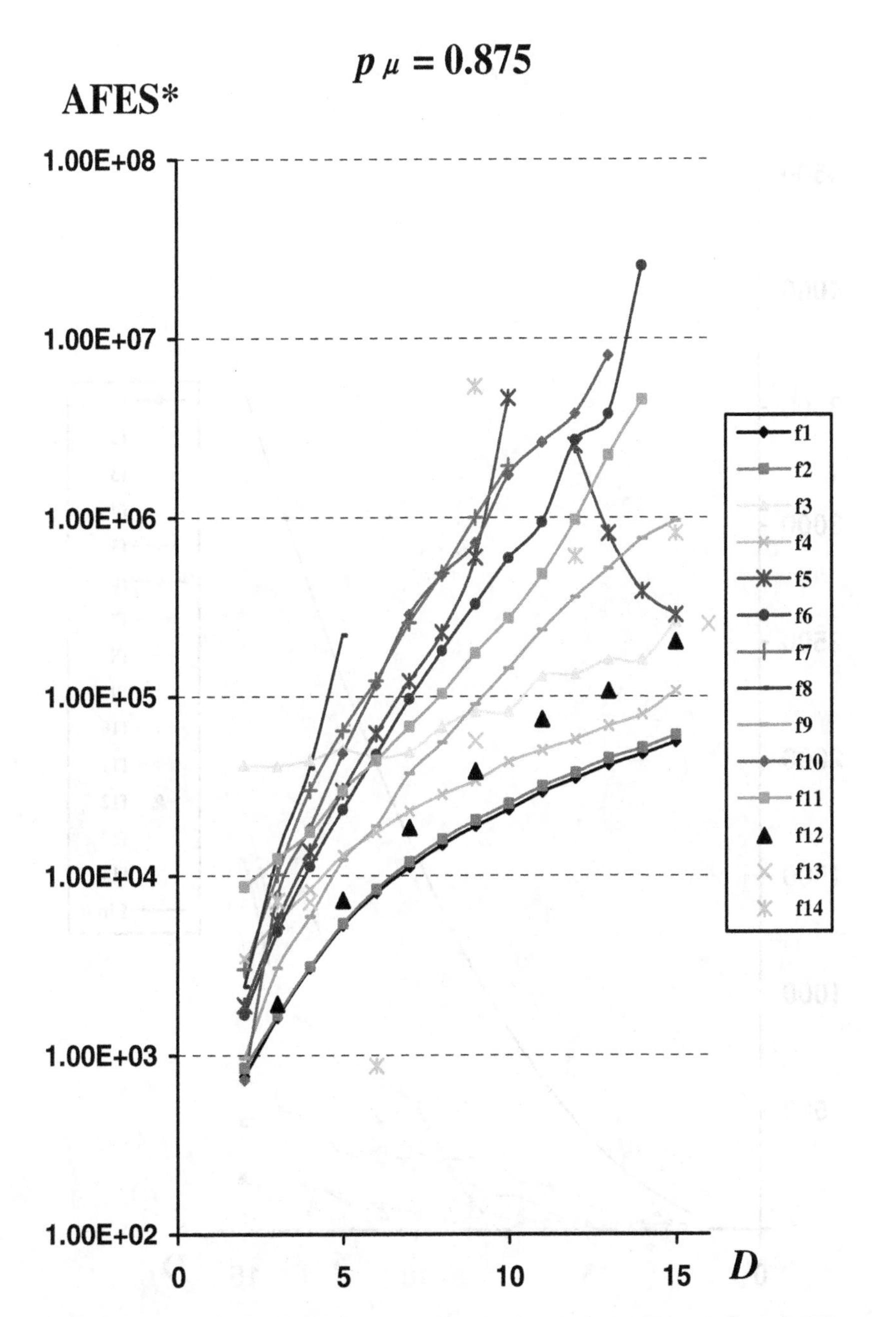

Fig. 15. A semi-log plot of AFES* *vs.* D for drift-free DE with $F = 1$ and $p_\mu = 0.875$

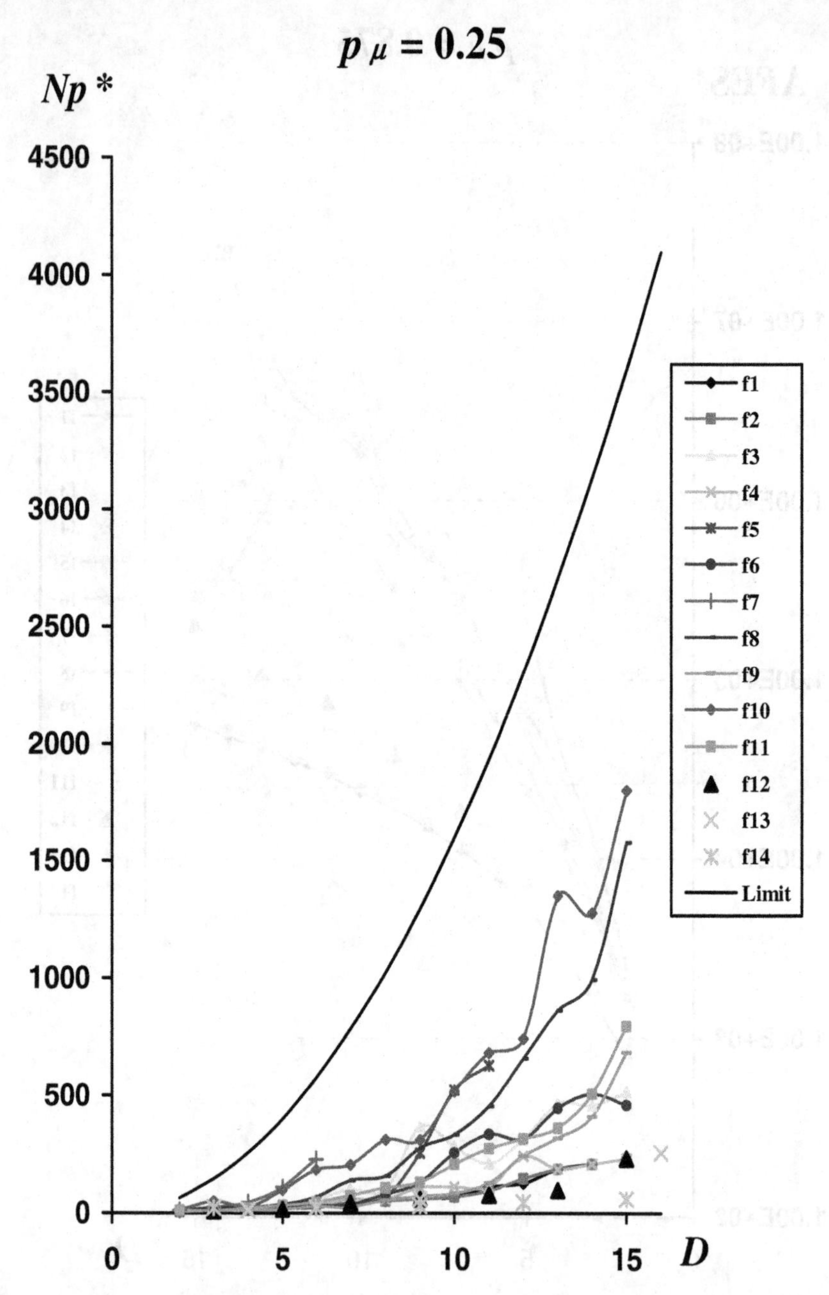

Fig. 16. *Np** *vs.* D for drift-free DE with $F = 1$ and $p_\mu = 0.25$. The limit line is the maximum allowed population for this experiment: $16D^2$.

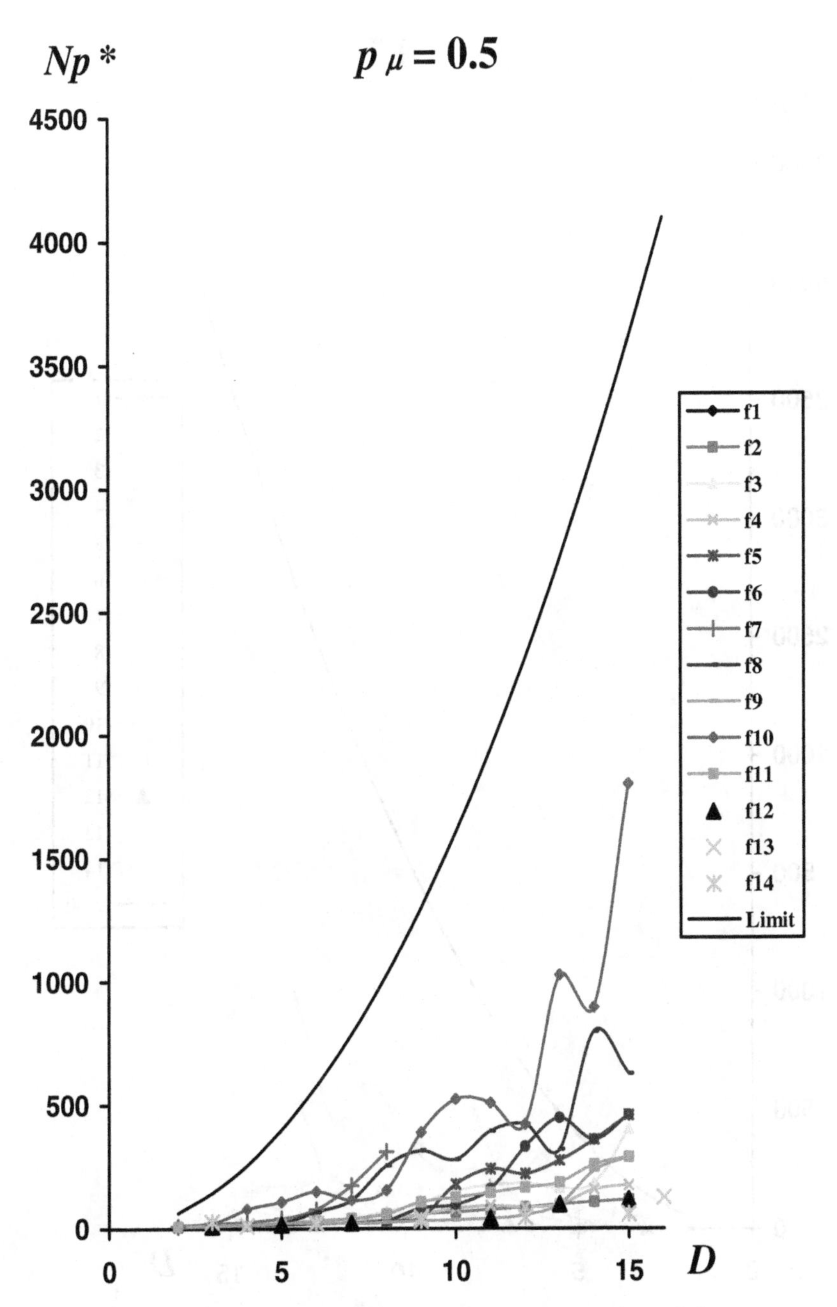

Fig. 17. Np^* *vs.* D for drift-free DE with $F = 1$ and $p_\mu = 0.5$

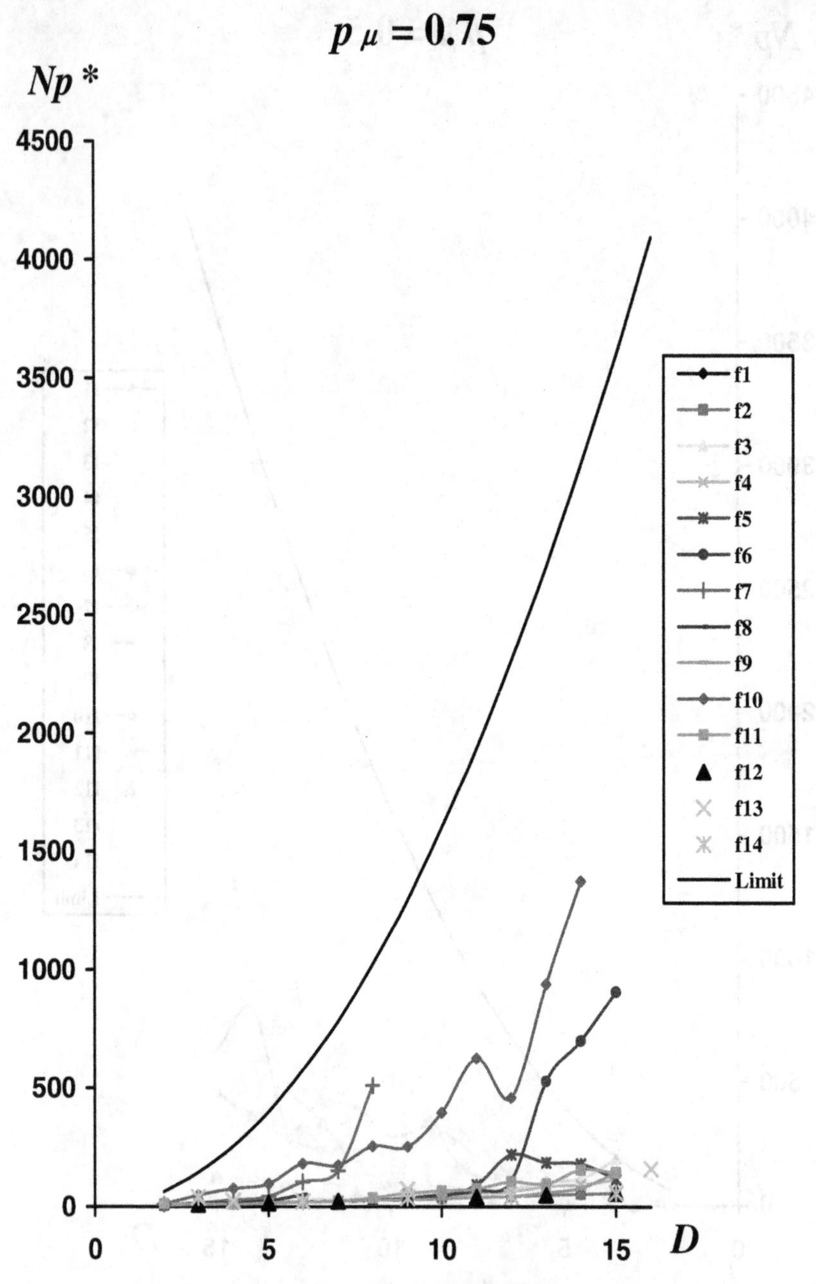

Fig. 18. *Np** *vs.* *D* for drift-free DE with $F = 1$ and $p_\mu = 0.75$

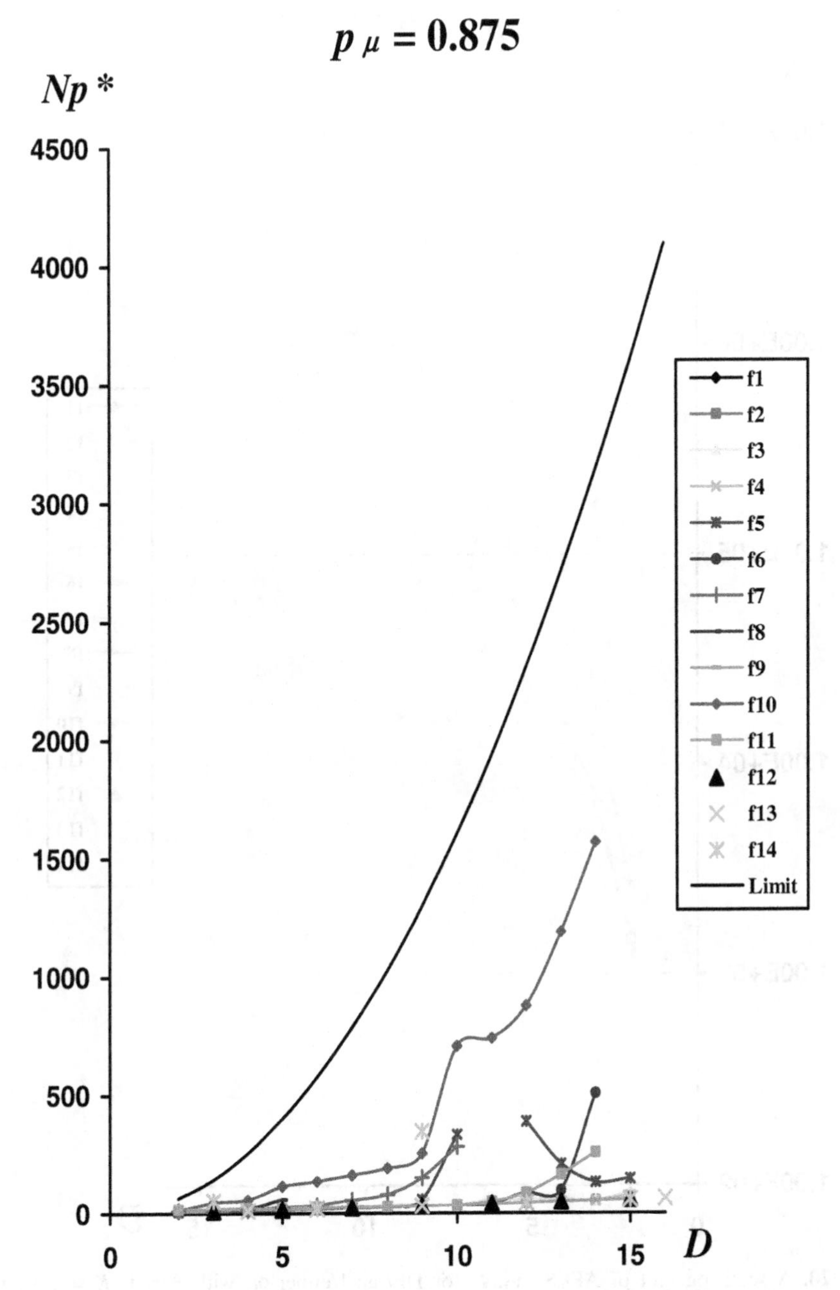

Fig. 19. Np^* *vs.* D for drift-free DE with $F = 1$ and $p_\mu = 0.875$

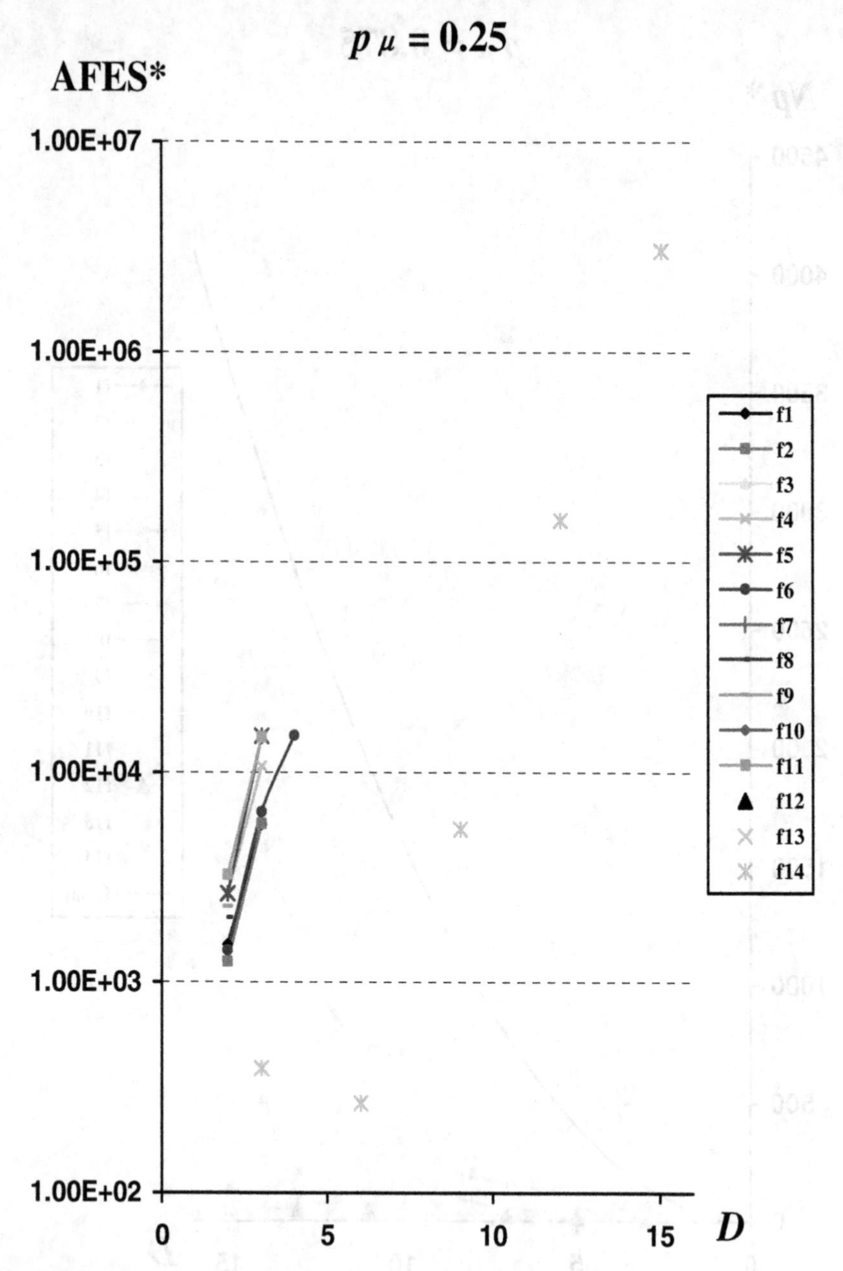

Fig. 20. A semi-log plot of AFES* *vs.* D for DE/ran/1/either-or, with F = 1, K = 1/3 and $p_\mu = 0.25$

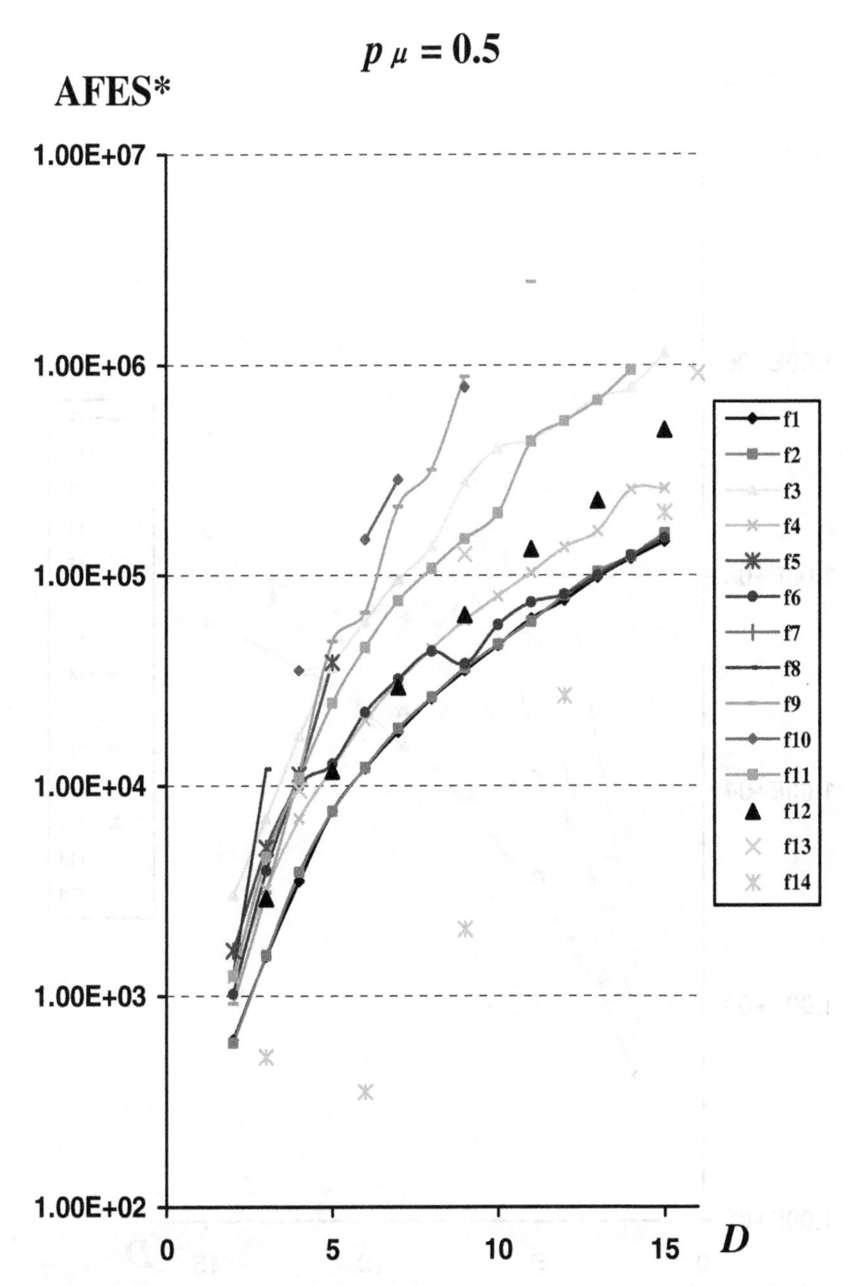

Fig. 21. A semi-log plot of AFES* *vs.* D for DE/ran/1/either-or, with $F = 1$, $K = 1/3$, $p_\mu = 0.5$

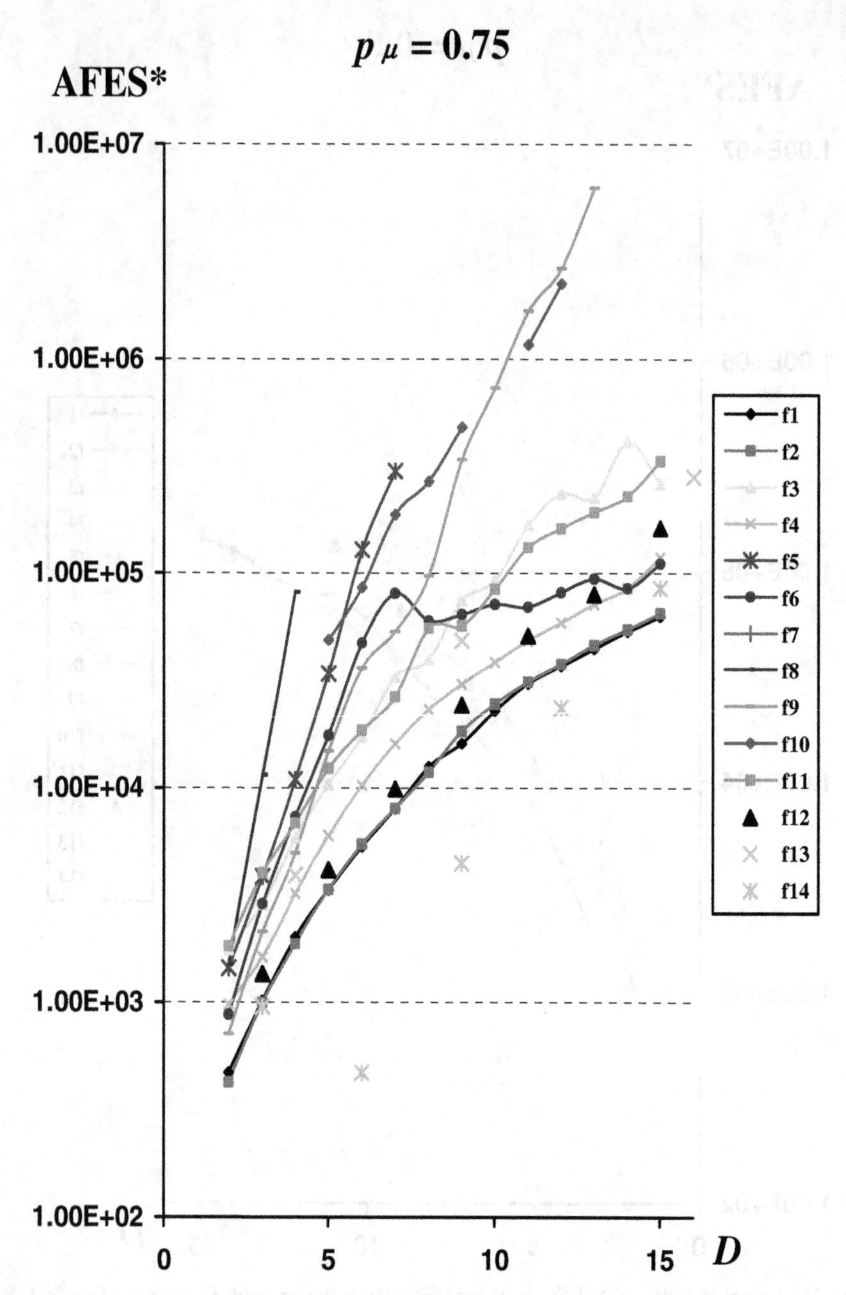

Fig. 22. A semi-log plot of AFES* *vs.* D for DE/ran/1/either-or, with $F = 1$, $K = 1/3$, $p_{\mu} = 0.75$

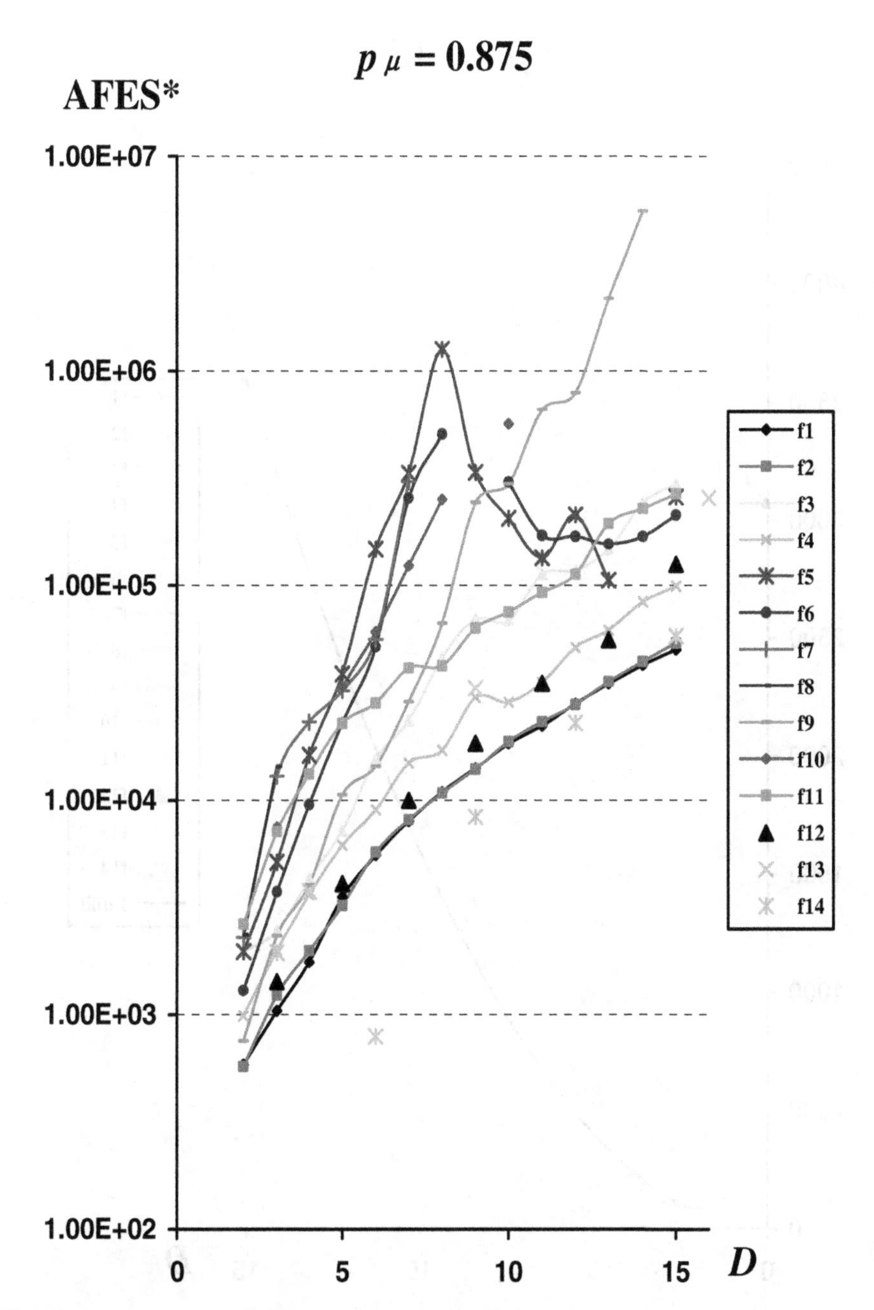

Fig. 23. A semi-log plot of AFES* *vs.* D for DE/ran/1/either-or, with $F = 1$, $K = 1/3$, $p_\mu = 0.875$

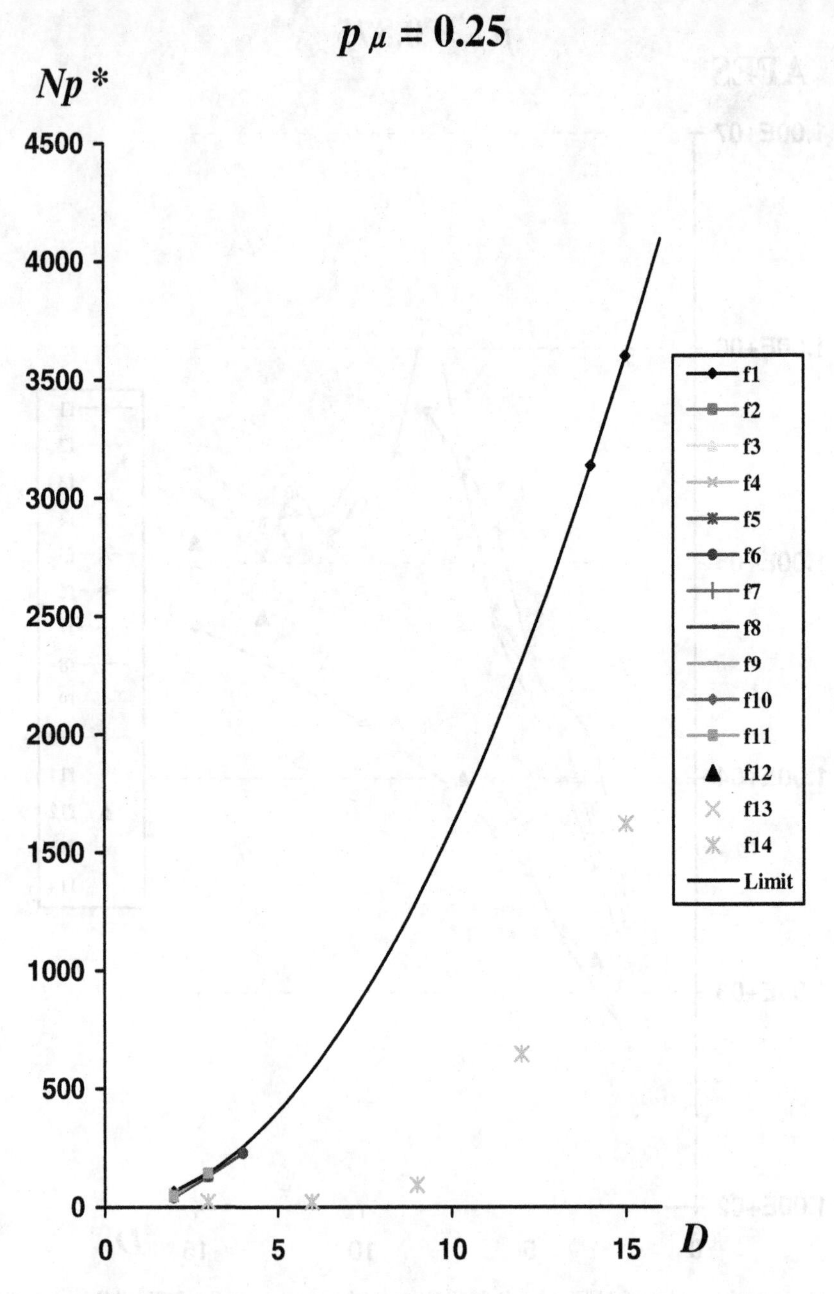

Fig. 24. *Np* vs. D* for DE/ran/1/either-or with $F = 1$, $K = 1/3$ and $p_\mu = 0.25$

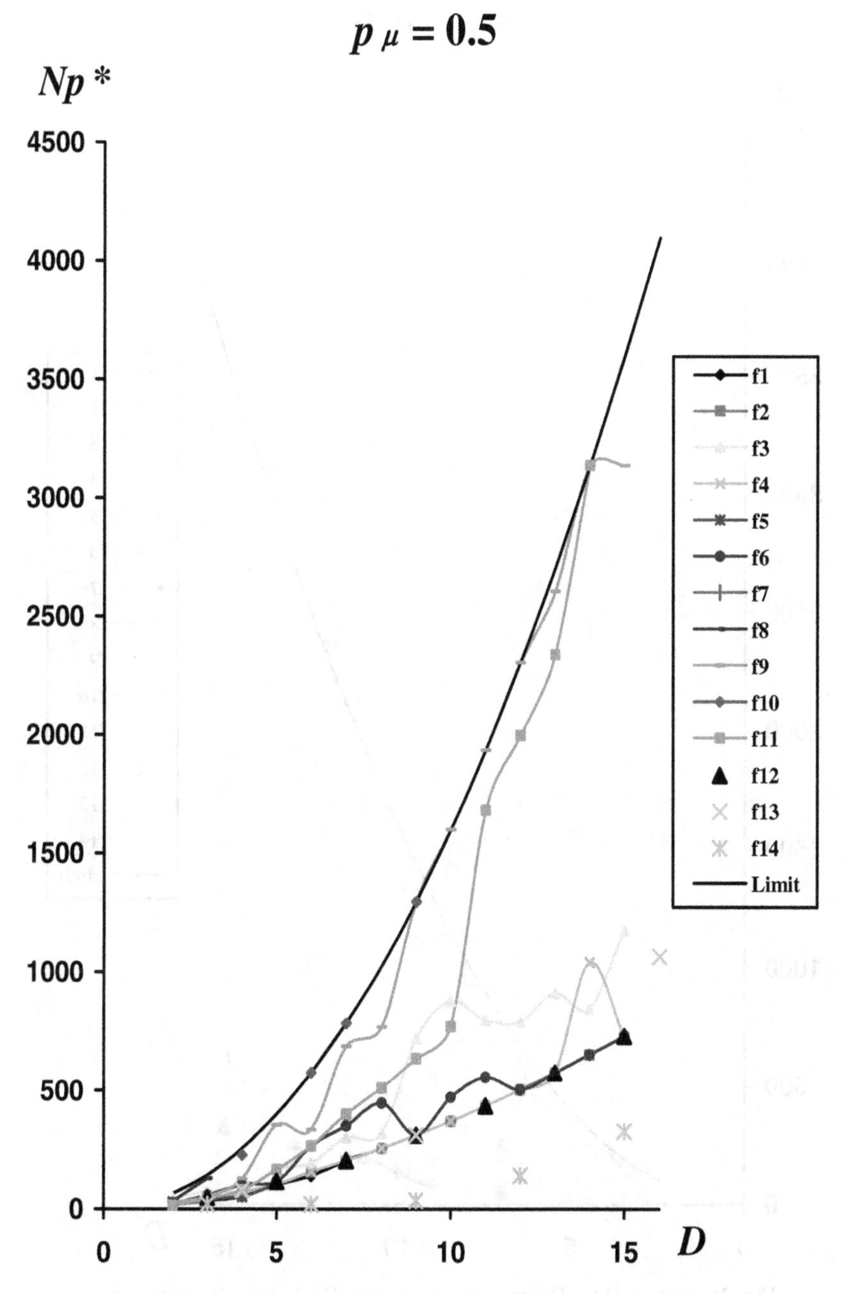

Fig. 25. Np^* *vs.* D for the DE/ran/1/either-or $F = 1$, $K = 1/3$ and $p_\mu = 0.5$

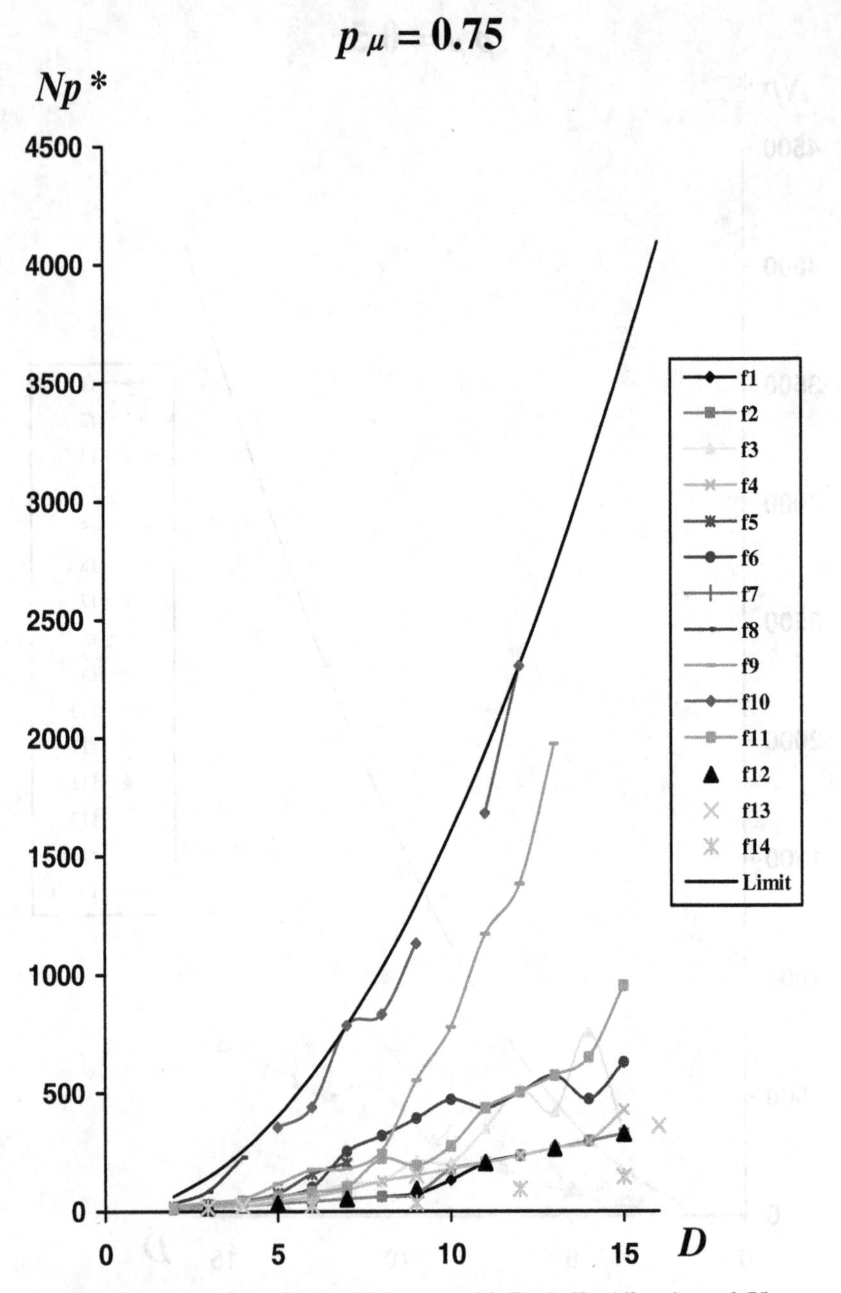

Fig. 26. Np^* *vs.* D for DE/ran/1/either-or with $F = 1$, $K = 1/3$ and $p_\mu = 0.75$

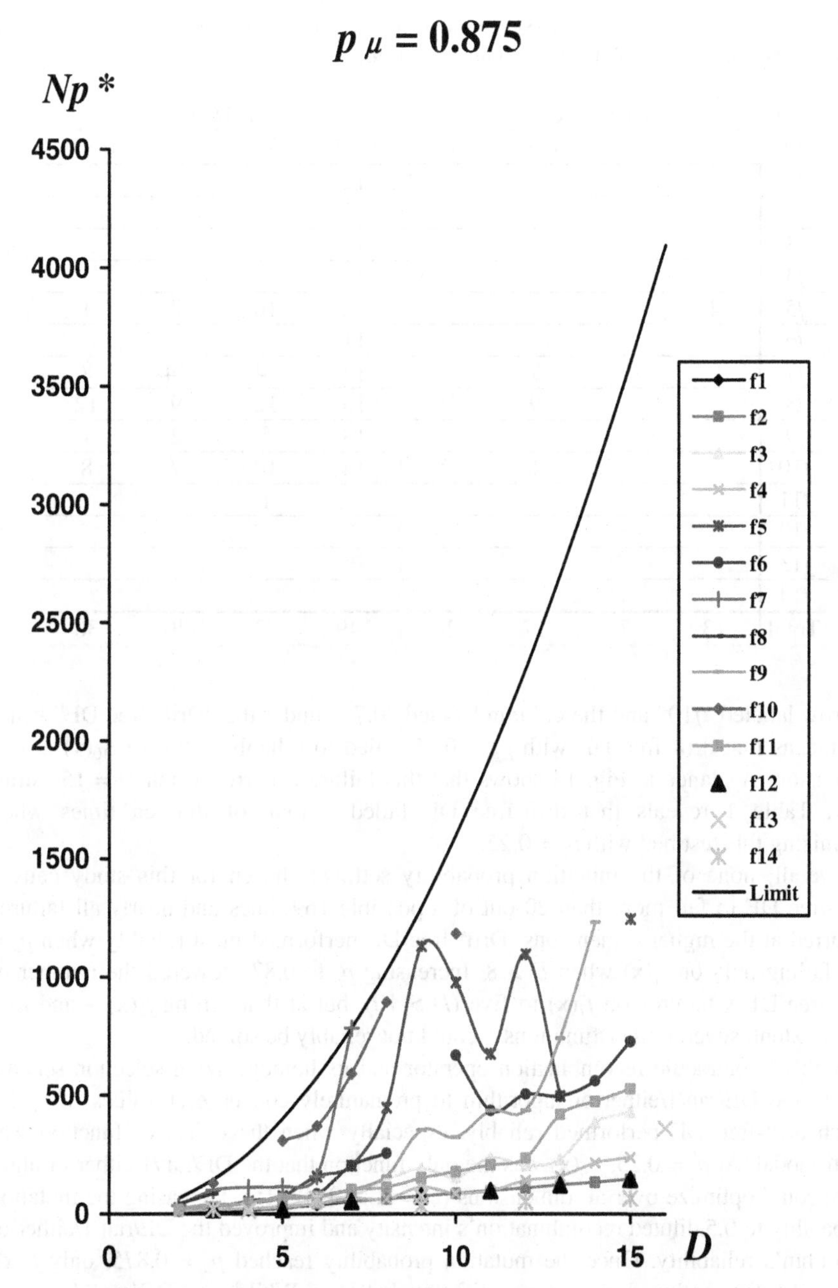

Fig. 27. Np^* *vs.* D for DE/ran/1/either-or with $F = 1$, $K = 1$ and $p_\mu = 0.875$

Reliability

Table1 records the number of times that drift-free DE and the DE/ran/1/either-or algorithm failed to optimize ten consecutive trials in all four runs. For example, the "1" in

Table 1. The number of failures for each combination of function, mutation probability and algorithm. The maximum total number of failures is 169.

p_μ / $f(\mathbf{x})$	Drift-free DE				DE/ran/1/either-or			
	0.25	0.5	0.75	0.875	0.25	0.5	0.75	0.875
*f*1					12			
*f*2					12			
*f*3					14			
*f*4					12			
*f*5	4			1	12	10	8	1
*f*6				1	11			
*f*7	9	7	7	5	14	14	14	8
*f*8			9	10	13	12	9	12
*f*9					13	5	2	1
*f*10			1	2	14	10	7	8
*f*11				1	12	1		
*f*12					7			
*f*13					3			
*f*14								
Total	13	7	17	20	149	52	40	30

the row labeled "*f*10" and the column labeled "0.75" under the "Drift-free DE" heading means that drift-free DE with p_μ = 0.75 failed to reliably optimize $f_{10}(\mathbf{x})$ at one dimension. A glance at Fig. 14 shows that this failure occurred when D = 15. Similarly, Table 1 reveals that drift-free DE failed a total of thirteen times when optimizing this test bed with p_μ = 0.25.

Overall, none of the mutation probability settings chosen for this study caused drift-free DE to fail more than 20 out of a possible 169 times and nearly all failures occurred at the higher dimensions. Drift-free DE performed most reliably when p_μ = 0.5, failing only on $f_7(\mathbf{x})$ when $D > 8$. Increasing p_μ to 0.875 lowered the number of drift-free DE's failures on $f_7(\mathbf{x})$ to five ($D > 10$), but at that setting $f_8(\mathbf{x})$ – and to a lesser extent, several other functions – could not reliably be solved.

Both its contractile recombination operator and its homogenizing selection scheme caused the DE/ran/1/either-or algorithm to prematurely converge at values of p_μ for which drift-free DE performed reliably, especially when the objective function was multi-modal. At p_μ = 0.25, $f_{14}(\mathbf{x})$ was the only function that the DE/ran/1/either-or algorithm could optimize over all dimensions (D = 3, 6, 9, 12, 15). Increasing the mutation probability to 0.5 diluted recombination's intensity and improved the DE/ran/1/either-or algorithm's reliability. Once the mutation probability reached p_μ = 0.875, only $f_7(\mathbf{x})$, $f_8(\mathbf{x})$ and $f_{10}(\mathbf{x})$ continued to pose a significant challenge. While most DE/ran/1/either-or failures occurred at the highest dimensions, there were cases – most notably $f_{10}(\mathbf{x})$ – in which failures occurred even when D was small (see Figs. 21–23).

It can be argued that arbitrarily choosing K = 1/3 and $p_\mu \leq 0.875$ penalizes the DE/ran/1/either-or algorithm and biases the empirical comparison. While extending p_μ

beyond 0.875 would probably improve the reliability of both algorithms on $f_7(\mathbf{x})$, it might also cause them both to fail more often on $f_8(\mathbf{x})$ and $f_{10}(\mathbf{x})$ – as the trend for both algorithms indicates – implying that p_μ = 0.875 is probably close to the DE/ran/1/either-or algorithm's most reliable setting when K = 1/3. Because of the control parameter dependence that exists between K and p_μ in the DE/ran/1/either-or algorithm, increasing K beyond 1/3 would reduce the magnitude of recombination's contractile effect and allow the DE/ran/1/either-or algorithm to perform more reliably when $p_\mu < 0.875$ than it does when K = 1/3, but to match drift-free DE's low failure rate, any adjustment to K would have to improve the DE ran/1/either-or algorithm's failure rate on $f_7(\mathbf{x})$ and several other functions without causing any additional failures. Thus, tuning K might modestly improve the DE/ran/1/either-or algorithm's reliability compared to the value chosen for this study, but allowing this extra degree of freedom would invalidate the fair comparison with drift-free DE, which does not explicitly depend on K.

Speed

To fairly measure algorithm speed, this study computed the AFES* at a constant probability of success. For the five-dimensional versions of functions $f_1(\mathbf{x})$–$f_{12}(\mathbf{x})$, Table 2 lists, for both algorithms, the lowest AFES* along with both the optimal mutation probability p_μ* and population size Np* at which it occurred (D = 4 for $f_{13}(\mathbf{x})$ and D = 6 for $f_{14}(\mathbf{x})$). Table 3 compiles the same information for D = 15 (D = 16 for $f_{13}(\mathbf{x})$). In both tables, Np* has been rounded to the nearest integer.

For functions $f_1(\mathbf{x})$, $f_2(\mathbf{x})$, $f_9(\mathbf{x})$, $f_{12}(\mathbf{x})$ and to a lesser extent $f_{11}(\mathbf{x})$ and $f_{13}(\mathbf{x})$, the difference between the two algorithm's AFES* performance was too small to be significant given the coarse sampling of control parameter combinations. The DE/ran/1/either-or

Table 2. Best AFES* (in **bold**) and its corresponding control parameter settings when D = 5 (D = 4 for $f_{13}(\mathbf{x})$ and D = 6 for $f_{14}(\mathbf{x})$)

	Drift-free DE			DE/ran/1/either-or		
	p_μ*	Np*	AFES*	p_μ*	Np*	AFES*
f1	0.5	16	**3252**	0.75	34	3349
f2	0.5	16	3426.8	0.875	22	**3233.45**
f3	0.5	22	12285.5	0.875	34	**7225**
f4	0.5	19	7711.73	0.75	41	**5976.15**
f5	0.5	16	**26536.8**	0.75	76	34078.1
f6	0.5	16	16651.2	0.5	116	**12664.3**
f7	0.75	44	51021.3	0.875	116	**32384.3**
f8	0.5	24	**60050.3**	–	–	–
f9	0.25	16	**10226.4**	0.875	76	10641.9
f10	0.25	97	**15132.8**	0.875	308	33717
f11	0.5	22	15319.3	0.75	63	**12333.8**
f12	0.5	16	4371.2	0.875	18	**4075.65**
f13	0.5	12	4311	0.875	14	**3765.3**
f14	0.25	24	650.7	0.25	22	**265.65**

Table 3. Best AFES* (in **bold**) and its corresponding control parameter settings when $D = 15$ ($D = 16$ for $f_{13}(\mathbf{x})$)

	Drift-free DE			DE/ran/1/either-or		
	p_μ*	Np*	AFES*	p_μ*	Np*	AFES*
*f*1	0.75	59	51353.6	0.875	147	**50321.8**
*f*2	0.75	59	54238.7	0.875	147	**54125.4**
*f*3	0.875	88	263678	0.75	328	**262384**
*f*4	0.875	74	108318	0.875	238	**99463.1**
*f*5	0.75	117	**143068**	0.875	1246	259134
*f*6	0.25	459	268526	0.75	629	**111072**
*f*7	–	–	–	–	–	–
*f*8	0.5	628	**3.90E+06**	–	–	–
*f*9	0.875	59	**959688**	–	–	–
*f*10	0.5	1800	**5.77E+06**	–	–	–
*f*11	0.75	146	**239698**	0.875	528	267679
*f*12	0.75	59	**90910.1**	0.875	147	125490
*f*13	0.875	64	**255682**	0.75	362	256350
*f*14	0.25	59	69764.6	0.875	66	**58276.3**

algorithm was less than twice as fast as drift-free DE on functions $f_3(\mathbf{x})$, $f_4(\mathbf{x})$, $f_6(\mathbf{x})$ and $f_7(\mathbf{x})$, whereas drift-free was less than twice as the DE/ran/1/either-or algorithm only on $f_5(\mathbf{x})$, but a little more than twice as fast on $f_{10}(\mathbf{x})$. In addition, the DE/ran/1/either-or algorithm was more than twice as fast as drift-free DE on $f_{14}(\mathbf{x})$, but only drift-free DE solved $f_8(\mathbf{x})$ when $D = 5$.

For every case in Table 2 except $f_{14}(\mathbf{x})$, Np* was larger for the DE/ran/1/either-or algorithm – sometimes dramatically so – than for drift-free DE. With the exception of $f_6(\mathbf{x})$ and $f_{14}(\mathbf{x})$ for which both algorithms posted p_μ* = 0.5 and p_μ* = 0.25, respectively, the DE/ran/1/either-or algorithm's mutation probabilities were also larger than were those registered by drift-free DE.

Table 3 reveals little difference between the two algorithm's AFES* performance on the fifteen-dimensional versions of functions $f_1(\mathbf{x})$, $f_2(\mathbf{x})$, $f_3(\mathbf{x})$, $f_4(\mathbf{x})$ $f_{11}(\mathbf{x})$, $f_{13}(\mathbf{x})$ and to a lesser extent on $f_{12}(\mathbf{x})$ and $f_{14}(\mathbf{x})$. Although the DE/ran/1/either-or algorithm was a little more than twice as fast as drift-free DE on $f_6(\mathbf{x})$, drift-free DE was twice as fast on $f_5(\mathbf{x})$. By default, drift-free DE was faster on the multimodal functions that DE/ran/1/either-or algorithm could not solve when $D = 15$, i.e., for $f_8(\mathbf{x})$, $f_9(\mathbf{x})$ and $f_{10}(\mathbf{x})$.

Although $p_\mu = 0.5$ was the most reliable setting for drift-free DE and its most common optimal mutation rate when $D = 5$, once $D = 15$, $p_\mu = 0.5$ produced the fastest convergence in only four of the fourteen cases listed in Table 3. Similarly, no value for p_μ* in Table 3 under the DE/ran/1/either-or heading was less than 0.75. Comparing the values of p_μ* and Np* listed in Tables 2 and 3 reveals that for a given function, the optimal mutation probability – like the optimal population size – tends to increase along with the function's dimension.

8 Control Parameter Guidelines for Drift-Free DE

8.1 Population Size

The optimal population size depends on the objective function, its dimension and the mutation probability. Because it is so simple, $f_1(\mathbf{x})$ provides a baseline for the minimum reliable Np^* for any D-dimensional function. When $p_\mu = 1$ and $F = 1.3/\sqrt{D}$, the authors of (Price and Rönkkönen 2006) found that $Np^* = 2D$ for both $f_1(\mathbf{x})$ and $f_2(\mathbf{x})$. Fifteen dimensions is too few to clearly indicate Np^*'s asymptotic behavior with respect to D, so to better understand how the *minimum* Np^* grows as a function of dimension, Fig. 28 plots Np^* *vs.* D for $f_1(\mathbf{x})$ over the range $D = 5n$, $n = 1,2,\ldots,8$ for several combinations of F and p_μ. The uppermost curve corresponds to $p_\mu = 0$, while the second and third lines from the top plot Np^* as a function of D for $p_\mu = 0.5$ and $p_\mu = 0.98$, respectively. In all three cases, $F = 1$. The lowest curve was generated with $p_\mu = 0.98$ and $F = 1.3/\sqrt{D}$. The mutation probability $p_\mu = 0.98$ was chosen to represent the population requirements for high rates of mutation because $p_\mu = 1$ in conjunction with $F = 1$ did not reliably converge.

Figure 28 includes trend lines that document Np^*'s rate of growth. For the case $p_\mu = 0$, a slightly flatter curve than the best-fit trend line in the figure would better model Np^*'s growth at high D, so the minimum population size when optimizing with recombination alone is likely to be very close to Np^2. The trend line is a good fit when $p_\mu = 0.5$, so the minimum Np^* when $p_\mu = 0.5$ is probably $0.5D^2$.

The lowest line in Fig. 28 represents the linear growth that Np displayed when mutation alone drives drift-free DE and the scale factor $F = 1.3/\sqrt{D}$ is tuned for optimizing convex quadratic basins. The trend line $Np^* = 2D$ confirms that the linear growth reported in (Price and Rönkkönen 2006) extends to 40 dimensions. When $F = 1$, and $p_\mu = 0.98$, growth in Np is initially linear, with $Np^* = 2D$ for all $D \leq 20$, but once $D \geq 25$, nonlinear growth begins to appear. Equation 63 sums up these results with a simple *rule of thumb* for predicting the minimum Np required for reliable convergence given p_μ and D.

$$Np^*_{\min} \approx \left(1 - p_\mu\right) \cdot D^2 + p_\mu \cdot 2D \,. \tag{63}$$

For example, this study found that $Np^* \approx 4D$ when $F = 1$ and $p_\mu = 0.875$, not only for $f_1(\mathbf{x})$ and $f_2(\mathbf{x})$, but also for $f_4(\mathbf{x})$, $f_9(\mathbf{x})$, $f_{12}(\mathbf{x})$, $f_{13}(\mathbf{x})$, $f_{14}(\mathbf{x})$ and to a lesser extent, $f_3(\mathbf{x})$. Equation 63 predicts that when $p_\mu = 0.875$ and $D = 10$, $Np^*_{\min} = 30$ (i.e., $3D$) and $Np^*_{\min} = 54$ (i.e., $3.6D$) when $D = 15$ – a slight under estimation of the actual $Np^* = 4D$, but a good estimate for $Np^*_{\min}$. Function $f_{14}(\mathbf{x})$ best illustrates that this rule of thumb is not always valid, since Np^* (= $4D$) was independent of p_μ.

While the sparse and noisy data makes it difficult to accurately estimate the maximum population size as a function of both D and p_μ, the highest populations required in this study were never more than D^3 and the dependence of the maximum Np^* on p_μ at a given dimension was roughly similar to that in Eq. 63. This meager data suggests that the maximum population needed to solve a D-dimensional multimodal function is no more than D times the minimum Np^*, i.e.,

$$Np^*_{\max} \approx D \cdot Np^*_{\min} \,. \tag{64}$$

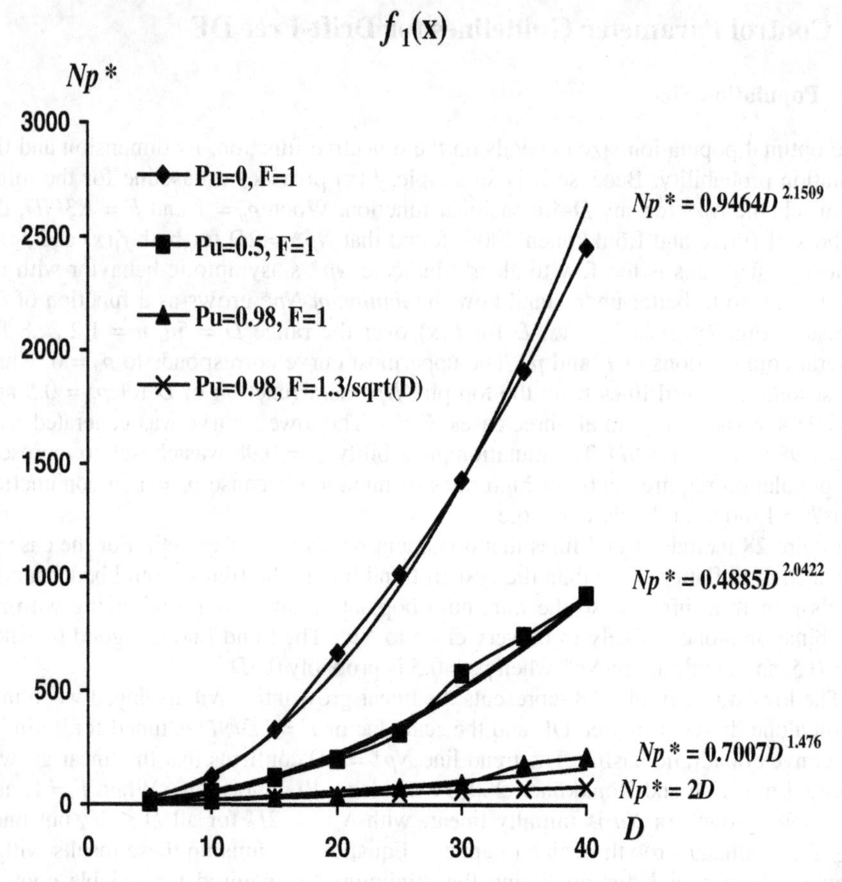

Fig. 28. Np^* for $f_1(\mathbf{x})$ as a function of $D = 5n$, $n = 1,2,\ldots,8$ and $p_\mu = 0.0$, 0.5 and 0.98. The upper three plots are with $F = 1$, while the bottom curve was generated with $F = 1.3/\sqrt{D}$.

Equations 63 and 64 are not meant to imply a theoretically derived relationship; they are only intended to establish convenient default settings for Np. While it is unlikely that populations smaller than Eq. 63 predicts will be useful, exceptional cases may require larger populations than Eq. 64 predicts.

8.2 Mutation Probability

Based on the results in Sect. 7, $p_\mu = 0.5$ was the most successful mutation probability, but the value that produced the fastest success was usually higher. As a comparison of Tables 2 and 3 shows, the optimal mutation probability tends to increase with dimension. Whereas $p_\mu = 0.5$ is both reliable and fast when $D = 5$, once $D = 15$, p_μ had increased for every function except $f_6(\mathbf{x})$, $f_8(\mathbf{x})$ and $f_{14}(\mathbf{x})$. Although weak, this trend suggests a possible rule of thumb for selecting the mutation probability:

$$p_\mu \approx 1 - \frac{1}{\sqrt{D}}. \tag{65}$$

Functions like $f_7(\mathbf{x})$ demonstrate that in lieu of a very large population, success may depend on p_μ being closer to 1.0 than Eq. 65 predicts. Conversely, some high-dimensional functions may require substantially more recombination than Eq. 65 specifies.

8.3 Mutation Scale Factor

If the results for drift-free DE in Sect. 7 are any guide, F will probably not have to be adjusted away from 1.0. On the other hand, $F \neq 1$ generates a more diverse population of potential trial vectors, so $F = 1$ risks failing if it cannot generate a sufficiently robust trial vector population – especially when used in conjunction with $p_\mu = 1$. If several combinations of Np and p_μ fail to satisfactorily optimize an objective function, tuning F should be considered. Setting $F = 0.98$ capitalizes on the ability of $F = 1$ to exploit functions with regularly spaced local minima without diminishing the pool of potential trial vectors.

The most obvious alternative value to $F \sim 1$ is $F = 1.3/\sqrt{D}$, which is optimal for convex quadratic functions like $f_1(\mathbf{x})$ and $f_2(\mathbf{x})$. This value for F is often too small to be effective when the objective function is multimodal, but it occasionally improves optimization speed. Although it is unusual for very small values for F to be effective, auxiliary experiments with $f_{14}(\mathbf{x})$ and published results (Chakraborti 2005) show that the most effective mutation scale factors for cluster optimization are on the order of $F = 0.001$. With such a small scale factor, mutation corresponds to a local search of the target vector's immediate neighborhood. Unless the optimization problem is analogous to clustering, tiny values for F should only be tried as a last resort.

8.4 Default Control Parameter Settings

Below is a default setting for drift-free DE's control parameters and a prescription for resetting them if the default settings produce unsatisfactory results.

- Default setting: $F = 1$; $p_\mu = 0.5$; $Np = D^2$ $(= 2Np^*_{\text{min}})$
- If results are unsatisfactory: $Np = k \cdot Np^*_{\text{min}}$, $k = 4, 8, 16, \ldots, D$
- If still unsatisfactory: $p_\mu = 1 - 1/\sqrt{D}$, 0.98, 0.25
- If all else fails: $F = 1.3/\sqrt{D}$, 0.98, 0.85, 0.5, 0.001

If the default setting fails to produce a satisfactory result, then double Np. Continue doubling Np until $Np = Np^*_{\text{max}} = D \cdot Np^*_{\text{min}}$. If increasing the population size does not improve the result, then experiment with different values for p_μ. Start by changing p_μ from 0.5 to $1 - 1/\sqrt{D}$, compute Np^*_{min} and reset Np to $2Np^*_{\text{min}}$. If results at the new settings are unsatisfactory, double Np until $Np = D \cdot Np^*_{\text{min}}$. If results are still unsatisfactory after repeating this procedure for the values of p_μ listed above, then experiment by tuning F.

9 Conclusion

9.1 Synopsis

This chapter describes two prior DE algorithms, discusses their limitations, proposes a new algorithm in response and offers a preliminary evaluation of its comparative effectiveness. More specifically, Sect. 2.2 showed that classic DE degenerates into a mutation-only algorithm at the one setting for which its performance is rotationally invariant (Cr = 1). In response, the DE/ran/1/either-or algorithm replaced uniform crossover with the rotationally invariant three-vector arithmetic recombination operator, showing it to be natural counterpart to differential mutation. By interleaving mutation and arithmetic recombination, the DE/ran/1/either-or algorithm reduced the control parameter dependence between F and K, albeit at the expense of adding a fourth control variable, p_F ($= p_\mu$).

Section 2.3 proved that the DE/ran/1/either-or algorithm generates a distribution of three-vector recombinants that is biased toward or away from the population's centroid whenever K ($\neq 0$) is held constant. Section 2.4 described how to eliminate recombination drift bias and the control variable K by projecting randomly chosen mutation differentials onto randomly chosen three-vector recombination differentials. By generating not only mutants but also recombinants with perturbations sampled from the population's own distribution of vector differences, the projection method unifies DE's approach to mutation and recombination.

Section 2.5 proved that the DE/ran/1/either-or algorithm's selection operator is also biased with respect to the population's centroid. Drift-free DE eliminates this selection bias by distributing trial vectors center-symmetrically around the vector with which they compete, i.e., the target vector. As a result, only recombination can homogenize the population in drift-free DE. Section 2.6 presented the drift-free DE algorithm, showing that the direction and distance from the target to a recombinant are respectively determined by recombination and mutation differentials.

Section 2.7 demonstrated that drift-free DE was more reliable than the DE/ran/1/either-or algorithm when optimizing a test bed of fourteen scalable benchmark functions. The speed of the two algorithms measured at a constant probability of success was comparable, with the DE/ran/1/either-or algorithm prevailing at low dimension and drift-free DE gaining the edge as D increased. For both algorithms, decreasing p_μ strongly correlated with increasing Np^*, showing that the population must be enlarged to offset the higher rate of homogenization due to recombination. The homogenizing effect of the DE/ran/1/either-or algorithm's selection bias was constant, so compared to drift-free DE, Np^* was large even when the recombination rate was low. The optimal population size also depended on the objective function's dimension. With $f_1(\mathbf{x})$ as a baseline, Sect. 2.8 developed guidelines for setting drift-free DE's control parameters.

9.2 Drift-Free DE's Benefits

Like earlier DE algorithms, drift-free DE is easy to understand and requires little effort to apply. Drift-free DE is also robust inasmuch as it reliably optimized most test bed functions for all four values of p_μ and for more than one value of Np. In most cases, $p_\mu = 0.5$ was a good default value and increasing Np until the solution ceased improving was the only tuning involved. Since both drift-free DE and the DE/ran/1/either-or algorithm were roughly equal in speed when compared at the same probability of success, drift-free DE's superior performance on high-dimensional multimodal functions, control parameter robustness and smaller system requirements (i.e., smaller Np^*) make it the better choice.

Apart from its reliable performance and robust control parameters, the drift-free DE algorithm has considerable esthetic appeal. Expressing trial vectors as linear combinations of three population vectors is both simple and fundamental, as are the symmetry conditions not met by earlier DE algorithms to which drift-free DE conforms. Furthermore, consistently implementing both recombination and mutation with randomly sampled vector differences improves the algorithm's logical coherence.

9.3 Future Research

Future research will employ an expanded test bed to compare drift-free DE's performance to that of one or more other EAs. The dependence of performance on control parameter settings will be more exhaustively sampled with the goal of understanding what general relationships exist between p_μ, Np (and F).

The delicate control that p_μ provides over the population's rate of homogenization potentially makes drift-free DE a valuable tool for multi-objective optimization where migration between evolutionary niches often must be restricted. Similarly, drift-free DE should be tested on both constrained problems and those with noisy objective functions, this last class of function being difficult for classic DE (Vesterstrom and Thomsen 2004). Finally, there may be an effective way to make p_μ adaptive, perhaps by modifying DE's selection rule.

Acknowledgment

The author wishes to thank Professor Uday Chakraborty at the University of Missouri not only for generously offering this opportunity to publish in his book, but also for the great patience he demonstrated while this manuscript was being prepared. Thanks are also extended to Dr. Rainer Storn and Professor Jouni Lampinen whose thoughtful comments have improved the quality of this effort.

References

Chakraborti, N.: Genetic algorithms and related techniques for optimizing Si–H clusters: A merit analysis for differential evolution. In: Price, K.V., Storn, R.M., Lampinen, J.A. (eds.) Differential evolution: A practical approach to global optimization. Springer, Heidelberg (2005)

Hassani, S.: Mathematical physics: A modern introduction to its foundations. Springer, New York (1999)

Nelder, J.A., Mead, R.: A simplex method for function minimization. Computer Journal 7, 308–313 (1965)

Price, K.V.: An introduction to differential evolution. In: Corne, D., Dorigo, M., Glover, F. (eds.) New ideas in optimization. McGraw-Hill, UK (1999)

Price, K., Storn, R.: Differential evolution. Dr. Dobb's Journal 78, 18–24 (1997)

Price, K.V., Storn, R.M., Lampinen, J.A.: Differential evolution: A practical approach to global optimization. Springer, Heidelberg (2005)

Price, K.V., Rönkkönen, J.I.: Comparing the unimodal scaling performance of global and local selection in mutation-only differential evolution algorithm. In: Proceedings of 2006 IEEE world congress on computational intelligence, Vancouver, Canada, July 16–21, pp. 7387–7394 (2006) ISBN 0-7803-9489-5

Rudolph, G.: Convergence of evolutionary algorithms in general search spaces. In: Proceedings of the third IEEE conference of evolutionary computation, pp. 50–54. IEEE Press, Los Alamitos (1996)

Salomon, R.: Re-evaluating genetic algorithm performance under coordinate rotation of benchmark functions: A survey of some practical and theoretical aspects of genetic algorithms. Biosystems 39(3), 263–278 (1996)

Schwefel, H.-P.: Evolution and optimum seeking. John Wiley and Sons, New York (1995)

Shang, Y.-W., Qiu, Y.-H.: A note on the extended Rosenbrock function. Evolutionary Computation 14(1), 119–126 (2006)

Storn, R., Price, K.: Differential evolution: A simple and efficient heuristic for global optimization over continuous spaces. Journal of Global Optimization 11, 341–359 (1997)

Vesterstrom, J., Thomsen, R.: A comparative study of differential evolution, particle swarm optimization and evolutionary algorithms on numerical benchmark problems. In: Proceedings of the 2004 congress on evolutionary computation, June 2004, vol. 2(19-23), pp. 1980–1987 (2004)

Whitley, D., Mathias, K., Rana, S., Dzubera, J.: Evaluating evolutionary algorithms. Artificial Intelligence 85, 1–32 (1996)

Yao, X., Liu, Y.: Fast evolution strategies. In: Angeline, P.J., Reynolds, R.G., McDonnell, J.R., Eberhardt, R. (eds.) Evolutionary Programming IV: Proceedings of the sixth international conference, Indianapolis, Indiana, USA, April 1997, pp. 151–161. Springer, Heidelberg (1997)

An Analysis of the Control Parameters' Adaptation in DE

Janez Brest, Aleš Zamuda, Borko Boškoviff, Sašo Greiner, and Viljem Žumer

Institute of Computer Science,
Faculty of Electrical Engineering and Computer Science,
University of Maribor, Smetanova 17, SI-2000 Maribor, Slovenia
janez.brest@uni-mb.si

Summary. The main goal of this chapter is to present an analysis of how self-adaptive control parameters are being changed during the current evolutionary process. We present a comparison of two distinct self-adaptive control parameters' mechanisms, both using Differential Evolution (DE). The first mechanism has recently been proposed in the jDE algorithm, which uses self-adaptation for F and CR control parameters. In the second one, we integrated the well known self-adaptive mechanism from Evolution Strategies (ES) into the original DE algorithm, also for the F and CR control parameters. Both mechanisms keep the third DE control parameter NP fixed during the optimization process. They both use the same DE strategy, same mutation, crossover, and selection operations, even the same initial population, and they both use self-adaptation at individual level.

1 Introduction

The Differential Evolution (DE) [13, 17, 21] algorithm was proposed by Storn and Price, and since then it has been used during many practical cases. The original DE was modified and many new versions have been proposed [13, 16, 17].

The original DE algorithm keeps all three control parameters fixed during the optimization process. However, there still exists a lack of knowledge on how to obtain reasonably good values for the control parameters of DE, over a given function [16, 22]. The necessity for changing control parameters during the optimization process was confirmed, based on the experiment in [8].

Self-adaptation has proved to be highly beneficial when automatically and dynamically adjusting control parameters. Self-adaptation is usually used in Evolution Strategies [4, 5, 6]. Self-adaptation allows an evolution strategy to adapt itself to any general class of problem, by reconfiguring itself accordingly without any user interaction [2, 3, 12]. DE with self-adaptive control parameters has already been presented in [8, 22].

In this analysis the unconstrained benchmark functions will be used. There are many studies that use DE algorithm in different research areas but, based on our knowledge, there is no current study, regarding the analyses of self-adaptive control parameters in DE algorithm.

U.K. Chakraborty (Ed.): Advances in Differential Evolution, SCI 143, pp. 89–110, 2008.
springerlink.com

This chapter makes the following contributions: (1) the application of a self-adaptive mechanism from evolution strategies to the original DE algorithm to construct a new version of the self-adaptive DE algorithm; (2) comparative study of the proposed DE algorithm with self-adaptive F and CR control parameters, the jDE algorithm, and the original DE algorithm; (3) analysis of how the control parameters are being changed during the evolutionary process.

The chapter is structured as follows. Section 2 gives an overview of work dealing with DE. Section 3 gives a brief background of the original differential evolution algorithm. Section 4 describes those differential evolution algorithms, which use self-adaptive adjusting control parameters. Two different self-adaptive mechanisms are described. Section 5 presents experimental results on the benchmark functions and gives performance comparisons for the self-adaptive and original DE algorithms. Discussion of the obtained results is given in Section 6. Section 7 concludes the chapter with some final remarks.

2 Work Related to Adaptation in Differential Evolution

Ali and Törn in [1] proposed new versions of the DE algorithm, and also suggested some modifications to the classical DE in order to improve its efficiency and robustness. They introduced an auxiliary population of NP individuals alongside the original population (noted in [1], a notation using sets is used – population set-based methods). Next they proposed a rule for calculating the control parameter F, automatically. Liu and Lampinen [16] proposed a version of DE, where the mutation control parameter and the crossover control parameter are adaptive. Teo in [22] proposed an attempt at self-adapting the population size parameter, in addition to self-adapting crossover and mutation rates. Brest et al. in [8] proposed a DE algorithm, using a self-adapting mechanism on the F and CR control parameters. The performance of the self-adaptive differential evolution algorithm was evaluated on the set of benchmark functions provided for constrained real parameter optimization [10]. In [18] Qin and Suganthan proposed the Self-adaptive Differential Evolution algorithm (SaDE), where the choice of learning strategy and the two control parameters F and CR do not require pre-defining. During evolution, suitable learning strategy and parameter settings are gradually self-adapted, according to the learning experience. Brest et al. [7] reported the performance comparison of certain selected DE algorithms, which use different self-adaptive or adaptive control parameter mechanisms.

In our paper [11] we presented experimental results on how control parameters are being changed during the evolutionary process on the constrained real parameter optimization benchmark functions (CEC2006 [10, 14]).

3 The Original DE Algorithm

In this section we give some background on the DE algorithm [19, 20, 21] that is important for understanding the rest of this chapter.

Differential Evolution (DE) is a floating-point encoding evolutionary algorithm for global optimization over continuous spaces [13, 16, 17, 21], which can also work with discrete variables. DE creates new candidate solutions by combining the parent individual and several other individuals of the same population. A candidate replaces the parent only if it has better fitness value. DE has three control parameters: the amplification factor of the difference vector – F, crossover control parameter – CR, and population size – NP.

The general problem an optimization algorithm is concerned with is to find a vector $\mathbf{x}$ so as to optimize $f(\mathbf{x}); \mathbf{x} = (x_1, x_2, ..., x_D)$. D is the dimensionality of the function f. The variables' domains are defined by their lower and upper bounds: $x_{j,low}, x_{j,upp}$; $j \in \{1, ..., D\}$. The initial population is selected uniform randomly between the lower ($x_{j,low}$) and upper ($x_{j,upp}$) bounds defined for each variable x_j. These bounds are specified by the user according to the nature of the problem.

DE is a population-based algorithm and vector $\mathbf{x}_{i,G}$, $i = 1, 2, ..., NP$ is an individual in the population. NP denotes population size and G the generation. During one generation for each vector, DE employs mutation, crossover and selection operations to produce a trial vector (offspring) and to select one of those vectors with the best fitness value.

By mutation for each population vector a mutant vector $\mathbf{v}_{i,G}$ is created. One of the most popular DE mutation strategy is *'rand/1/bin'* [17, 21]:

$$\mathbf{v}_{i,G} = \mathbf{x}_{r_1,G} + F \times (\mathbf{x}_{r_2,G} - \mathbf{x}_{r_3,G}) \tag{1}$$

where the indexes r_1, r_2, r_3 represent the random and mutually different integers generated within the range $[1, NP]$ and also different from index i. F is an amplification factor of the difference vector within the range $[0, 2]$, but usually less than 1.

The original DE algorithm is described very well in literature [17, 21], and, therefore, we will skip a detailed description of the whole DE algorithm.

4 Self-Adaptive DE Algorithms

In this section we describe two different self-adaptive mechanisms of control parameters in the DE algorithm. Both mechanisms use self-adaptation of control parameters at the individual level. The first mechanism uses uniform distribution for changing the values of the control parameter, while the second uses a self-adaptive mechanism found in evolution strategies.

4.1 The Self-Adaptive Control Parameters in a jDE Algorithm

Self-Adaptive DE refers to the self-adaptive mechanism on the control parameters, as proposed by Brest et al. [8]. This self-adapting mechanism uses the already exposed *'rand/1/bin'* strategy (see formula (1)).

In [8] a self-adaptive control mechanism was used to change the control parameters F and CR during the evolutionary process. The third control parameter NP was kept unchanged.

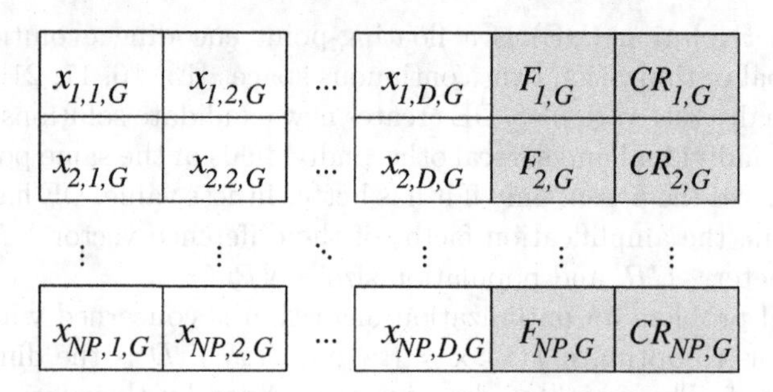

Fig. 1. Self-adapting control parameters F and CR are encoded into the individual. The vector of each individual $\mathbf{x}_{i,G}$ is extended by the values of two control parameters: $F_{i,G}$ and $CR_{i,G}$.

Each individual in the population was extended using the values of these two control parameters (see Figure 1). Both of them were applied at an individual level. Better values for these (encoded) control parameters lead to better individuals which, in turn, are more likely to survive and produce offspring and, hence, propagate these better parameter values.

New control parameters $F_{i,G+1}$ and $CR_{i,G+1}$ were calculated as follows:

$$F_{i,G+1} = \begin{cases} F_l + rand_1 \times F_u & \text{if } rand_2 < \tau_1, \\ F_{i,G} & \text{otherwise}, \end{cases} \tag{2}$$

$$CR_{i,G+1} = \begin{cases} rand_3 & \text{if } rand_4 < \tau_2, \\ CR_{i,G} & \text{otherwise}. \end{cases} \tag{3}$$

They produce control parameters F and CR in a new vector. $rand_j, j \in \{1, 2, 3, 4\}$ are uniform random values $\in [0, 1]$. τ_1 and τ_2 represent the probabilities of adjusting control parameters F and CR, respectively. τ_1, τ_2, F_l, F_u were taken fixed values $0.1, 0.1, 0.1, 0.9$, respectively. The new F takes a value from $[0.1, 1.0]$ in a random manner. The new CR takes a value from $[0, 1]$. $F_{i,G+1}$ and $CR_{i,G+1}$ are obtained before the mutation is performed, so they influence the mutation, crossover, and selection operations of the new vector $\mathbf{x}_{i,G+1}$.

4.2 The SA-DE Algorithm

As mentioned earlier, evolution strategies [6] are well-known for including a self-adaptive mechanism, encoded directly in each individual of the population. An evolution strategy (ES) has a notation $\mu/\rho, \lambda$-ES, where μ is parent population size, ρ is the number of parents for each new individual, and λ is child population size. An individual is denoted as $\mathbf{a} = (\mathbf{x}, \mathbf{s}, F(\mathbf{x}))$, where $\mathbf{x}$ are search parameters, $\mathbf{s}$ are control parameters, and $F(\mathbf{x})$ is the evaluation of the individual.

We used the idea of self-adaptive mechanism from evolution strategies and applied this idea to the original DE. We shall name the new constructed version of DE, the SA-DE algorithm.

Each individual (see Figure 1) of the SA-DE algorithm is extended to include self-adaptive F and CR control parameters in a similar way as in the jDE algorithm.

A trial vector is composed by mutation and recombination for each individual in population. The mutation procedure is different in the SA-DE algorithm in comparison to the original DE. For adapting the amplification factor of the difference vector F_i for trial individual i, from parent generation G into child generation $G+1$ for the trial vector, the following formula is used:

$$F_{i,G+1} = \langle F_G \rangle_i \times e^{\tau N(0,1)}, \tag{4}$$

where τ denotes the learning factor and is equal to $\tau = 1/\sqrt{2D}$, D being the dimension of the problem. $N(0,1)$ is a random number with a Gauss distribution. The $\langle F_G \rangle_i$ denotes the averaging of the parameters F of individuals i, r_1, r_2, and r_3 from generation G:

$$\langle F_G \rangle_i = \frac{F_{i,G} + F_{r_1,G} + F_{r_2,G} + F_{r_3,G}}{4}. \tag{5}$$

An analogous formula is used for CR of the trial individual i:

$$CR_{i,G+1} = \langle CR_G \rangle_i \times e^{\tau N(0,1)}, \tag{6}$$

where the τ used here is the same as for the adaptation of the F parameter. The $\langle CR_G \rangle_i$ denotes the averaging of the parameters again:

$$\langle CR_G \rangle_i = \frac{CR_{i,G} + CR_{r_1,G} + CR_{r_2,G} + CR_{r_3,G}}{4}. \tag{7}$$

The recombination process is not affected by our strategy, but rather taken from the strategy *'rand/1/bin'* (see Eq. (1)) of the original DE, and the adapted CR_i is used for each individual. The selection principle also helps in adapting F and CR, because only the individuals adapting good parameters can survive.

During the experiments, the following parameter settings were used for the SA-DE algorithm: the global lower and upper bounds for control parameter F were $0.3 \leq F \leq 1.1$, and for control parameter CR were $1/D \leq CR \leq 1$.

5 Experimental Results

The benchmark function test suite used in the experiments for this work, is presented in Table 4. A detailed description about test functions is given in [23]. All the included functions are to be minimised and have the same number of parameters, but they are tested with different number of function evaluations (FES), have different search space domains, and test various optimizer characteristics. Based on these functions, a performance evaluation of the listed algorithms is applied here.

Parameters settings for the jDE and SA-DE algorithms were presented in the previous section. Population size ($NP = 100$) was fixed for all algorithms in all

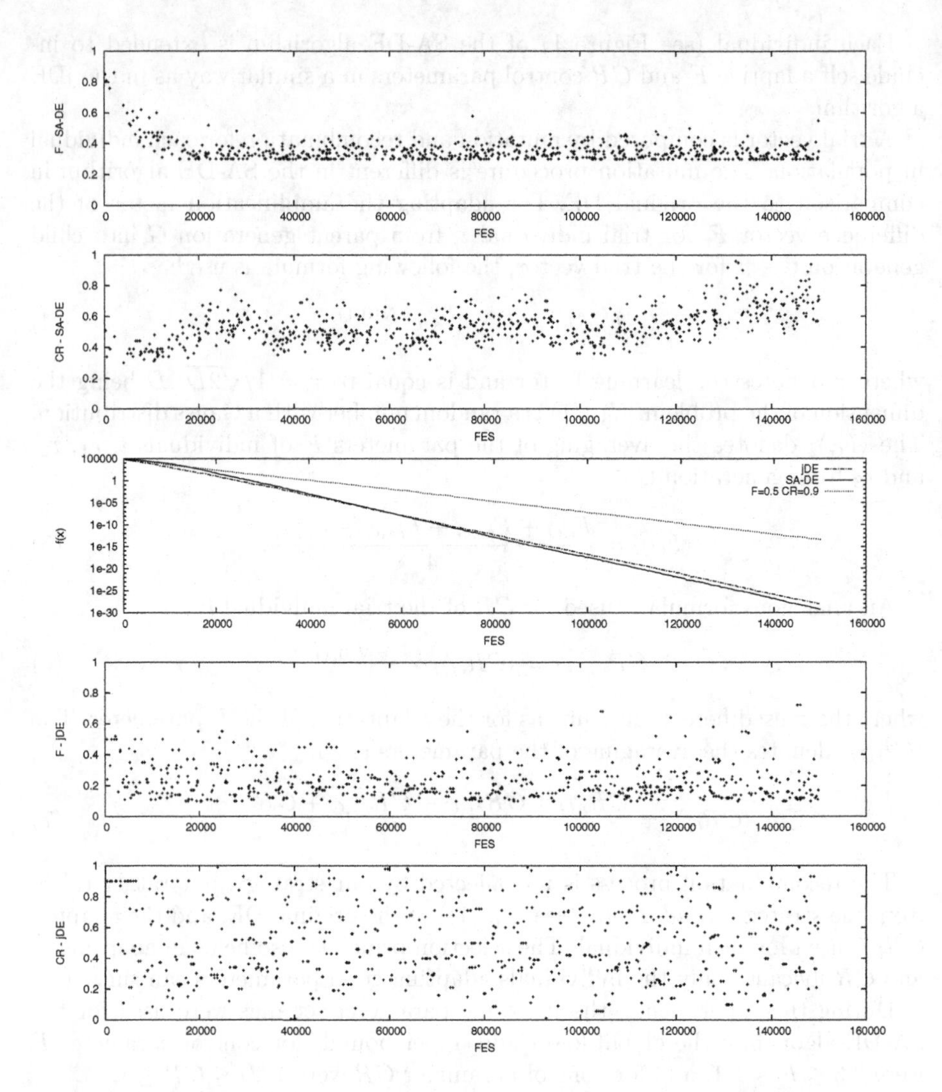

Fig. 2. Values of F and CR control parameters and fitness values of algorithms over one run for function f_1

experiments. As already mentioned, it was of particular interest how the control parameters are being changed during the evolutionary process.

Figures 2–13 show the values for initial parameters F and CR, and the convergence graphs for benchmark functions. Each figure has five sub-figures. Let us describe the sub-figures from the top to the bottom: the first two sub-figures represent the values for the SA-DE algorithm, the first one representing the values of control parameter F and the second the values for the CR control parameter. The fourth and fifth sub-figures represent the same control parameters for

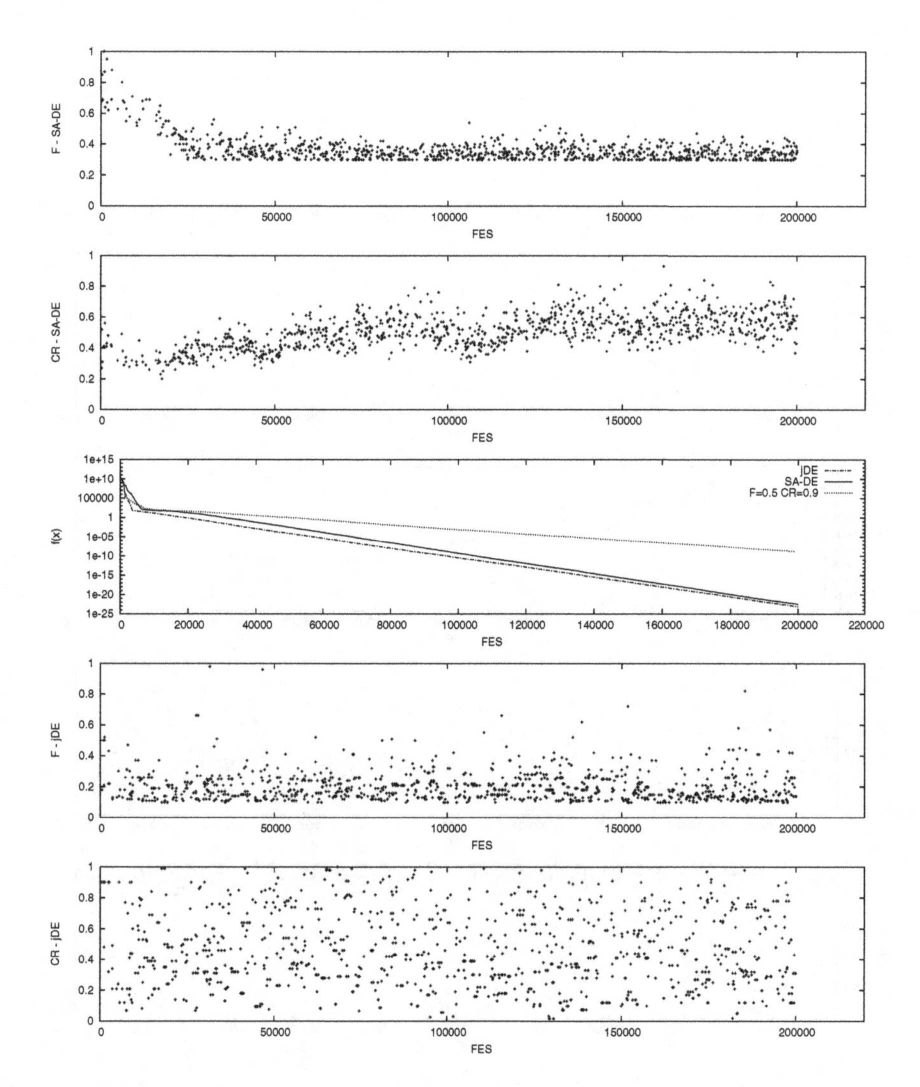

Fig. 3. Values of F and CR control parameters and fitness values of algorithms over one run for function f_2

the jDE algorithm as the first two sub-figures for the SA-DE algorithm. Third sub-figure presents convergence graphs of the fitness values for the jDE, SA-DE, and original DE algorithms. The original DE algorithm used fixed values for control parameters $F = 0.5$ and $CR = 0.9$. All algorithms used the same initial population (same seed for random generator).

Figure 2 shows the results obtained by typical evolutionary run, for function f_1. Both self-adaptive algorithms obtained similar results on the convergence

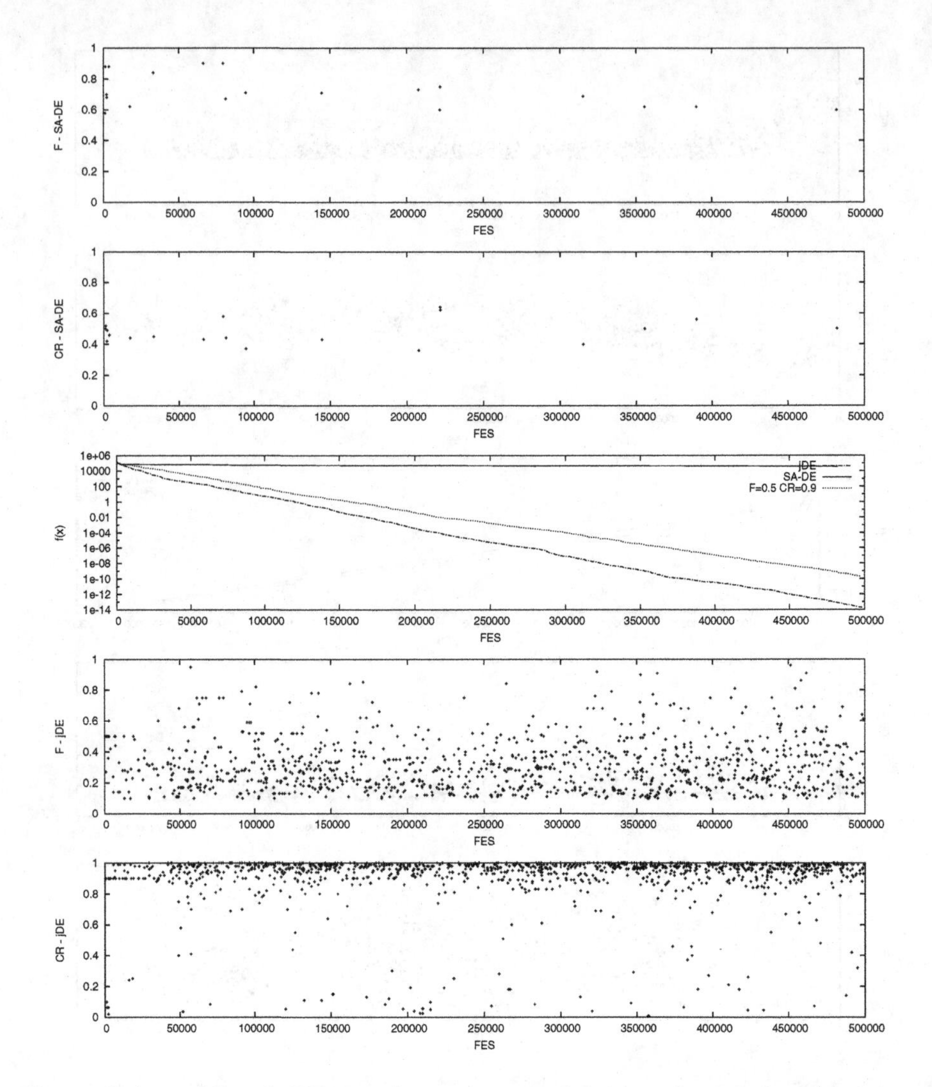

Fig. 4. Values of F and CR control parameters and fitness values of algorithms over one run for function f_3

graph, but the graphs for control parameters F and CR differ. The original DE algorithm obtained the worst results on the fitness convergence graph.

It can be noticed from Figure 3, that convergence graphs for the fitness values regarding the SA-DE and jDE algorithms are very similar (overlapped). The values for control parameter F are, in most cases, less than 0.5 for both algorithms. The original DE algorithm obtained the worst results regarding the fitness convergence graph.

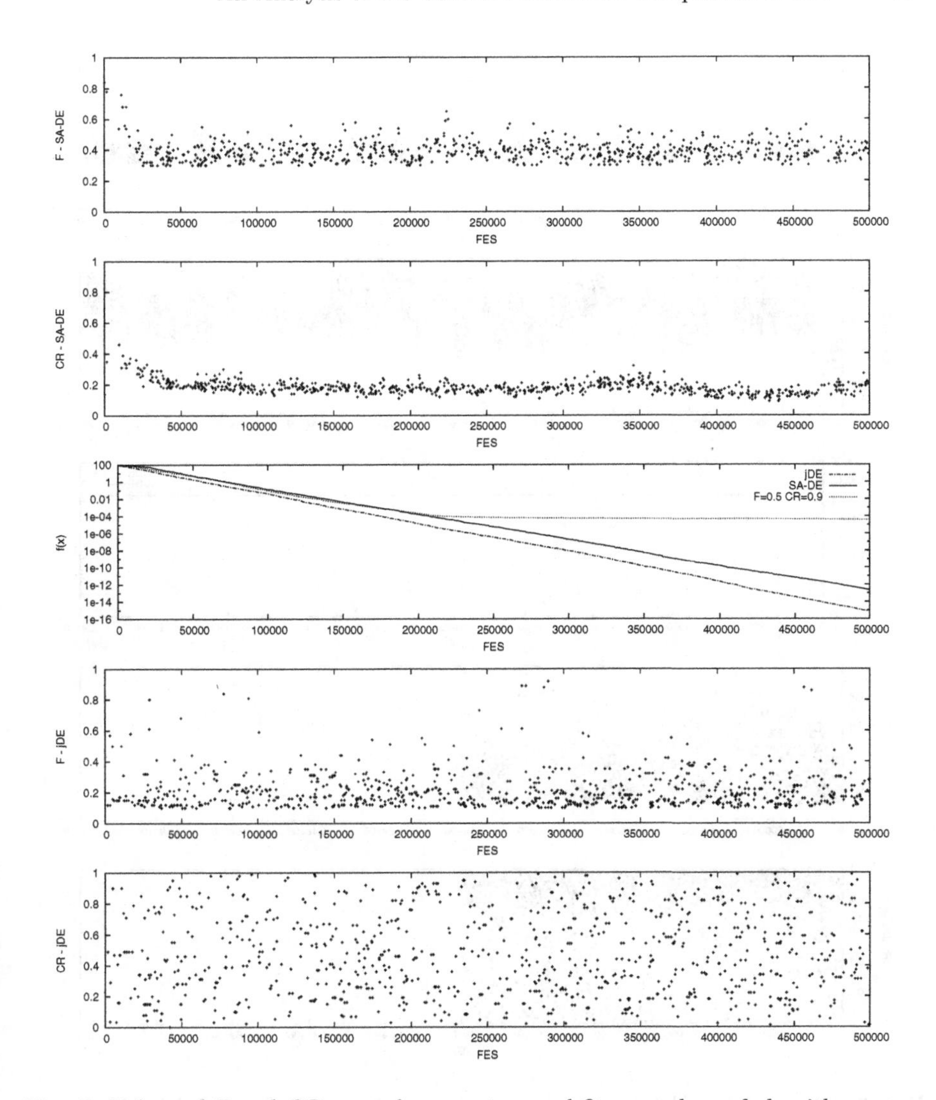

Fig. 5. Values of F and CR control parameters and fitness values of algorithms over one run for function f_4

Figure 4 shows the obtained results for function f_3, where the jDE algorithm outperformed the SA-DE algorithm. A very small number of the currently best fitness value's improvement occurred by the SA-DE algorithm. This algorithm obtained the worst results on the convergence graph.

If we compare the values for the control parameters F and CR of the jDE algorithm in Figures 2–4, for functions f_1, f_2, and f_3, a similarity to the control parameter F can be noticed: there are more values for F less than 0.5. Values for control parameter CR are equally distributed from 0 to 1 for functions f_1

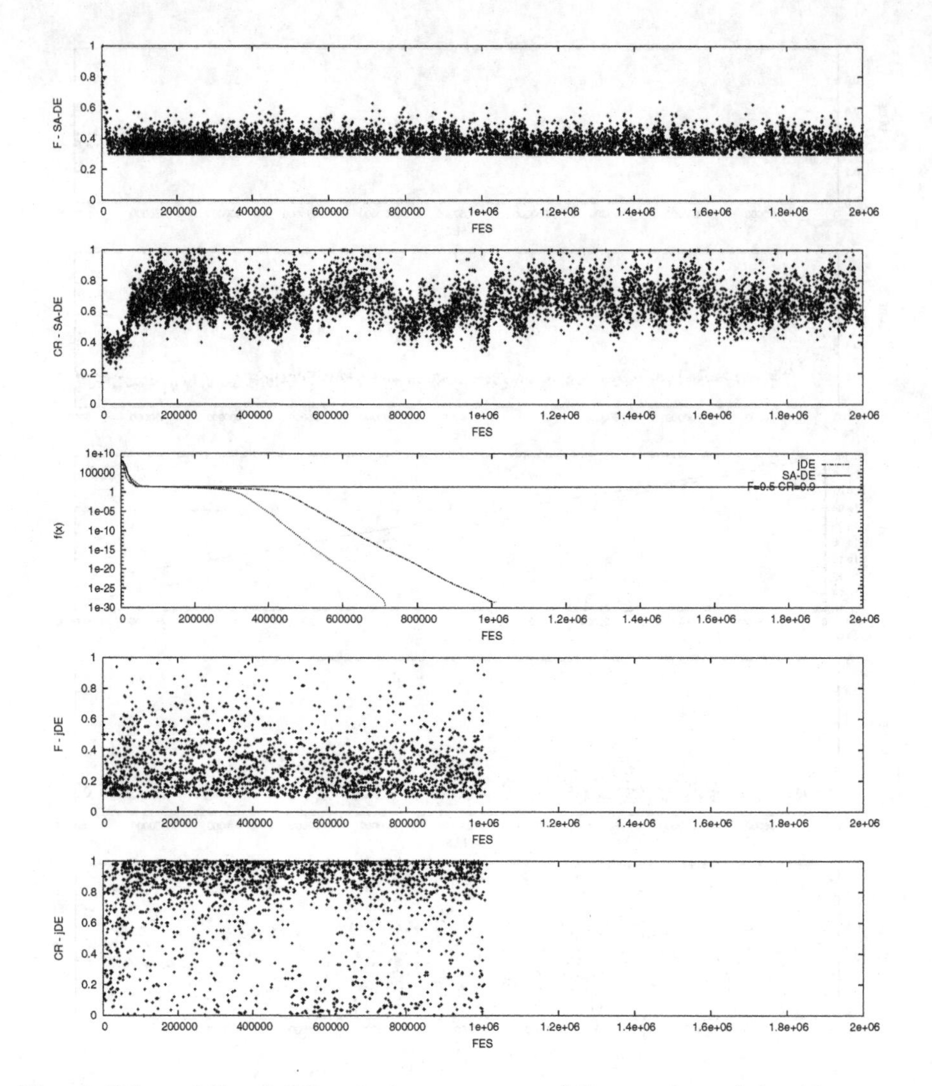

Fig. 6. Values of F and CR control parameters and fitness values of algorithms over one run for function f_5

and f_2, while the values of CR are very high (CR is greater than 0.8) in the case of function f_3.

The jDE and SA-DE algorithms obtained quite similar results on convergence graphs for function f_4 (see Figure 5). The jDE algorithm performed slightly better. The original DE algorithm obtained the worst results on the fitness convergence graph. It did not obtain much improvement in the fitness values after a half of the predefined maximum number of function evaluations was reached.

Figure 6 shows the obtained results for function f_5. In this case the best results ware obtained by the original DE algorithm, followed by the jDE algorithm.

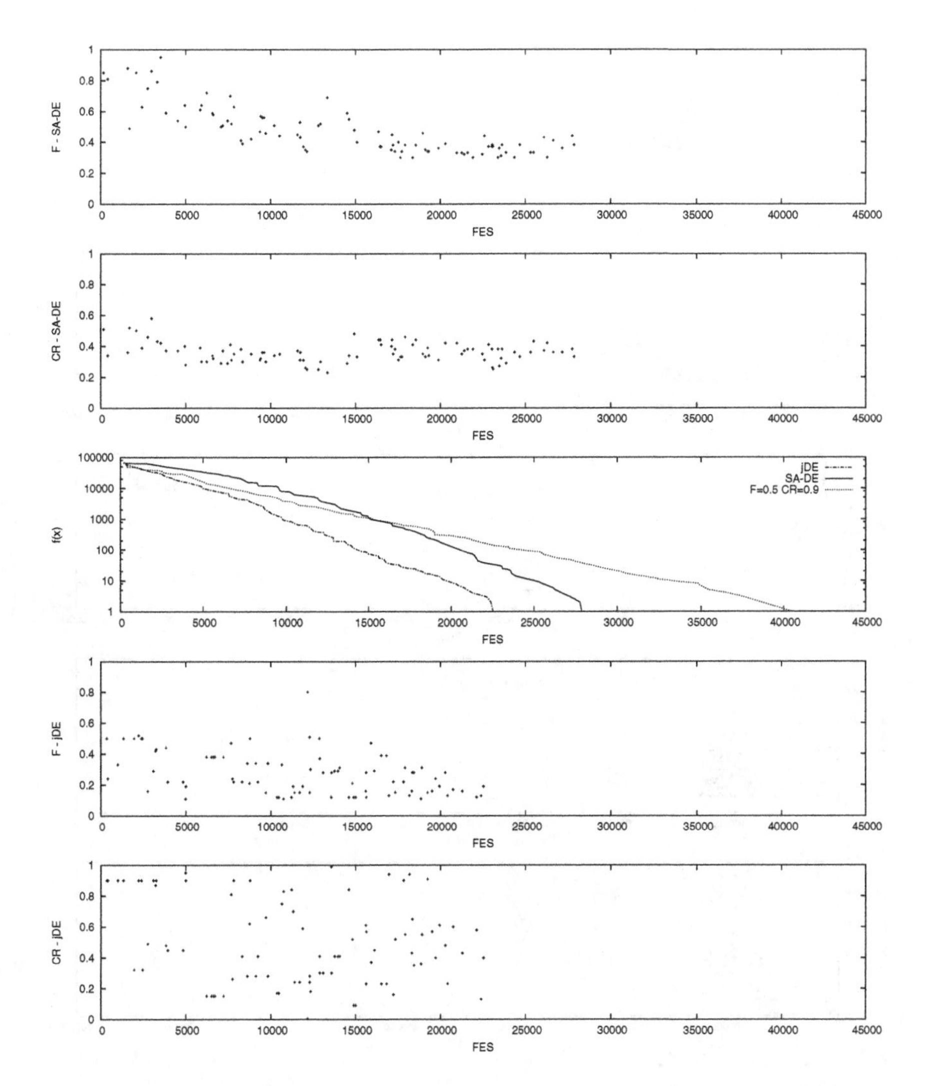

Fig. 7. Values of F and CR control parameters and fitness values of algorithms over one run for function f_6

The SA-DE algorithm got trapped in local optimum and, therefore obtained the worst result on the convergence graph. It can be noticed that both SA-DE and jDE algorithms conducted a great number of improvements of the currently best individual fitness value during the evolutionary process. CR values are usually high ($CR > 0.7$) for the jDE algorithm.

Figure 7 shows that all algorithms succeeded in solving function f_6. A slightly better performance was obtained by the jDE algorithm, followed by the SA-DE algorithm.

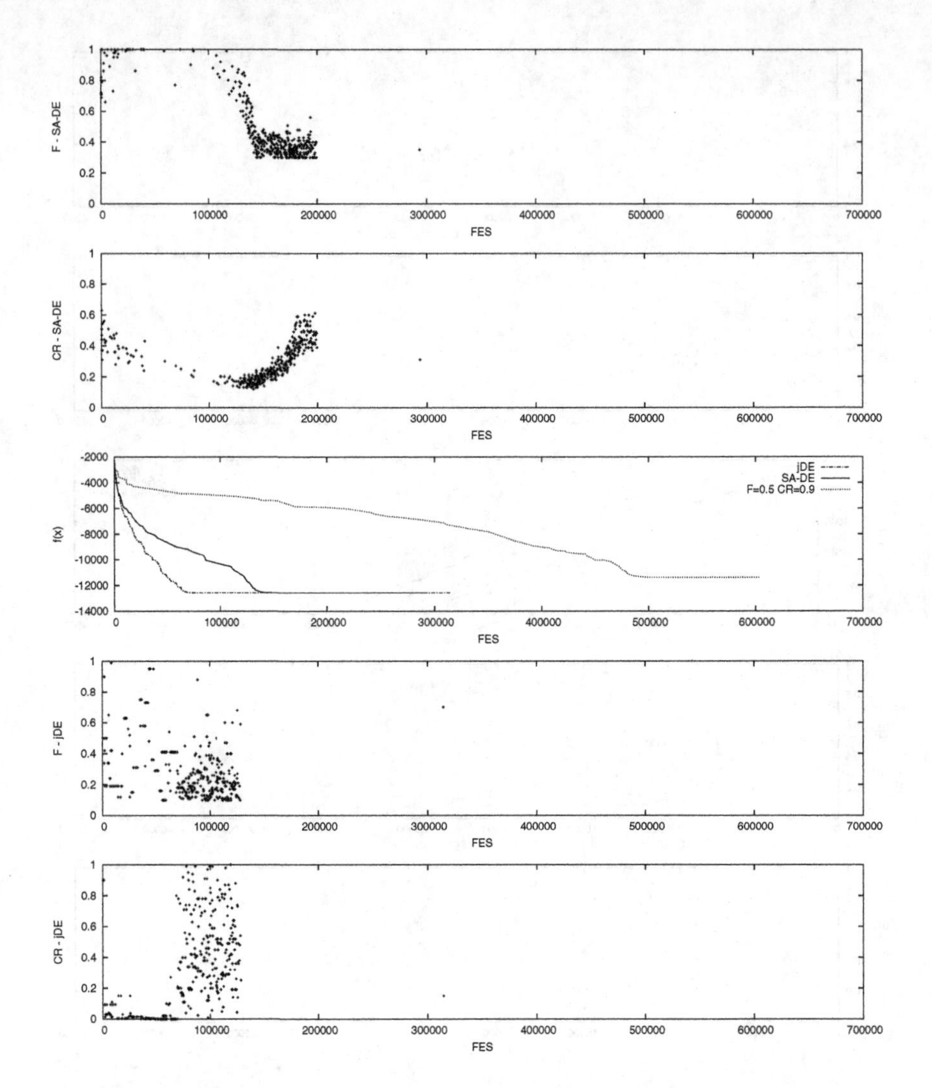

Fig. 8. Values of F and CR control parameters and fitness values of algorithms over one run for function f_8

The self-adaptive algorithms jDE and SA-DE performed much better than the original DE algorithm regarding function f_8 (see Figure 8). The original DE algorithm did not even get close to global optimum. It found the fitness value −11382.06.

Figure 9 shows the results for function f_9. The best performance results for function f_9 were obtained by the jDE algorithm. The worst performance was obtained by the original DE algorithm. When comparing sub-figures with F and CR for self-adaptive algorithms, it can seen that CR values by both algorithms are very low ($CR < 0.2$).

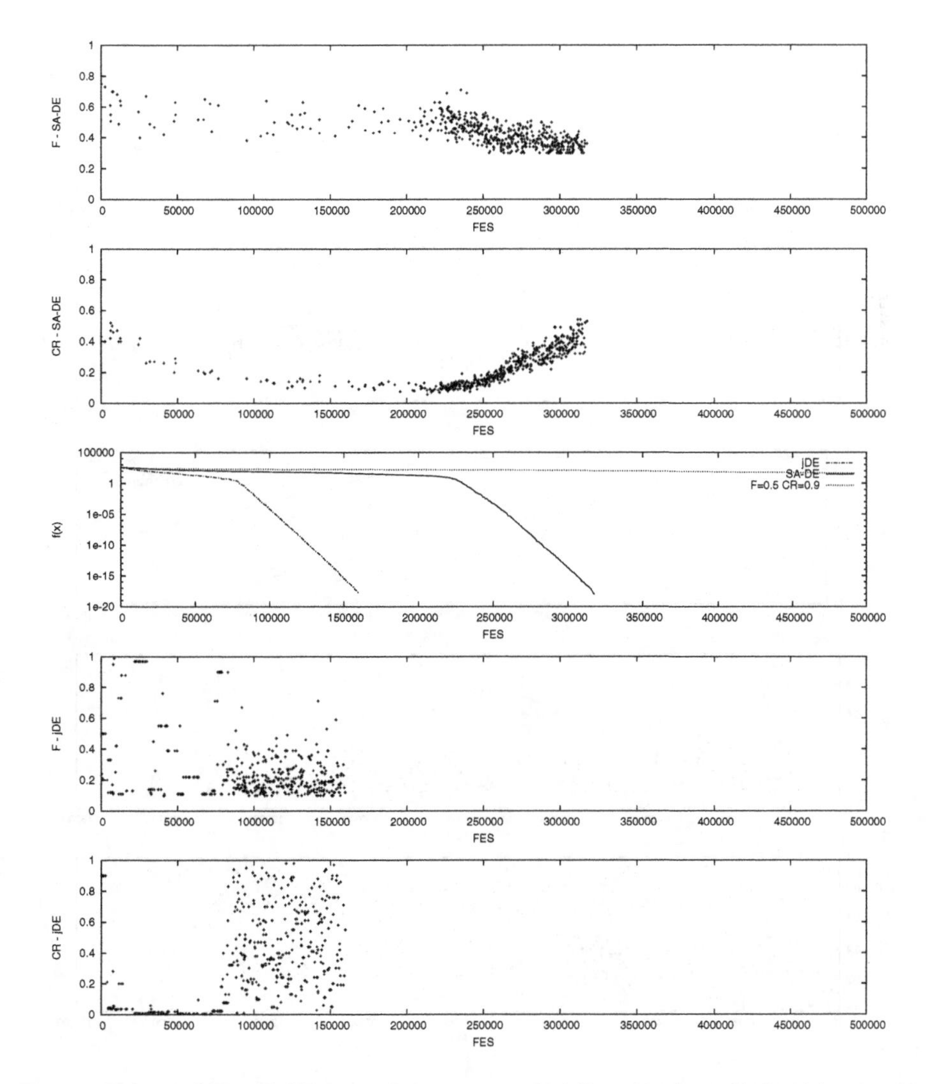

Fig. 9. Values of F and CR control parameters and fitness values of algorithms over one run for function f_9

Figure 9 shows that the jDE algorithm performed best for function f_{10}. If we look at Figures 8–10, it can be noticed that more improvements to the currently best individual occurred by the SA-DE algorithm after approximately 100 000, 200 000 and 70 000 FES for functions f_8, f_9, and f_{10}, respectively.

Figures 11–13 show the results for functions f_{11}, f_{12}, and f_{13}, respectively. For those functions both self-adaptive algorithms obtained better performance than the original DE algorithm. The values for the F control parameter were less than 0.5, while the CR values were equally distributed between 0 and 1 for

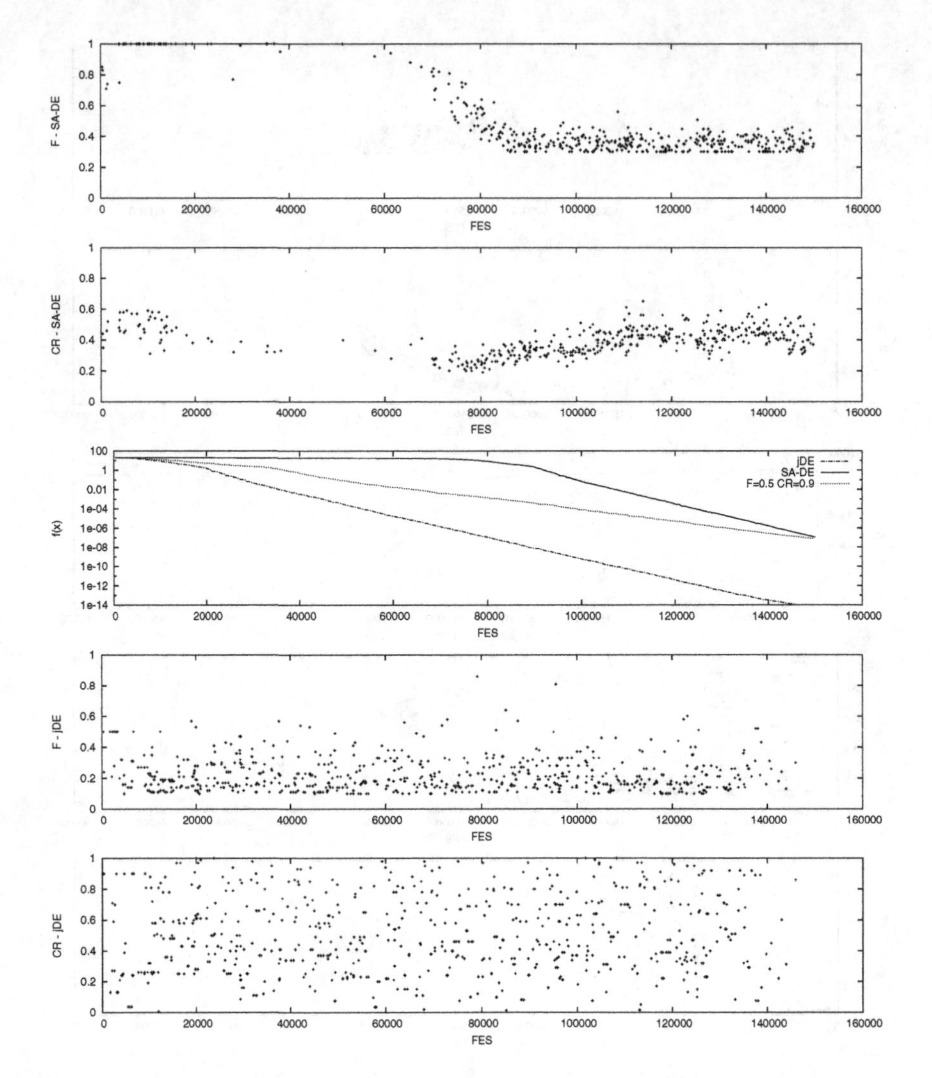

Fig. 10. Values of F and CR control parameters and fitness values of algorithms over one run for function f_{10}

functions f_{11}–f_{13}. For these functions the F values were usually less than 0.5 for the SA-DE algorithm, while the CR values were between 0.3 and 0.7.

The most important conclusion based on the results from Figures 2–13 is that the F and CR values obtained by the self-adaptive jDE and SA-DE algorithms differ. Actually, they also differ for functions, where the convergence graph shows almost equal algorithm performances (f_1, f_2, f_4, f_{11}, f_{12}, and f_{13}).

Figures 2–13 show the obtained results for one typical run of algorithms. Our description of the obtained results is only one point of view on how the control parameters of self-adaptive DE algorithms are being changed during the

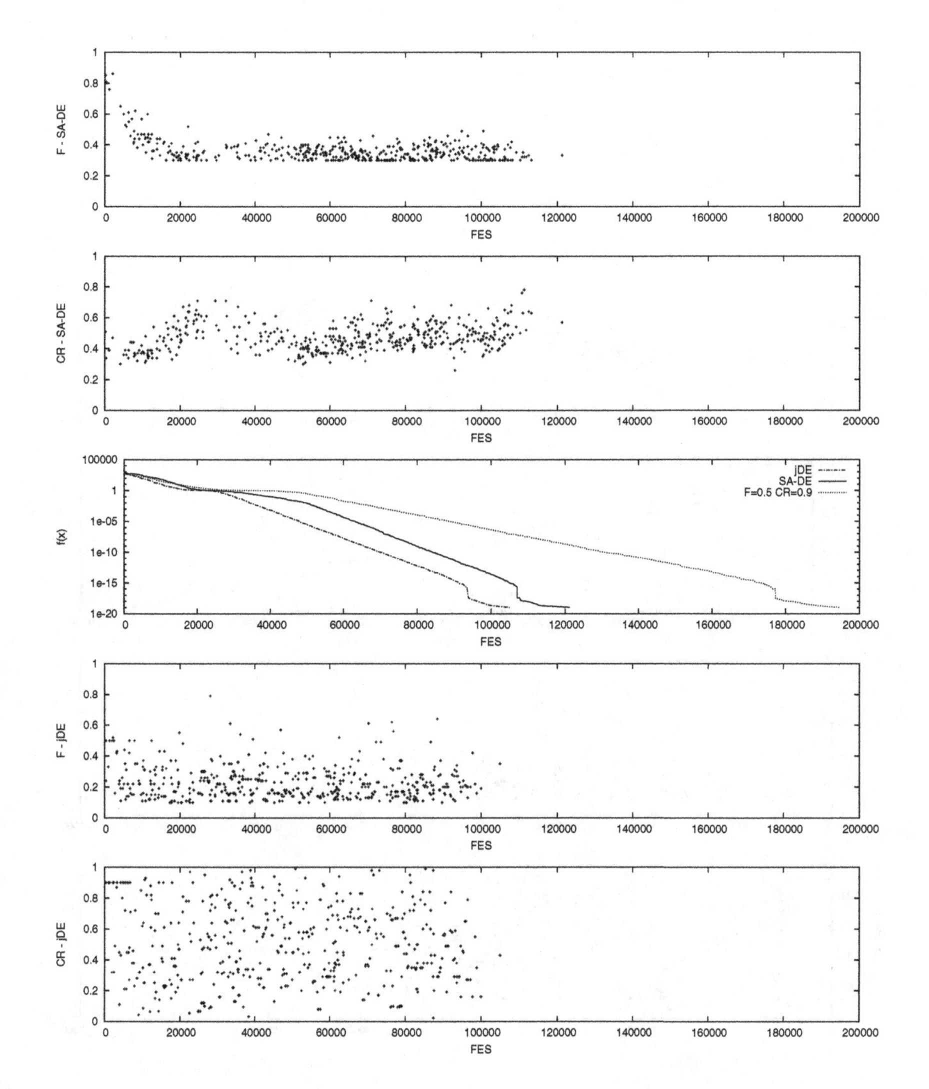

Fig. 11. Values of F and CR control parameters and fitness values of algorithms over one run for function f_{11}

optimization process, and on how changes in control parameters F and CR have an influence on the fitness value of the convergence graph.

Table 1 shows the obtained results for the three algorithms. The jDE algorithm performed well, on average. It obtained the best results for some benchmark functions, but it did not optimize the function f_5 so well, because it got trapped in a local optimum once. Similar observations were gathered for function f_{12}. The SA-DE algorithm gets the best results for the functions f_1, f_{12}, and f_{13}, while it has the worst results for functions f_3, f_8 (only a small number of missed global

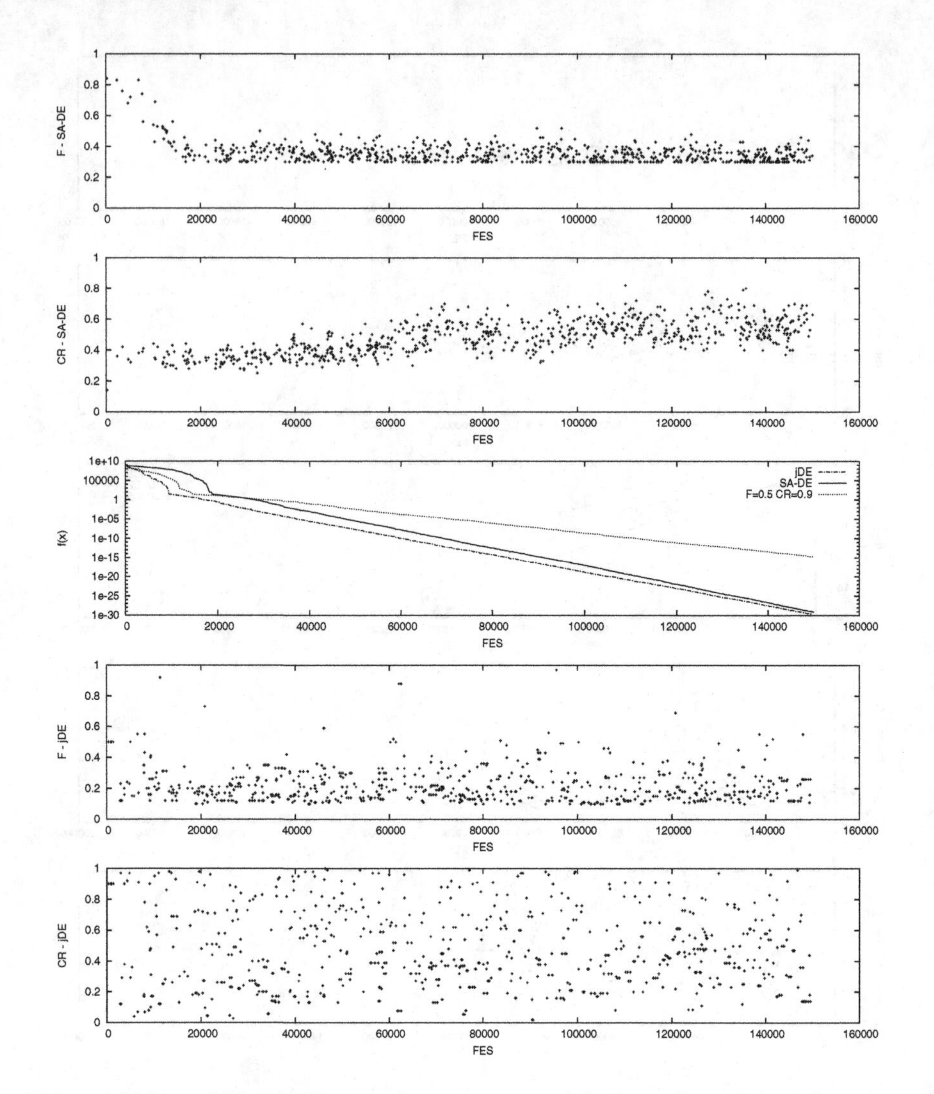

Fig. 12. Values of F and CR control parameters and fitness values of algorithms over one run for function f_{12}

optima), and f_5. The original DE algorithm performs well, but convergence speed is lower than for the self-adaptive algorithms on some benchmark functions.

Table 2 shows the average values for the control parameters F and CR, obtained in the experiment during one evolutionary run. The average values for F using the jDE algorithm are between 0.2 and 0.35, while CR values are more evenly distributed over the $[0, 1]$ interval. For the functions f_8 and f_9, the values are both around 0.35. For the function f_3, the value of the CR control parameter is approximately 0.9, and for the other functions, the CR values are around

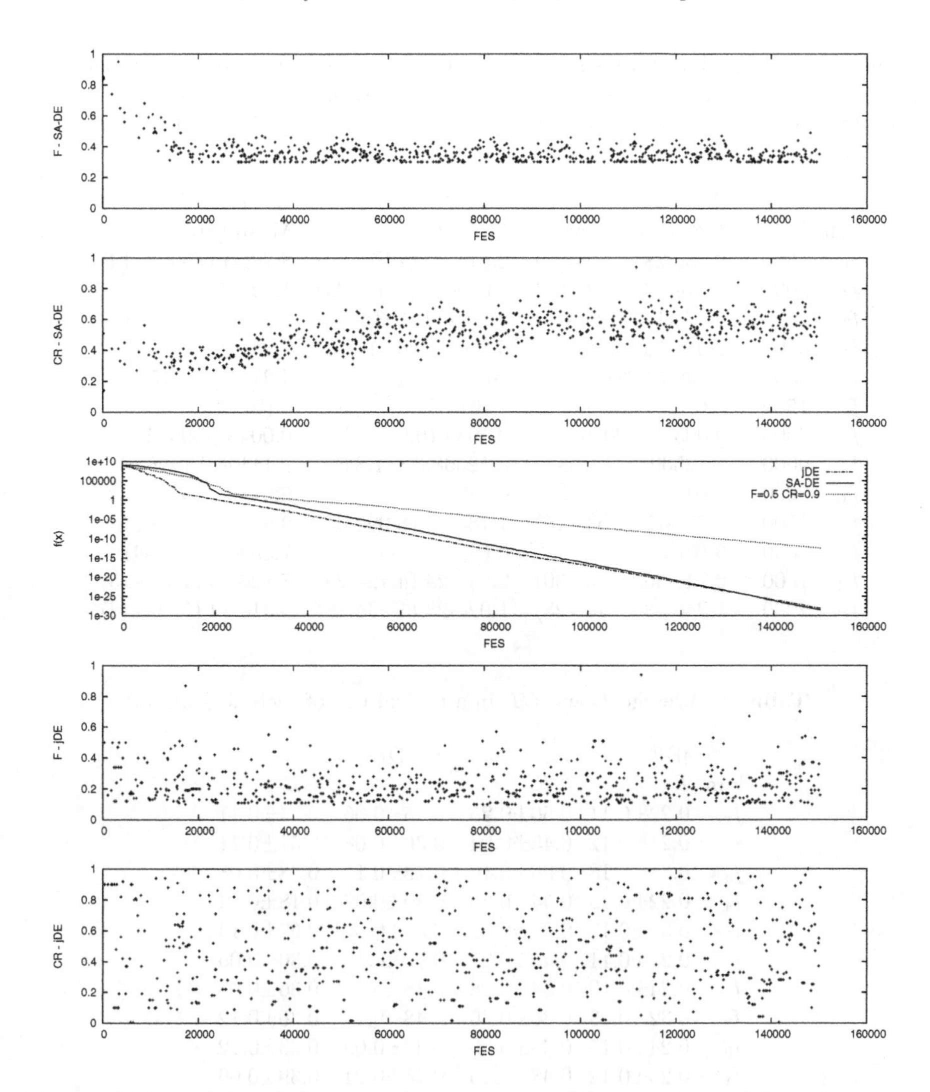

Fig. 13. Values of F and CR control parameters and fitness values of algorithms over one run for function f_{13}

0.5. Remember that the initial values for the control parameters using the jDE algorithms were $F = 0.5$ and $CR = 0.9$.

The next algorithm in Table 2 is the SA-DE algorithm. After evolutionary process, the obtained average F and CR were quite different for each function. For functions f_1, f_2, f_4, f_5, f_{11}, f_{12}, and f_{13}, the obtained average F was between $[0.35, 0.39]$. On these functions, SA-DE was quite successful, with the exception of f_5. For these functions, an average CR parameter was also quite similar, between $[0.47, 0, 54]$, with f_5 again being an exception. A greater CR indicates that many parameters still have to be changed to reach global optimum. In

Table 1. The experimental results, averaged over 100 independent runs, of the jDE algorithm, SA-DE algorithm, and the original DE algorithm ($F = 0.5$, $CR = 0.9$ 'Mean' indicates the average of the minimum best values obtained and 'Std.Dev' stands for the standard deviation

		jDE	SA-DE	DE $F_{0.5}CR_{0.9}$
Fun.	Gen.	Mean (Std.Dev)	Mean (Std.Dev)	Mean (Std.Dev)
f_1	1500	2.83e-28 (2.54e-28)	2.61e-29 (2.97e-29)	8.79e-14 (5.83e-14)
f_2	2000	1.51e-23 (9.13e-24)	4.08e-23 (4.02e-23)	1.42e-9 (9.95e-9)
f_3	5000	6.47e-14 (1.25e-13)	21645 (16103)	6.25e-11 (6.64e-11)
f_4	5000	2.08e-15 (3.18e-15)	5.32e-13 (6.31e-13)	7.35e-2 (1.17e-1)
f_5	20000	0.039 (0.02)	26.46 (7.24)	4.21e-31 (2.27e-30)
f_6	1500	0 (0)	0 (0)	0 (0)
f_7	3000	0.0031 (0.0009)	0.0038 (0.00086)	0.0046 (0.0014)
f_8	9000	−12569.5 (1.07e-11)	−12568.3 (11.84)	−11148.5 (496.6)
f_9	5000	0 (0)	0 (0)	68.18 (33.67)
f_{10}	1500	8.73e-15 (2.54e-15)	1.18e-05 (8.09e-05)	9.97e-8 (4.13e-8)
f_{11}	2000	0 (0)	0 (0)	7.39e-5 (7.39e-4)
f_{12}	1500	6.74e-30 (8.15e-30)	2.84e-29 (4.12e-29)	7.82e-15 (7.79e-15)
f_{13}	2000	1.24e-28 (1.44e-28)	1.02e-28 (1.33e-28)	5.31e-14 (5.76e-14)

Table 2. Average F and CR in a typical run of each algorithm

	jDE		SA-DE	
Fun.	F	CR	F	CR
f_1	0.22±0.11	0.50±0.25	0.35±0.06	0.54±0.11
f_2	0.21±0.12	0.45±0.24	0.36±0.08	0.51±0.11
f_3	0.31±0.18	0.90±0.19	0.72±0.11	0.48±0.08
f_4	0.22±0.13	0.44±0.25	0.39±0.07	0.18±0.04
f_5	0.30±0.17	0.69±0.32	0.38±0.06	0.67±0.13
f_6	0.21±0.11	0.49±0.27	0.47±0.16	0.36±0.06
f_7	0.24±0.10	0.54±0.28	0.48±0.13	0.36±0.05
f_8	0.32±0.23	0.36±0.30	0.48±0.21	0.30±0.12
f_9	0.24±0.15	0.35±0.29	0.43±0.09	0.23±0.12
f_{10}	0.23±0.12	0.48±0.25	0.45±0.21	0.39±0.09
f_{11}	0.23±0.11	0.47±0.25	0.36±0.07	0.47±0.09
f_{12}	0.22±0.11	0.48±0.25	0.36±0.06	0.48±0.11
f_{13}	0.22±0.11	0.47±0.25	0.36±0.07	0.51±0.11

the case of f_4, a CR drop allows a much more precise selection, which makes the algorithm perform better than the original DE on this function, because the overall function evaluation improvement is achieved by diminishing each x_i component. Another pattern can be observed with the functions f_6–f_{10}. The average F is between $[0.43, 0.48]$ here, and CR is smaller and between $[0.23, 0.39]$. For all these functions, the performance of the SA-DE algorithm is the same or better than the original DE, except for the f_8, where the convergence is not as rapid as with the other two algorithms.

Table 3. Number of improvements

Fun.	jDE	SA-DE	DE $F_{0.5}$ $CR_{0.9}$
f_1	805	726	353
f_2	1024	985	411
f_3	1224	19	387
f_4	898	729	870
f_5	2833	8106	1561
f_6	106	93	80
f_7	53	56	39
f_8	415	402	364
f_9	516	529	43
f_{10}	680	440	325
f_{11}	525	449	406
f_{12}	763	734	287
f_{13}	690	709	308

For the function f_3, which is the worst case for SA-DE, F and CR still remained quite high, keeping the algorithm from converging into a local optimum, thus promising a potential global optimum convergence.

Table 3 shows how many times the currently best individual was changed. The results were obtained during the same experiment used to obtain the results for control parameters F and CR, as reported in Table 2. From Table 3 it can be noticed that the original DE usually improved the currently best individual fewer times than for other algorithms. If we compare only both self-adaptive algorithms, the number of improvements is quite similar except for functions f_3 and f_5, where the SA-DE algorithm performed either little or high numbers of improvements, respectively. In both cases the jDE algorithm obtained better results (see Table 1).

The best setting for control parameters is problem dependent. Self-adaptation may help an algorithm to have higher convergence speed to global optimum. The results in this section show that no algorithm performed superiorly better than any other algorithm for all optimization problems.

6 Discussion

When introducing the SA-DE algorithm, we did not make fine-tunings of the τ learning parameter, which is still open for further research. The τ could have been separately defined for F and CR, or it could even be self-adapted, projecting several new experimental combinations to try out. Another constraint that could be changed or alleviated is our initialization phase and the bounds, mostly for F, in SA-DE, where the bounds could be extended or dynamically adapted. As has been confirmed in [15], the lower bound of F has indeed a strong impact on the algorithm convergence.

The SA-DE control parameters crossover process could also be changed to include all the members (in ES, called global intermediate – GI) in the population or the few random best ones.

Other possibilities for self-adaptation would be at the component level, where each component x_i of each individual would have its own control parameters. Another possibility is to encode control parameters at the population level, where same parameters are used for one generation – similarly, but not the same as the GI approach. We have indeed tried many combinations of these proposals, confirming that many research opportunities are still open.

The presented control parameters analysis did not include the NP parameter adaptation, which is also a candidate for future research. Readers are referred to our recent work [9].

7 Conclusion

The chapter presents two self-adaptive mechanisms in the DE. Both mechanisms are implemented at the individual level. Self-adaptation may help an algorithm to perform better for convergence speed, and an algorithm with self-adaptation may have greater robustness, on average.

Our goal in this work was not to make fine tunning of each self-adaptive mechanism to obtain the best result for a particular optimization problem, but rather to give some ideas on how to apply self-adaptive control parameters (F and CR) in a DE algorithm, in order to achieve better performance, in general.

References

[1] Ali, M.M., Törn, A.: Population Set-Based Global Optimization Algorithms: Some Modifications and Numerical Studies. Computers & Operations Research 31(10), 1703–1725 (2004)

[2] Bäck, T.: Adaptive Business Intelligence Based on Evolution Strategies: Some Application Examples of Self-Adaptive Software. Information Sciences 148, 113–121 (2002)

[3] Bäck, T., Fogel, D.B., Michalewicz, Z. (eds.): Handbook of Evolutionary Computation. Institute of Physics Publishing and Oxford University Press (1997)

[4] Bäck, T.: Evolutionary Algorithms in Theory and Practice: Evolution Strategies, Evolutionary Programming, Genetic Algorithms. Oxford University Press, New York (1996)

[5] Beyer, H.-G.: The theory of Evolution Strategies. Springer, Berlin (2001)

[6] Beyer, H.-G., Schwefel, H.-P.: Evolution strategies: a comprehensive introduction. Natural Computing 1, 3–52 (2002)

[7] Brest, J., Bošković, B., Greiner, S., Žumer, V., Sepesy Maučec, M.: Performance comparison of self-adaptive and adaptive differential evolution algorithms. Soft Computing - A Fusion of Foundations, Methodologies and Applications 11(7), 617–629 (2007)

[8] Brest, J., Greiner, S., Bošković, B., Mernik, M., Žumer, V.: Self-Adapting Control Parameters in Differential Evolution: A Comparative Study on Numerical Benchmark Problems. IEEE Transactions on Evolutionary Computation 10(6), 646–657 (2006)
[9] Brest, J., Sepesy Maučec, M.: Population Size Reduction for the Differential Evolution Algorithm. Applied Intelligence (accepted) DOI: 10.1007/s10489-007-0091-x
[10] Brest, J., Žumer, V., Sepesy Maučec, M.: Self-adaptive Differential Evolution Algorithm in Constrained Real-Parameter Optimization. In: The 2006 IEEE Congress on Evolutionary Computation CEC 2006, pp. 919–926. IEEE Press, Los Alamitos (2006)
[11] Brest, J., Žumer, V., Maučec, M.S.: Control Parameters in Self-Adaptive Differential Evolution. In: Filipič, B., Šilc, J. (eds.) Bioinspired Optimization Methods and Their Applications, Ljubljana, Slovenia, October 2006, pp. 35–44. Jožef Stefan Institute (2006)
[12] Eiben, A.E., Smith, J.E.: Introduction to Evolutionary Computing. In: Natural Computing. Springer, Berlin (2003)
[13] Feoktistov, V.: Differential Evolution: In Search of Solutions. Springer, New York (2006)
[14] Liang, J.J., Runarsson, T.P., Mezura-Montes, E., Clerc, M., Suganthan, N., Coello, C.A.C., Deb, K.: Problem Definitions and Evaluation Criteria for the CEC 2006 Special Session on Constrained Real-Parameter Optimization. Technical Report Report #2006005, Nanyang Technological University, Singapore and et al. (December 2005), http://www.ntu.edu.sg/home/EPNSugan
[15] Liang, K.-H., Yao, X., Newton, C.S.: Adapting self-adaptive parameters in evolutionary algorithms. Appl. Intell. 15(3), 171–180 (2001)
[16] Liu, J., Lampinen, J.: A Fuzzy Adaptive Differential Evolution Algorithm. Soft Computing - A Fusion of Foundations, Methodologies and Applications 9(6), 448–462 (2005)
[17] Price, K.V., Storn, R.M., Lampinen, J.A.: Differential Evolution, A Practical Approach to Global Optimization. Springer, Heidelberg (2005)
[18] Qin, A.K., Suganthan, P.N.: Self-adaptive Differential Evolution Algorithm for Numerical Optimization. In: The 2005 IEEE Congress on Evolutionary Computation CEC 2005, vol. 2, pp. 1785–1791. IEEE Press, Los Alamitos (2005)
[19] Rönkkönen, J., Kukkonen, S., Price, K.V.: Real-Parameter Optimization with Differential Evolution. In: The 2005 IEEE Congress on Evolutionary Computation CEC 2005, vol. 1, pp. 506–513. IEEE Press, Los Alamitos (2005)
[20] Storn, R., Price, K.: Differential Evolution - a simple and efficient adaptive scheme for global optimization over continuous spaces. Technical Report TR-95-012, Berkeley, CA (1995)
[21] Storn, R., Price, K.: Differential Evolution – A Simple and Efficient Heuristic for Global Optimization over Continuous Spaces. Journal of Global Optimization 11, 341–359 (1997)
[22] Teo, J.: Exploring dynamic self-adaptive populations in differential evolution. Soft Computing - A Fusion of Foundations, Methodologies and Applications 10(8), 673–686 (2006)
[23] Yao, X., Liu, Y., Lin, G.: Evolutionary Programming Made Faster. IEEE Transactions on Evolutionary Computation 3(2), 82–102 (1999)

Appendix

Table 4. Benchmark functions used in this study

Test function	D	S	f_{min}
$f_1(x) = \sum_{i=1}^{D} x_i^2$	30	$[-100, 100]^D$	0
$f_2(x) = \sum_{i=1}^{D} \lvert x_i \rvert + \prod_{i=1}^{D} \lvert x_i \rvert$	30	$[-10, 10]^D$	0
$f_3(x) = \sum_{i=1}^{D} (\sum_{j=1}^{i} x_j)^2$	30	$[-100, 100]^D$	0
$f_4(x) = \max_i \{\lvert x_i \rvert, 1 \leq i \leq D\}$	30	$[-100, 100]^D$	0
$f_5(x) = \sum_{i=1}^{D-1} [100(x_{i+1} - x_i^2)^2 + (x_i - 1)^2]$	30	$[-30, 30]^D$	0
$f_6(x) = \sum_{i=1}^{D} (\lfloor x_i + 0.5 \rfloor)^2$	30	$[-100, 100]^D$	0
$f_7(x) = \sum_{i=1}^{D} i x_i^4 + random[0, 1)$	30	$[-1.28, 1.28]^D$	0
$f_8(x) = \sum_{i=1}^{D} -x_i \sin(\sqrt{\lvert x_i \rvert})$	30	$[-500, 500]^D$	-12569.5
$f_9(x) = \sum_{i=1}^{D} [x_i^2 - 10\cos(2\pi x_i) + 10]$	30	$[-5.12, 5.12]^D$	0
$f_{10}(x) = -20\exp\left(-0.2\sqrt{\frac{1}{D}\sum_{i=1}^{D} x_i^2}\right) - \exp\left(\frac{1}{D}\sum_{i=1}^{D}\cos 2\pi x_i\right)$ $+20 + e$	30	$[-32, 32]^D$	0
$f_{11}(x) = \frac{1}{4000}\sum_{i=1}^{D} x_i^2 - \prod_{i=1}^{D}\cos\left(\frac{x_i}{\sqrt{i}}\right) + 1$	30	$[-600, 600]^D$	0
$f_{12}(x) = \frac{\pi}{D}\{10\sin^2(\pi y_i) + \sum_{i=1}^{D-1}(y_i - 1)^2[1 + 10\sin^2(\pi y_{i+1})]$ $+(y_D - 1)^2\} + \sum_{i=1}^{D} u(x_i, 10, 100, 4)$, $y_i = 1 + \frac{1}{4}(x_i + 1)$, $u(x_i, a, k, m) = \begin{cases} k(x_i - a)^m, & x_i > a, \\ 0, & -a \leq x_i \leq a, \\ k(-x_i - a)^m, & x_i < -a. \end{cases}$	30	$[-50, 50]^D$	0
$f_{13}(x) = 0.1\{\sin^2(3\pi x_1) + \sum_{i=1}^{D-1}(x_i - 1)^2[1 + \sin^2(3\pi x_{i+1})]$ $+(x_D - 1)^2[1 + \sin^2(2\pi x_D)]\} + \sum_{i=1}^{D} u(x_i, 5, 100, 4)$	30	$[-50, 50]^D$	0

Stopping Criteria for Differential Evolution in Constrained Single-Objective Optimization

Karin Zielinski and Rainer Laur

Institute for Electromagnetic Theory and Microelectronics, University of Bremen, P.O. Box 330440, 28334 Bremen, Germany
{zielinski,rlaur}@item.uni-bremen.de

Summary. Because real-world problems generally include computationally expensive objective and constraint functions, an optimization run should be terminated as soon as convergence to the optimum has been obtained. However, detection of this condition is not a trivial task. Because the global optimum is usually unknown, distance measures cannot be applied for this purpose. Stopping after a predefined number of function evaluations has not only the disadvantage that trial-and-error methods have to be applied for determining a suitable number of function evaluations, but the number of function evaluations at which convergence occurs may also be subject to large fluctuations due to the randomness involved in evolutionary algorithms. Therefore, stopping criteria should be applied which react adaptively to the state of the optimization run. In this work several stopping criteria are introduced that consider the improvement, movement or distribution of population members to derive a suitable time for terminating the Differential Evolution algorithm. Their application for other evolutionary algorithms is also discussed. Based on an extensive test set the criteria are evaluated using Differential Evolution, and it is shown that a distribution-based criterion considering objective space yields the best results concerning the convergence rate as well as the additional computational effort.

1 Introduction

Since the development of Differential Evolution (DE) in 1995 [1], considerable effort has been spend to improve its convergence characteristics e.g. by varying operators [2, 3] or changing the handling of constraints [4, 5]. As a consequence, several enhancements have been found during the last years that have led to successful applications in many different fields [6]. However, even the performance of a very good algorithm may be bad for practical purposes when it is not stopped at a proper time. For theoretical work about convergence properties or a comparison of different implementations of DE this aspect is generally not important because for this purpose usually test functions are employed for which the optimum is known. In that case, the execution of the algorithm can be terminated if the optimum is found with a given accuracy, and the involved computational effort can be used to analyze the performance of different DE implementations. An alternative is to terminate after a defined number of function evaluations (FEs) and to evaluate the distance of the best individual to the

U.K. Chakraborty (Ed.): Advances in Differential Evolution, SCI 143, pp. 111–138, 2008.
springerlink.com

optimum. This approach works well for theoretical work when algorithm variants are tested against each other but for real-world problems the situation is different because the optimum is usually unknown.

A stopping rule for problems with unknown optimum that is widely used in the literature is to terminate the execution of an algorithm after a given maximum number of function evaluations FE_{max} (this stopping criterion will be called *LimFuncEval* in the following). This approach is associated with two problems: Suitable settings for FE_{max} vary considerably for different optimization problems, so usually FE_{max} has to be figured out by trial-and-error methods. A further problem is that the number of objective function evaluations FE_{conv} that is needed for convergence for one specific optimization problem may also be subject to large variations due to the stochastic nature of DE. This statement holds for many different implementations of DE as can be seen in [3, 4, 5, 7, 8, 9, 10]. Because real-world problems usually contain computationally expensive objective and constraint functions it is imperative that unnecessary function evaluations are avoided. Therefore, it is important to examine other alternatives for stopping the execution of the DE algorithm besides termination after a fixed number of function evaluations. In order to deal with the problem that is caused by fluctuations of FE_{conv}, the stopping criteria have to be able to detect when convergence is reached. Thus, they have to react adaptively to the current state of an optimization run. The stopping criteria have to ensure that the algorithm is executed long enough to obtain convergence to the global optimum but without wasting of computational resources.

Different mechanisms can be used for deriving conclusions about the current state of an optimization run. In principle any phenomenon can be used that exhibits a definite trend from the beginning to the end of an optimization run. For instance both the improvement as well as the movement of individuals are typically large in the beginning of an optimization run and both become small when approaching convergence. Another example is the distribution of population members as they are scattered throughout the search space initially but usually converge to one point towards the end of an optimization run. Consequently, each of these properties is basically usable for detecting convergence.

Any of the before-mentioned population characteristics like improvement, movement and distribution can be used in various implementations for the creation of stopping conditions. Because the performance of different implementations is not necessarily similar, an extensive study analyzing their abilities will be presented in this chapter. Conclusions will be derived about which mechanisms are best suited to meet the demands of reliable stopping after convergence to the optimum has been obtained without wasting of computational resources.

Besides the problem with fluctuations of FE_{conv}, terminating after a fixed number of function evaluations is also connected with the problem that a suitable setting for parameter FE_{max} has to be found. Apparently, this kind of problem is inherent to all stopping criteria because up to now no parameter-free stopping criterion is known. The adaptive stopping criteria which will be presented in this chapter are also associated with parameters which have to be chosen by the user. Interestingly, it will be shown in this work that standard settings which can be used for a large range of optimization problems exist for some of them. As a consequence, the application of these stopping criteria is easy for the user.

One problem in the field of optimization is that authors often use different sets of test functions or different accuracies (for defining convergence to the optimum as well as for the allowed constraint violation of equality constraints). Hence, it is generally difficult to compare results. In contrast, a subset of a standardized well-defined test set that was specified in [11] for the Special Session on Constrained Real Parameter Optimization at the Congress on Evolutionary Computation 2006 (CEC06) is used in this work. The reason for using only a subset of the mentioned test set is that for the examination of stopping criteria it is reasonable to use optimization problems for which the employed algorithm is able to converge reliably, meaning that convergence is obtained in every optimization run. In that case, the evaluation of stopping criteria is simplified because a convergence rate of less than 100% can be considered to be a result of unsuitable stopping conditions. The performance of the DE variant that is used in the present examination has already been analyzed adhering to the demands of [11] in a former study [10]. Based on this study, 16 out of 24 test functions have been selected for which a reliable convergence behavior has been found in [10].

Based on the previous considerations, the remainder of this chapter is organized as follows: In Sect. 2 the specification of the Differential Evolution variant that is used for the present examination is given. In Sect. 3 an overview about stopping criteria is provided, including a discussion if they can also possibly be used for other evolutionary algorithms besides DE. The description of experimental settings in Sect. 4 specifies parameter settings of DE, parameter settings of the stopping criteria and the performance measures that are applied in this work. Results are discussed in Sect. 5, and Sect. 6 ends with conclusions about the suitability of the presented stopping criteria for Differential Evolution.

2 Differential Evolution

In this section first the general process of Differential Evolution is described before giving details about the variant that is used here. For DE the positions of individuals are represented as real-coded vectors which are randomly initialized inside the limits of the given search space in the beginning of an optimization run (see Fig. 1). The individuals are evolved during the optimization run by applying mutation, recombination and selection to each individual in every generation. A stopping criterion determines after the building of every new generation if the optimization run should be terminated.

In this work the Differential Evolution algorithm is used in the variant DE/rand/1/bin [12]. This notation means that in the mutation process a randomly chosen population member $\mathbf{x}_{r_1}$ is added to one vector difference (also built from two randomly chosen members $\mathbf{x}_{r_2}$ and $\mathbf{x}_{r_3}$ of the current population) where $\mathbf{x}_{r_1}$, $\mathbf{x}_{r_2}$, $\mathbf{x}_{r_3}$ and the so-called target vector $\mathbf{x}_i$ are mutually different:

$$\mathbf{v}_i = \mathbf{x}_{r_1} + F \cdot (\mathbf{x}_{r_2} - \mathbf{x}_{r_3}) \tag{1}$$

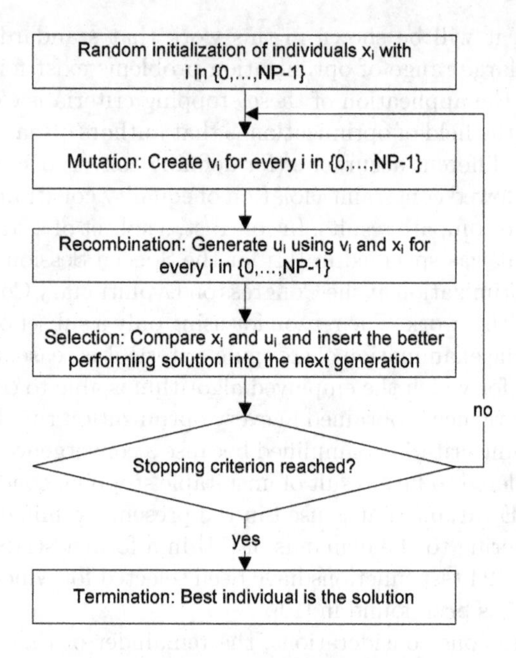

Fig. 1. Flowchart of Differential Evolution

F is a control parameter of DE that is usually chosen from the interval $[0, 1]$. Best values are usually in the range $[0.5, 0.9]$ as indicated in [13, 14, 15, 16].

Furthermore, the notation of the variant DE/rand/1/bin specifies that a binomial recombination process is used that can be written as follows:

$$u_{i,j} = \begin{cases} v_{i,j} & \text{if } rand_j \leq CR \text{ or } j = k \\ x_{i,j} & \text{otherwise} \end{cases} \tag{2}$$

Equation 2 generates the so-called trial vector $\mathbf{u}_i$ by copying components from the mutated vector $\mathbf{v}_i$ and the target vector $\mathbf{x}_i$ in dependence on a random number $rand_j \in [0, 1]$ that is compared to the control parameter CR (where $rand_j$ is chosen anew for each parameter in every generation). Good settings for CR are typically close to one but for some functions also small values have been reported to yield good results [13, 14, 15, 16]. Because during the selection process the target vector and the corresponding trial vector will compete for a place in the subsequent generation, it is ensured that $\mathbf{u}_i \neq \mathbf{x}_i$ by selecting at least one component from the mutated vector $\mathbf{v}_i$. For this purpose the variable $k \in \{0, \ldots, D-1\}$ (where D is the dimension of the optimization problem that is equal to the number of objective function parameters) is randomly chosen for every trial vector in each generation, and the k-th component of the trial vector is copied from the mutated vector.

Hence, four population members are involved in the creation of each trial vector which leads to an adaptive scaling of step sizes because the magnitude of the vector differences varies during the different stages of an optimization run.

Selection is a deterministic process in DE: The target vector and the trial vector are compared to each other, and the one with the lower objective function value (for minimization problems like in this work) is inserted into the next generation. Because this selection scheme allows only improvement but not deterioration of the objective function value, it is called greedy [14]. A further characteristic of the DE selection process is that the best objective function value cannot get lost when moving from one generation to the next. This property is called elitist, and it is usually associated with fast convergence behavior [6].

Mutation, recombination and selection is applied to every population member $\mathbf{x}_i$ with $i \in \{0, \ldots, NP-1\}$ in each generation where NP specifies the population size that has to be adjusted by the user. The fact that the evolutionary operators are applied to every population member is a property that distinguishes DE from several other evolutionary algorithms which often select only a subset of the population for mating [17]. In that case, individuals with better characteristics generally have better chances to reproduce, resulting in a possible increase of the convergence speed but also in loss of diversity. Because DE already generates enough convergence pressure by using an elitist selection procedure, diversity is emphasized by allowing each individual to generate offspring.

Differential Evolution has originally been developed for unconstrained single-objective optimization. Hence, a method for constraint-handling has to be added if constrained optimization problems should be solved like in this work. Several different constraint-handling approaches have been suggested in the literature [4, 5, 6]. In this work a method is employed that is widely used because it is simple but effective. It does not change the mutation and recombination processes of DE but only modifies selection in the following way:

- Feasible individuals (meaning individuals that fulfill all constraints) are favored over infeasible individuals.
- In the comparison of two infeasible individuals the one with the lower sum of constraint violation wins.
- The original selection method is used in the comparison of two feasible individuals.

Using this approach the search is guided to feasible regions of the search space by preferring individuals with lower or no constraint violation. This technique is easy to use because no additional parameters have to be set. With a small modification it can also be used for multi-objective optimization [18].

Boundary constraints are treated as a special case of constraint functions here because especially for real-world problems it may be crucial that individuals stay inside certain boundaries. A position that exceeds a limit is reset to the middle between old position and boundary, so the boundary is approached asymptotically [19]:

$$u_{i,j,G+1} = \begin{cases} \frac{1}{2}\left(x_{i,j,G} + X_{max,j}\right) & \text{if } u_{i,j,G+1} > X_{max,j} \\ \frac{1}{2}\left(x_{i,j,G} + X_{min,j}\right) & \text{if } u_{i,j,G+1} < X_{min,j} \\ u_{i,j,G+1} & \text{otherwise} \end{cases} \quad (3)$$

where $X_{max,j}$ is the upper limit and $X_{min,j}$ is the lower limit for the j-th parameter, and Eq. 3 is given for the i-th individual in generation G.

3 Stopping Criteria

The stopping criteria which are used in this work are grouped into three classes:

- Improvement-based criteria,
- movement-based criteria and
- distribution-based criteria.

In the following several implementations of stopping criteria based on monitoring improvement, movement and distribution are summarized. Most of the stopping criteria that are presented here have already been used for DE [20, 21]. Additionally, for one of them a generalization is newly introduced here because it has been indicated elsewhere [22] that the generalized criterion may exhibit improved behavior over the special case that was formerly used. Many of the stopping criteria that are examined in this work can also be employed for other evolutionary algorithms but it was shown that the performance is not necessarily equal [20, 22]. This conclusion is also reached in [23] where it is stated that the effectiveness of a stopping criterion is closely related to the procedure of a certain optimization strategy and not automatically transferable to other algorithms. Therefore, in the following description of stopping criteria references are added, if available, in which other context or for which other optimization algorithm the stopping criteria can be used also.

Every criterion that is presented here includes one or two specific parameters which have to be set by the user. This property seems to be inherent to all stopping criteria because even if a problem with known optimum is used and termination is done when the optimum is found, the accuracy has to be set by the user. Similarly, parameter FE_{max} has to be set when criterion *LimFuncEval* is employed. Usually, no general guidelines can be given for the setting of FE_{max} but the adaptive stopping criteria do not necessarily have this property as will be shown in this work.

3.1 Improvement-Based Criteria

If the improvement of the objective function value decreases to a small value for some time, it can be assumed that convergence has been obtained. Because improvement can be measured in different ways, three conditions are examined here:

- *ImpBest*: The improvement of the best objective function value of each generation is monitored. If it falls below a user-defined threshold t for a number of generations g, the optimization run will be terminated.

A similar approach is also discussed in [24] for Particle Swarm Optimization (PSO) and furthermore in [25] to determine a suitable switch-over point from a Genetic Algorithm to a local optimization technique.

- *ImpAv*: Because the best objective function value might not correctly reflect the state of the whole population, the average improvement computed from the whole population is examined for this criterion. Similar to *ImpBest*, an optimization run is terminated if the average improvement is below a given threshold t for g generations.

  This criterion is also used in [26] to stop a local search procedure that is embedded in a Genetic Algorithm. For the same purpose similar criteria as *ImpBest* and *ImpAv* are also employed in [27].
- *NoAcc*: Because DE incorporates a greedy selection scheme, the acceptance of trial vectors means that there is improvement in the population. Based on this fact, it is monitored if still trial vectors have been accepted in a specified number of generations g, and the optimization run is terminated if this condition is violated.

  NoAcc has the advantage that only one parameter has to be set whereas all other stopping conditions that are presented here require the setting of two parameters. However, in contrast to the other improvement-based criteria, it is specific to the functionality of DE and may not be assignable to other evolutionary algorithms. For PSO *NoAcc* can be adapted by observing if new personal best positions have been found in a predefined number of generations but in [22] the performance of this criterion was poor.

  NoAcc is also described for DE in [6], and it is recommended to set g not too low because long periods without improvement may occur during optimization runs.

3.2 Movement-Based Criteria

In the beginning of an optimization run the individuals are randomly scattered in the search space, and large step sizes are generated in mutation and recombination. Towards the end of an optimization run the population generally converges to one point in the search space. Thus, step sizes become small because of the adaptive scaling of DE. As a result, the movement of individuals in parameter space can also be used to derive a stopping criterion (parameter space means that the positions of the individuals are regarded while objective space refers to the objective function values of the individuals):

- *MovPar*: If the average movement of the population members is below a threshold t for a given number of generations g, the optimization run is terminated.

  MovPar is also usable for other evolutionary algorithms with real-coded variables. It might be possible to adapt it also for binary-coded individuals if a suitable distance measure can be found. Moreover, stopping criteria like this are used in classical optimization algorithms like hill climbing techniques [23].

Movement can also be measured in objective space but because of the greedy selection scheme of DE, the objective function value can only improve but not deteriorate. Therefore, a stopping criterion based on movement in objective space would be equal to an improvement-based criterion. In contrast, a criterion *MovObj* could be used for other evolutionary algorithms which permit deterioration of objective function values:

- *MovObj*: If the average movement of the population members in objective space is below a threshold t for g generations, the optimization run is stopped.

3.3 Distribution-Based Criteria

In single-objective optimization DE individuals usually converge to one point in the search space towards the end of an optimization run. As a result, the distribution of individuals can be used to derive conclusions about the state of an optimization run. Several possibilities exist to measure the distribution of individuals. One of the easiest alternatives is the following:

- *MaxDist*: The maximum distance of any population member to the individual with the best objective function value is monitored in parameter space. If it falls below a threshold m, the optimization run will be terminated.

  A similar criterion is also discussed in [24] for PSO.

If the positions of all population members should be regarded instead of observing only the maximum distance to the best individual, the following stopping criterion can be used:

- *StdDev*: The standard deviation of positions of all population members is examined. The optimization run is stopped if it drops below a given threshold m.

  In [28] a similar criterion is also used for DE.

Especially for Particle Swarm Optimization it has been shown that a generalization of *MaxDist* has advantages [20, 22]:

- *MaxDistQuick*: Instead of examining the maximum distance of all population members to the current best individual, only a subset of the current population is used. For this purpose, the population members are sorted due to their objective function value using a Quicksort algorithm, and only for the best p% of the population it is checked if their distance is below a threshold m. Because a feasible solution is wanted, it is also checked if the best p% of the individuals are feasible.

  MaxDist can be derived from *MaxDistQuick* by setting p to 100%.

In [22] it was concluded for PSO that it might be beneficial if a generalization of *StdDev* is also examined because it was shown that the generalized criterion *MaxDistQuick* has advantages over the special case *MaxDist*. *StdDev* and *MaxDist* rely on similar mechanisms, so it can be expected that the performance of a generalized criterion might also be better for *StdDev*. Consequently, the following criterion is newly introduced here:

- *StdDevQuick*: Similar to *MaxDistQuick*, the population is first sorted due to their objective function value using a Quicksort algorithm. The standard deviation of positions is then calculated for the best $p\%$ of the population and compared to the user-defined threshold m. Again, it is also examined if the best $p\%$ of the individuals are feasible.

  Similar to the relationship between *MaxDist* and *MaxDistQuick*, *StdDev* is a special case of *StdDevQuick* with $p = 100\%$.

Because *MaxDist* and *StdDev* are special cases of *MaxDistQuick* and *StdDev-Quick*, respectively, only the generalizations *MaxDistQuick* and *StdDevQuick* are regarded in the following.

All distribution-based criteria that have been mentioned so far are calculated in parameter space. Another possibility to evaluate the distribution of the population members is to regard objective space:

- *Diff*: The difference between best and worst objective function value in a generation is checked if it is below a given threshold d. Furthermore, it is demanded that at least $p\%$ of the individuals are feasible because otherwise *Diff* could lead to early termination of an optimization run if e.g. only two individuals are feasible and they are close to each other by chance but the population has not converged yet.

  A similar implementation of this criterion without parameter p is described in [6, 23] and also used in [29] (interestingly, it will be shown in the following that the results of the present examination indicate that the performance of *Diff* is independent from p so it may be omitted). It is recommended in [6] to set d to a value that is several orders of magnitude lower than the desired accuracy of the optimum.

No DE-specific information is used for the distribution-based criteria so in principle they can be used for other algorithms also. However, if another representation than real-coded vectors is used for the positions of the individuals, the distribution-based criteria in parameter space will have to be adapted.

3.4 Combined Criteria

Because functions have different features it can be concluded that a combination of different stopping criteria may result in good performance. For example an criterion like *Diff* that is easy to check can be tested first. Because the first criterion might fail for certain characteristics of the objective function (e.g. it was shown in former work [20] that *Diff* fails for functions with a flat surface), a second criterion that is based on another mechanism might be evaluated after the stopping condition of the first criterion has been fulfilled. In former work the following combined criteria were tested:

- *ComCrit*: First, the improvement-based criterion *ImpAv* is evaluated. If *ImpAv* indicates that the optimization run should be stopped, the distribution-based criterion *MaxDist* is regarded additionally. *ComCrit* was examined in [20, 21] for DE and also in [22] for PSO.

- *Diff_MaxDistQuick*: In this case, distribution-based criteria in objective and parameter space were joined (*Diff* and *MaxDistQuick*) so that *MaxDistQuick* is only checked if the stopping condition of *Diff* has been fulfilled. Up to now this criterion has only been applied for PSO in [22].

In all former examinations the combined criteria always needed more function evaluations for detecting convergence than the individual criteria. Furthermore, selecting appropriate parameter settings was complicated because three parameters have to be set for each stopping criterion in contrast to one or two parameters for the individual criteria, respectively. Moreover, the connection between parameter settings and problem features like desired accuracy was obliterated. Because of these disadvantages, combined criteria are not considered further in this work.

4 Experimental Settings

To be able to derive general conclusions about the suitability of stopping criteria for a broad range of optimization problems, 16 test functions are used here. They are chosen from the standardized test set that was used in the Special Session on Constrained Real Parameter Optimization at the Congress on Evolutionary Computation 2006. The test set originally consists of 24 constrained single-objective test functions (g01–g24) but this work concentrates on the functions for which a convergence rate of 100% has been found in former work [10] for the same algorithm with the same parameter settings ($F = 0.7$, $CR = 0.9$, $NP = 50$). In this case, the analysis of stopping criteria is simplified because performance variations concerning the convergence rate can be accredited to the unsuitability of stopping criteria. Because of these considerations, functions g02, g03, g13, g17, g20, g21, g22 and g23 are omitted here. Nevertheless, the remaining functions permits extensive testing of stopping criteria because a broad spectrum of different features is represented by them:

- Dimensionality,
- type of function,
- ratio of feasible space to the whole search space,
- linear and nonlinear inequality and equality constraints,
- active constraints at the optimum (meaning that the optimum is located at the boundary of one or more constraints) and
- disconnected feasible regions.

Because the exact definition of the test functions has been shown in several works, it is not repeated here as it takes a lot of space. Instead, the interested reader should refer to [11, 30, 31, 32], while some general information is also given in Table 1. In Table 1 $\rho = \frac{|\mathcal{F}|}{S}$ specifies the estimated ratio of feasible space to the whole search space. The following four columns of Table 1 give information about the number and type of constraints: LI is the number of linear inequality constraints, NI is the number of nonlinear inequality constraints, LE is the number of linear equality constraints and NE is the number of nonlinear

Table 1. Details of the test functions (from [10] and [11])

Problem	D	Type of function	ρ	LI	NI	LE	NE	a	Median FEs in [10]
g01	13	quadratic	0.0111%	9	0	0	0	6	32996
g04	5	quadratic	52.1230%	0	6	0	0	2	16166
g05	4	cubic	0.0000%	2	0	0	3	3	105780
g06	2	cubic	0.0066%	0	2	0	0	2	7198
g07	10	quadratic	0.0003%	3	5	0	0	6	93752
g08	2	nonlinear	0.8560%	0	2	0	0	0	1091
g09	7	polynomial	0.5121%	0	4	0	0	2	25602
g10	8	linear	0.0010%	3	3	0	0	6	120624
g11	2	quadratic	0.0000%	0	0	0	1	1	14993
g12	3	quadratic	4.7713%	0	1	0	0	0	5398
g14	10	nonlinear	0.0000%	0	0	3	0	3	68147
g15	3	quadratic	0.0000%	0	0	1	1	2	51619
g16	5	nonlinear	0.0204%	4	34	0	0	4	11522
g18	9	quadratic	0.0000%	0	13	0	0	6	80322
g19	15	nonlinear	33.4761%	0	5	0	0	0	176127
g24	2	linear	79.6556%	0	2	0	0	2	3067

equality constraints. Besides, the number of active constraints at the optimum is given by a. Table 1 also gives information about the median number of function evaluations that have been needed for convergence for the same algorithm in former work [10] because it will be shown that the behavior of some stopping criteria is dependent on it.

Each stopping criterion includes one or two parameters. The parameter settings that are examined in this work are given in Table 2. For every parameter combination and each test function 100 independent runs are conducted. In the CEC06 special session 500,000 FEs were allowed for solving each optimization problem [11], so a maximum number of $FE_{max} = 500,000$ is used in connection with each stopping criterion to terminate the optimization run if the stopping criterion is not able to do it. However, an optimization run that is stopped at 500,000 FEs is considered as unsuccessful. In successful runs the execution of the algorithm must be stopped before reaching 500,000 FEs, and the optimum must be located with an accuracy of 10^{-4} as it was required in [11] for the CEC06 special session (naturally, the solution must also be feasible where the allowed remaining constraint violation for equality constraints is 10^{-4} as in [11]).

Two aspects are important for the assessment of the performance of stopping criteria:

- Has convergence been achieved i.e. has the optimum been reached with the desired accuracy before the algorithm was terminated?
- How fast was the termination i.e. how many function evaluations were done after convergence has been reached?

Table 2. Parameter settings for the stopping criteria

Criterion	Parameter	Start value	Stop value	Modifier
ImpBest	t	1e-2	1e-6	· 1e-1
	g	5	20	+ 5
ImpAv	t	1e-2	1e-6	· 1e-1
	g	5	20	+ 5
NoAcc	g	1	5	+ 1
MovPar	t	1e-2	1e-6	· 1e-1
	g	5	20	+ 5
MaxDistQuick	m	1e-2	1e-5	· 1e-1
	p	0.1	1.0	+ 0.1
StdDevQuick	m	1e-2	1e-5	· 1e-1
	p	0.1	1.0	+ 0.1
Diff	d	1e-1	1e-6	· 1e-1
	p	0.1	1.0	+ 0.1

The first performance measure is evaluated by computing the percentage of successful runs out of 100 independent optimization runs. Clearly, the first performance measure is more important than the second here because fastness is irrelevant if convergence is not obtained. Given that a sufficient convergence rate has been achieved, the second performance measure also provides important information about the abilities of stopping criteria. It is examined by calculating the additional computational effort $\frac{FE_{stop}-FE_{conv}}{FE_{conv}}$: The difference between the number of function evaluations at which the execution of the algorithm is terminated (FE_{stop}) and the number of function evaluations at which convergence is achieved for the first time (FE_{conv}) is computed and the result is divided by FE_{conv}. Thereby, the additional computational effort is normalized because the test problems require very different amounts of FEs for convergence. For both performance measures the median is calculated for each function (the median is preferred instead of the average that is often used in the literature because it is more robust to outliers).

Due to the high amount of data that has been collected for this examination, only general results can be visualized instead of going into detail. Therefore, box plots of the number of successful runs and the additional computational effort will be shown that provide a concise overview over the performance for all 16 optimization problems. Hence, performance over a large range of functions can be easily evaluated for each combination of parameter settings of the stopping criteria from Table 2.

Naturally, different parameter settings of stopping criteria are required if the demanded accuracy of the result is varied. As a consequence, all examinations are repeated with an accuracy of $\epsilon = 10^{-2}$ in order to make comparisons with the formerly used accuracy of $\epsilon = 10^{-4}$ that was specified in [11]. Because the visualization of results takes a lot of space, the results for $\epsilon = 10^{-2}$ will only be summarized qualitatively in the text.

5 Results

In this section the results based on the experimental settings discussed in the previous section are shown for each stopping criterion.

5.1 Criterion *ImpBest*

Criterion *ImpBest* has a very bad performance (see Fig. 2). For many functions the DE algorithm is stopped too early so the convergence rate is low. Only for g08, g16 and g24 a rather high convergence rate has been achieved for certain parameter settings (visible in Fig. 2 as outliers in the box plots). When searching for commonalities of functions with a good performance, it is noticeable that function g08 for which the best results have been achieved needs the least amount of function evaluations for convergence of all functions (see Table 1). g16 and g24 also belong to the functions which can be optimized with a comparably low amount of function evaluations. With $\epsilon = 10^{-2}$ the convergence rate improves for several functions but again only for functions which need a low amount of function evaluations for convergence (e.g. g01, g06, g11, g12). Therefore, the conclusion can be derived that *ImpBest* can only be used for functions which can be optimized with low computational effort. However, obviously this property is not sufficient because a poor performance concerning convergence rate is yielded for several functions which can be optimized using few function evaluations.

Concerning the additional computational effort, a moderate result has been achieved when compared to the performance of other stopping criteria. Naturally, the additional computational effort is always higher if $\epsilon = 10^{-2}$ is used with the same parameter settings of stopping criteria, respectively, because convergence is obtained earlier.

When analyzing the dependence of the convergence rate and the additional computational effort on parameter settings, it can be noticed that both slightly increase with decreasing improvement threshold t and increasing number of generations g. In general, results are very different for different functions, so it is not obvious how parameter settings should be chosen.

Mainly because of its bad results concerning convergence rate, *ImpBest* cannot be regarded as a reliable stopping criterion. Moreover, the applicability of *ImpBest* is complicated because of the difficulty that is connected with choosing parameter settings. Former work on stopping criteria for DE also supports this conclusion [20, 21].

5.2 Criterion *ImpAv*

Criterion *ImpAv* shows a better performance than *ImpBest* but again the algorithm is constantly terminated too early for many functions (see Fig. 3). For

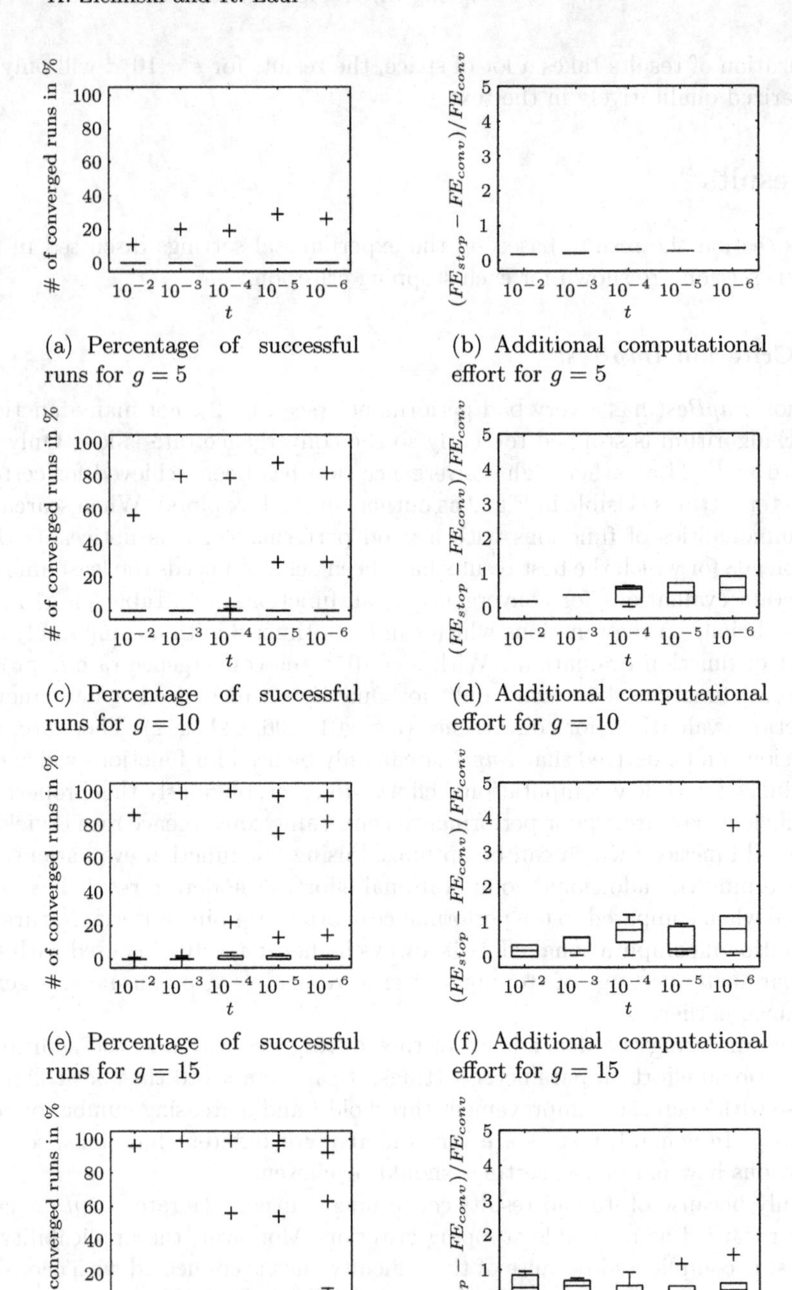

(a) Percentage of successful runs for $g = 5$

(b) Additional computational effort for $g = 5$

(c) Percentage of successful runs for $g = 10$

(d) Additional computational effort for $g = 10$

(e) Percentage of successful runs for $g = 15$

(f) Additional computational effort for $g = 15$

(g) Percentage of successful runs for $g = 20$

(h) Additional computational effort for $g = 20$

Fig. 2. Results for criterion *ImpBest*

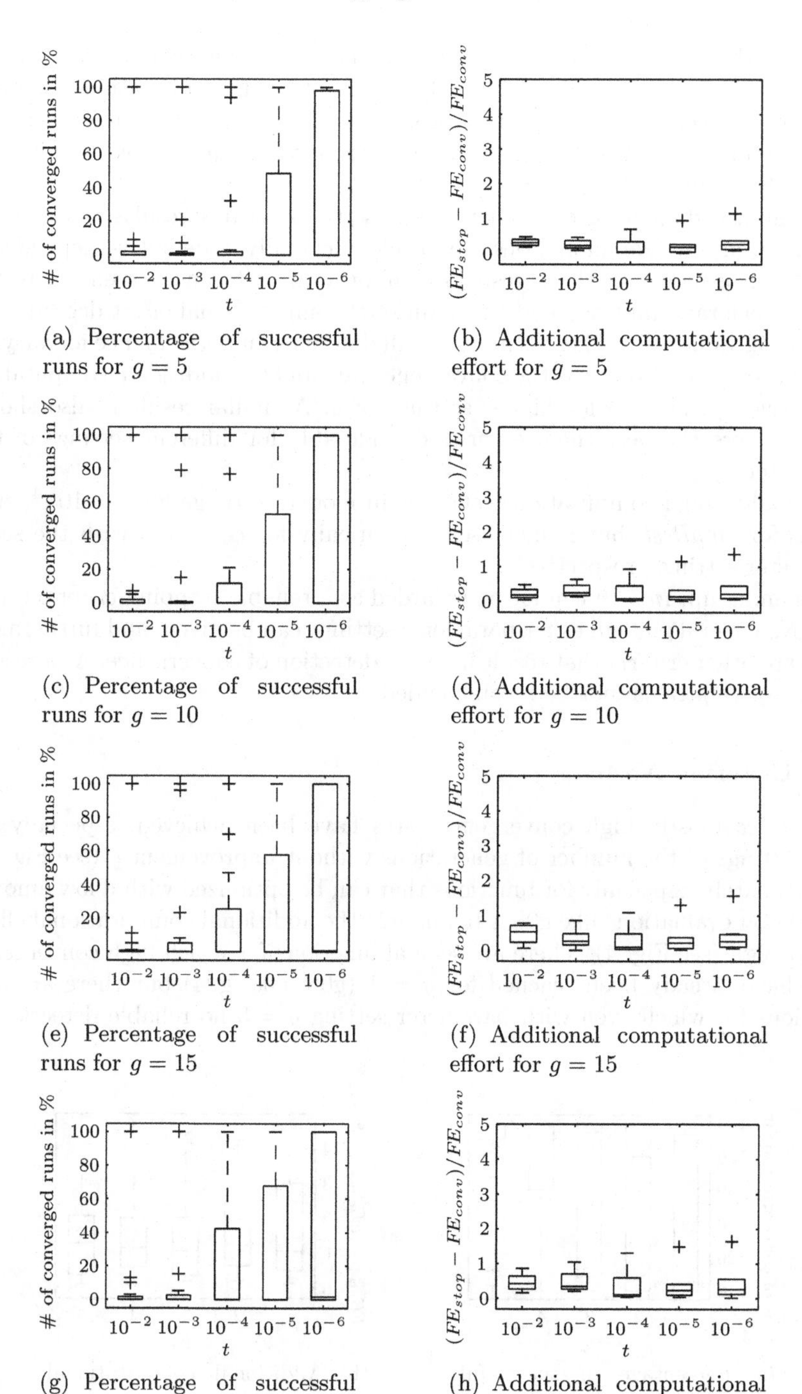

(a) Percentage of successful runs for $g = 5$

(b) Additional computational effort for $g = 5$

(c) Percentage of successful runs for $g = 10$

(d) Additional computational effort for $g = 10$

(e) Percentage of successful runs for $g = 15$

(f) Additional computational effort for $g = 15$

(g) Percentage of successful runs for $g = 20$

(h) Additional computational effort for $g = 20$

Fig. 3. Results for criterion *ImpAv*

$\epsilon = 10^{-4}$ the functions with better performance share the same property as for *ImpBest* which is that a relatively small number of function evaluations is needed for convergence. For $\epsilon = 10^{-2}$ this property is less pronounced, and there are only two functions left for which convergence is never reached before termination of the algorithm (g04 and g10).

Again, it is difficult to choose parameter settings because results differ for the functions, and no commonalities are visible. Generally, a definitive dependence of the performance on the threshold of improvement t can be seen where the convergence rate improves and the additional computational effort degrades for decreasing values of t. There is also a small difference in performance for varying numbers of generations g as the convergence rate and the additional computation effort slightly increase for higher settings of g. A similar result is also shown in [21] where the performance varies considerably for different settings of the parameters.

The additional computational effort is in moderate range for $\epsilon = 10^{-4}$, similar as for *ImpBest*, but it increases dramatically for $\epsilon = 10^{-2}$ with the same parameter settings, respectively.

Summing up, *ImpAv* cannot be regarded as a reliable stopping criterion, neither: No general guidelines for parameter settings can be given, and furthermore there are other criteria that result in faster detection of convergence. As a result, the use of *ImpAv* cannot be recommended.

5.3 Criterion *NoAcc*

For *NoAcc* mostly high convergence rates have been achieved, especially for high settings of the number of generations without improvement g (see Fig. 4). Unfortunately, especially for functions that can be optimized with a low amount of function evaluations like g08, g16 and g24 the additional computational effort is very high (see Fig. 4). There are several functions for which high convergence rates have already been reached for $g = 1$ (g06, g08, g24) but there are also functions for which even with parameter setting $g = 5$ no reliable detection of

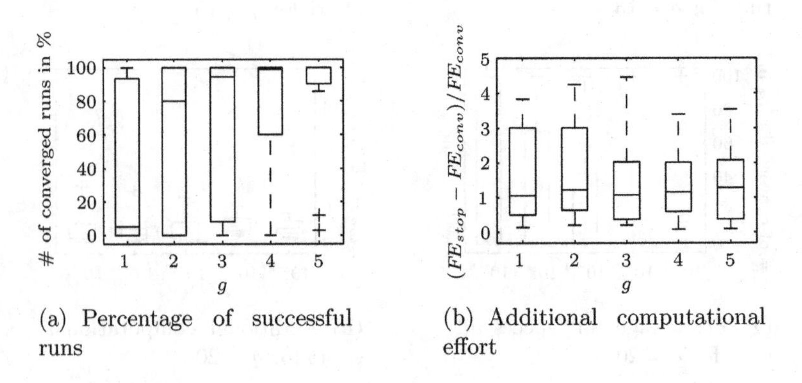

(a) Percentage of successful runs

(b) Additional computational effort

Fig. 4. Results for criterion *NoAcc*

convergence can be achieved. For $\epsilon = 10^{-2}$ the convergence rates become higher but there are still two functions for which the convergence rate is only moderate (g18 and g19).

Although *NoAcc* has the advantage that it only incorporates one parameter, its reliability is limited because it may not be easy to set parameter g as the results concerning convergence rate vary for different functions. The additional computational complexity also reaches considerable magnitude for *NoAcc*, especially for $\epsilon = 10^{-2}$. In former work *NoAcc* has shown a good performance [21] but failed for a function with a flat surface [20].

In summary, *NoAcc* has the advantages that it incorporates only one parameter, and moreover it is easy to check as only the number of accepted trial vectors has to be counted. Nevertheless, it cannot be recommended without hesitation because suitable settings of parameter g vary for different functions, and the additional computational effort is very high for several functions. Additionally, parameter g cannot take values smaller than 1 but even this setting leads to a high additional computational effort for several functions. The missing possibility of scaling to lower values which would lead to earlier termination of the algorithm may be unfavorable.

5.4 Criterion *MovPar*

The convergence rate of criterion *MovPar* is dependent on both the threshold of improvement t and the number of generations g (see Fig. 5). For small settings of t and large settings of g convergence rates of 100% have been found for most functions. Exceptions are g01, g04 and g06 but it is not clear which property is the determining factor because the characteristics of these functions are quite dissimilar. In contrast, for $\epsilon = 10^{-2}$ convergence rates of 100% are found for all functions if parameter settings $g = 20$ and $t \leq 10^{-5}$ are used.

Concerning the additional computational effort, mostly moderate performance is shown with $\epsilon = 10^{-4}$, but it reaches a considerable size for few outliers (g08 and g18). Again, it increases strongly for $\epsilon = 10^{-2}$ with the same parameter settings.

Similar as for *ImpBest* and *ImpAv*, determination of suitable settings for the parameters is not easy which can also be seen in former work [21]. In the present examination mostly settings of $t \leq 10^{-5}$ and $g \geq 10$ yielded good results.

To sum up, it can be stated that *MovPar* yields better results than *ImpBest* and *ImpAv* here, but there are still a few functions where no reliable convergence behavior was observed for $\epsilon = 10^{-4}$. It can be argued that different parameter settings may result in better convergence rates but it must also be considered that the additional computational effort becomes large for some functions. As a consequence, *MovPar* cannot be recommended as a stopping criterion for DE.

5.5 Criterion *MaxDistQuick*

MaxDistQuick achieved very good results in former work on stopping criteria for DE [20, 21] but with the broad set of functions that is used in this work the

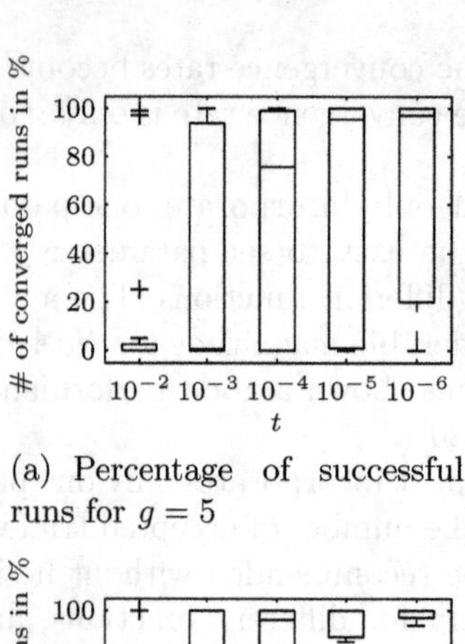

(a) Percentage of successful runs for $g = 5$

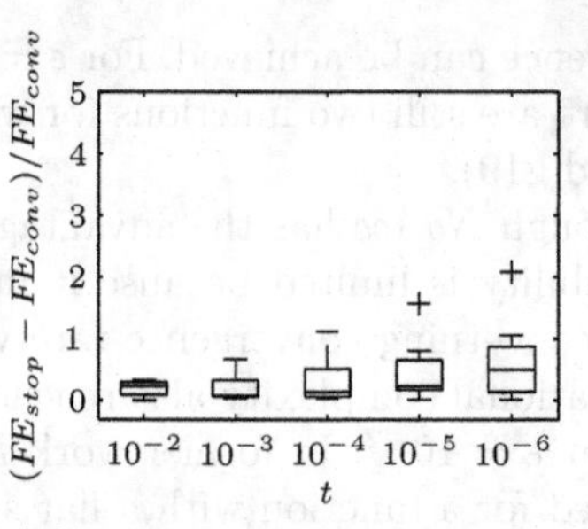

(b) Additional computational effort for $g = 5$

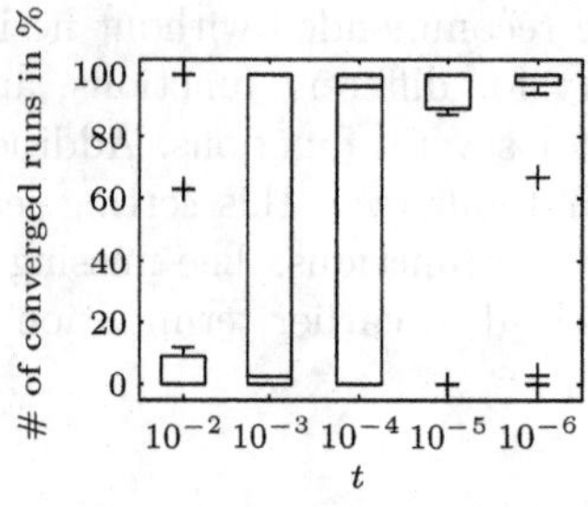

(c) Percentage of successful runs for $g = 10$

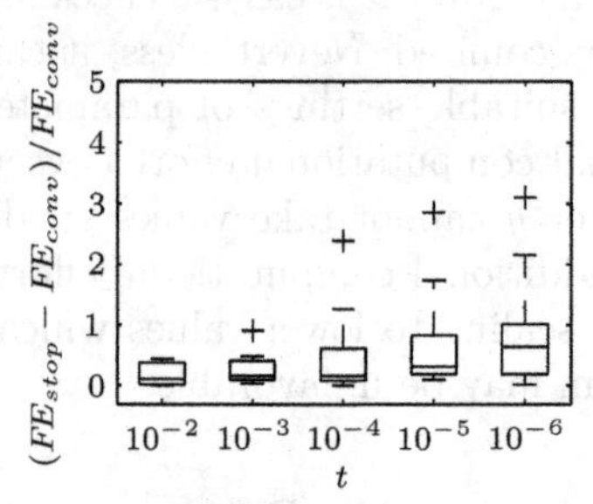

(d) Additional computational effort for $g = 10$

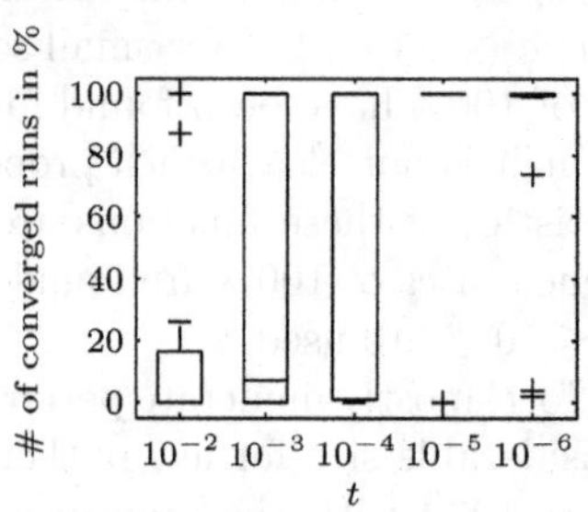

(e) Percentage of successful runs for $g = 15$

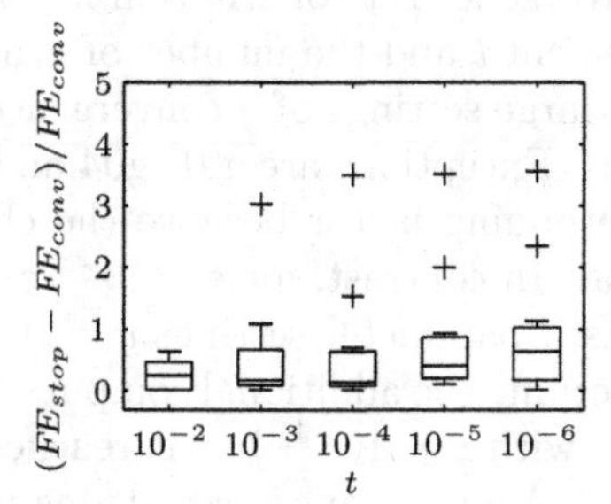

(f) Additional computational effort for $g = 15$

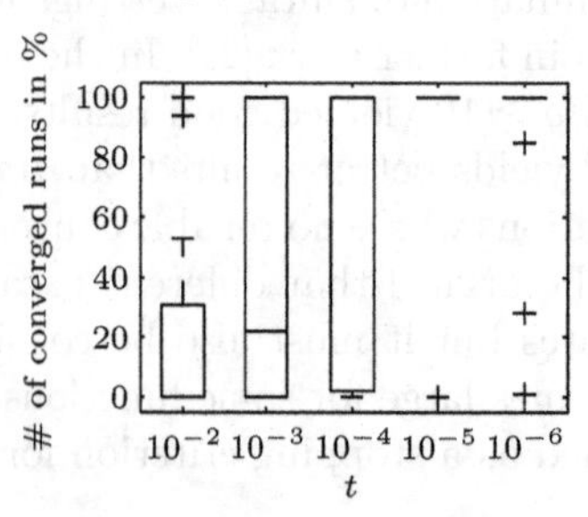

(g) Percentage of successful runs for $g = 20$

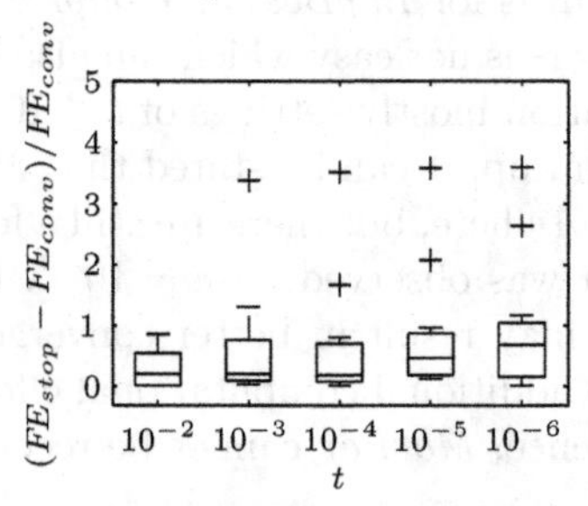

(h) Additional computational effort for $g = 20$

Fig. 5. Results for criterion *MovPar*

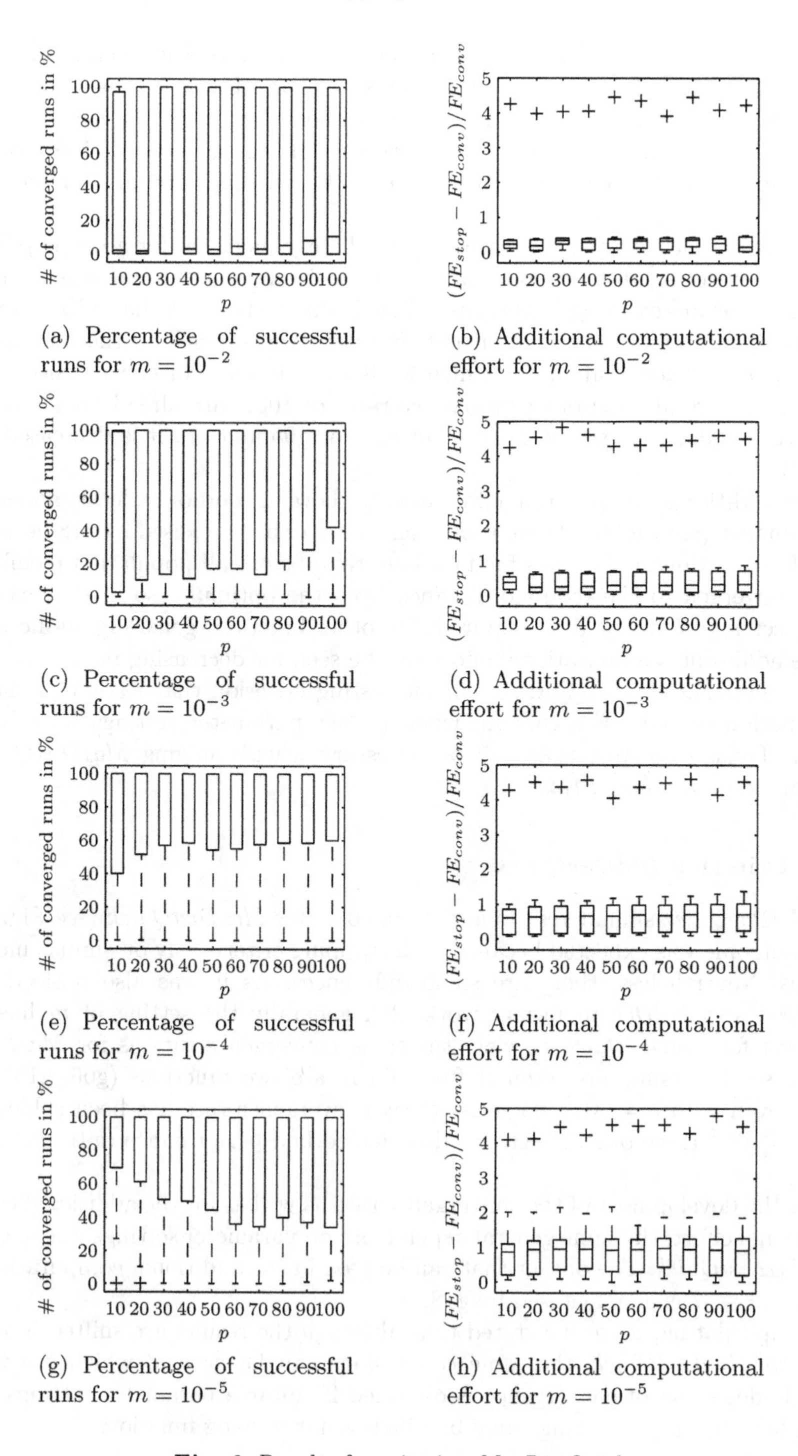

(a) Percentage of successful runs for $m = 10^{-2}$

(b) Additional computational effort for $m = 10^{-2}$

(c) Percentage of successful runs for $m = 10^{-3}$

(d) Additional computational effort for $m = 10^{-3}$

(e) Percentage of successful runs for $m = 10^{-4}$

(f) Additional computational effort for $m = 10^{-4}$

(g) Percentage of successful runs for $m = 10^{-5}$

(h) Additional computational effort for $m = 10^{-5}$

Fig. 6. Results for criterion *MaxDistQuick*

evaluation of its performance is not that clear (see Fig. 6). For several functions convergence rates of 100% have been reached but for other functions no parameter settings have resulted in good convergence behavior (g04, g06, g18). For g18 this is still the case for $\epsilon = 10^{-2}$. In contrast, convergence rates of 100% could be obtained for all other functions with $\epsilon = 10^{-2}$ if proper parameter settings are used.

For most functions the convergence rate slightly increases for growing p if m is large, but for small m the opposite effect can be seen as the convergence rate decreases for increasing p in this case. This is due to the fact that with a small setting of m convergence is often not detected before reaching the maximum number of function evaluations which leads to a decrease of convergence rate. Hence, for several functions convergence rates of 100% are already reached for $m = 10^{-2}$ (g05, g08, g10, g12, g16) but e.g. for g05 and g10 it is decreased for $m = 10^{-5}$.

The additional computation effort mostly shows a good or at least moderate performance (see Fig. 6). There is only one outlier that is caused by g18 because the global optimum is always found a long time before all population members have converged to the required distance from the optimum, so in that case a larger setting of m would be required. For other functions generally an increase of the additional computational effort can be seen for decreasing m.

In summary, *MaxDistQuick* is an interesting criterion that leads to reliable termination of the DE algorithm when proper parameter settings have been found. Thus, some test runs will be necessary when applying *MaxDistQuick*, similar as for *LimFuncEval*.

5.6 Criterion *StdDevQuick*

For *StdDevQuick* similar results are obtained as for *MaxDistQuick* (see Fig. 7). This outcome was expected because both stopping criteria rely on similar mechanisms. Nevertheless, there are some differences: As it was also noticed for *MaxDist* and *StdDev* in former work [21], generally the setting of m has to be lower for *StdDevQuick* to yield the same convergence rate as for *MaxDistQuick*, so the results are often shifted. There are two functions (g06, g18) for which even with $\epsilon = 10^{-2}$ no satisfactory convergence rate has been achieved. For $\epsilon = 10^{-4}$ three other functions also yielded bad performance (g01, g04 and g15).

For the development of the convergence rate as well as for the additional computational effort, the same general dependence on parameter settings was seen as for *MaxDistQuick*. The outlier that can be seen in the additional computational effort in Fig. 7 is again caused by g18.

Recapitulating, it can be stated that although the results are shifted in contrast to *MaxDistQuick*, the same conclusions can be derived which are that reliable detection of convergence is obtained if suitable parameter settings are used, but the proper settings may be different for varying functions.

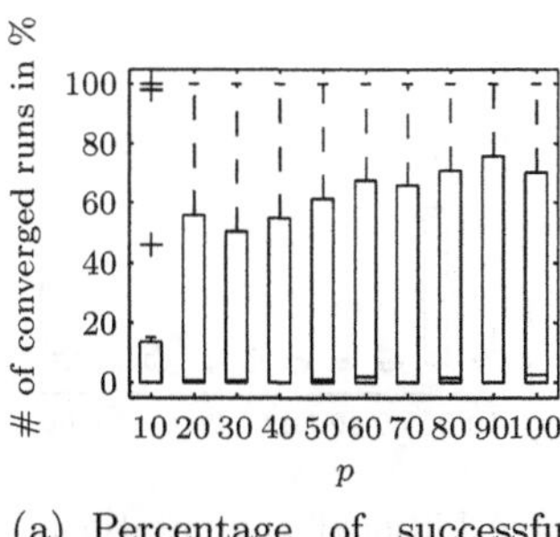

(a) Percentage of successful runs for $m = 10^{-2}$

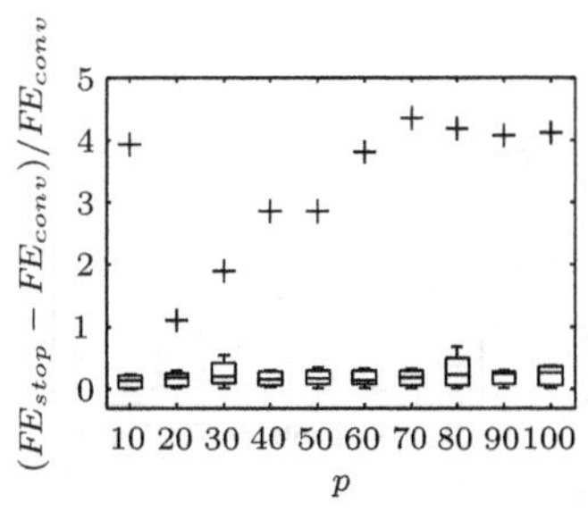

(b) Additional computational effort for $m = 10^{-2}$

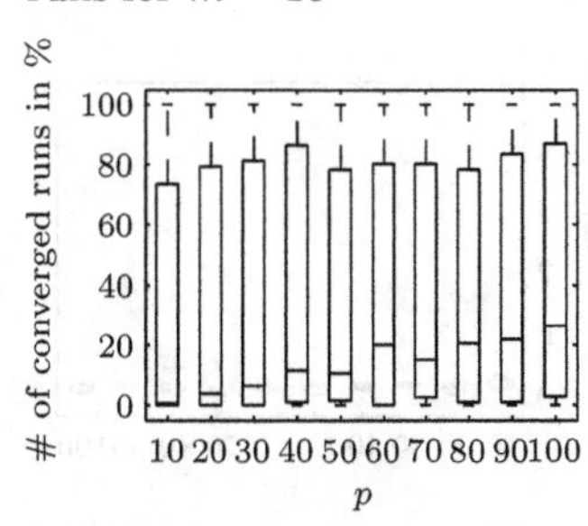

(c) Percentage of successful runs for $m = 10^{-3}$

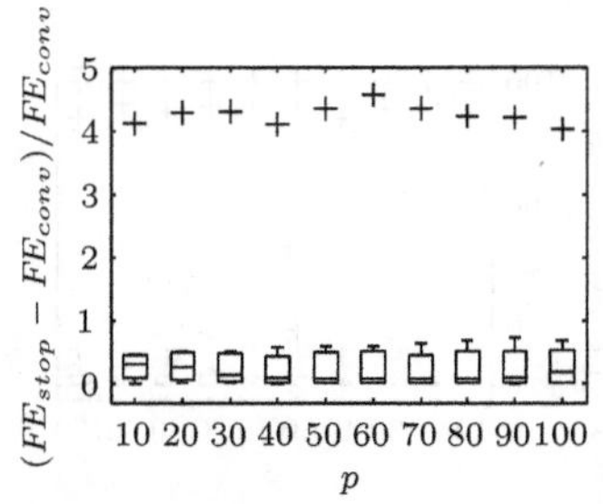

(d) Additional computational effort for $m = 10^{-3}$

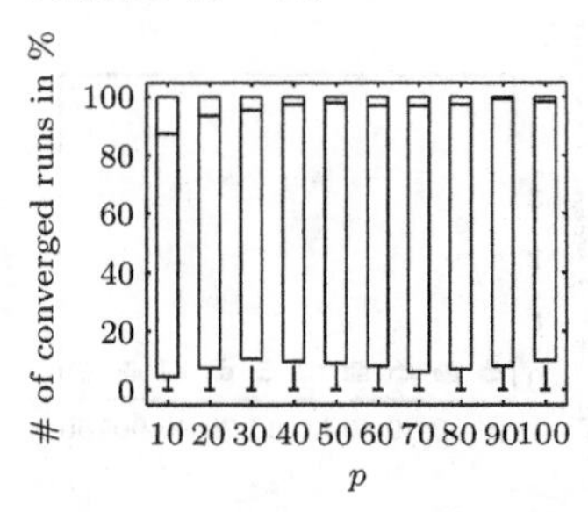

(e) Percentage of successful runs for $m = 10^{-4}$

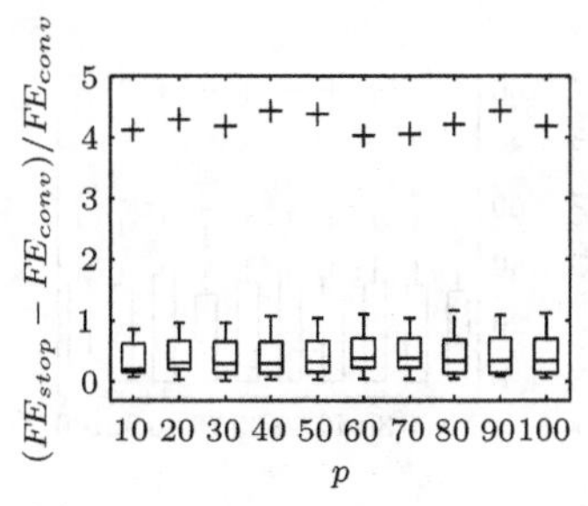

(f) Additional computational effort for $m = 10^{-4}$

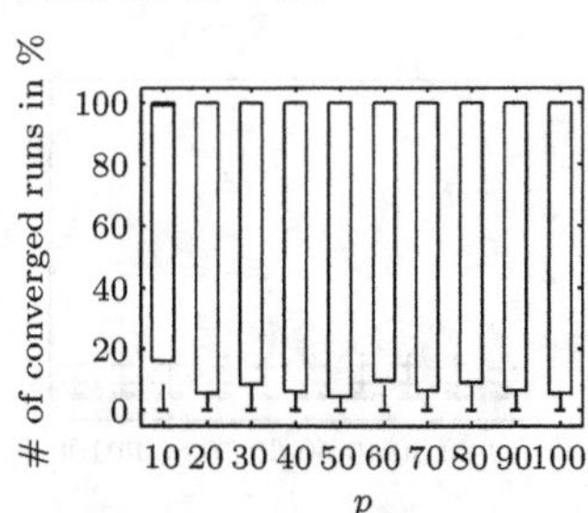

(g) Percentage of successful runs for $m = 10^{-5}$

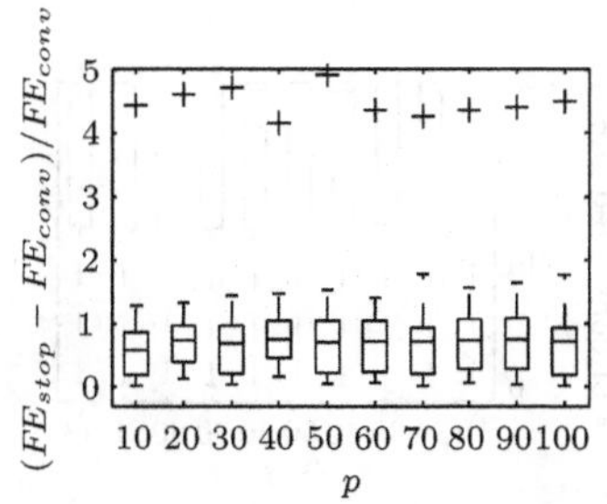

(h) Additional computational effort for $m = 10^{-5}$

Fig. 7. Results for criterion *StdDevQuick*

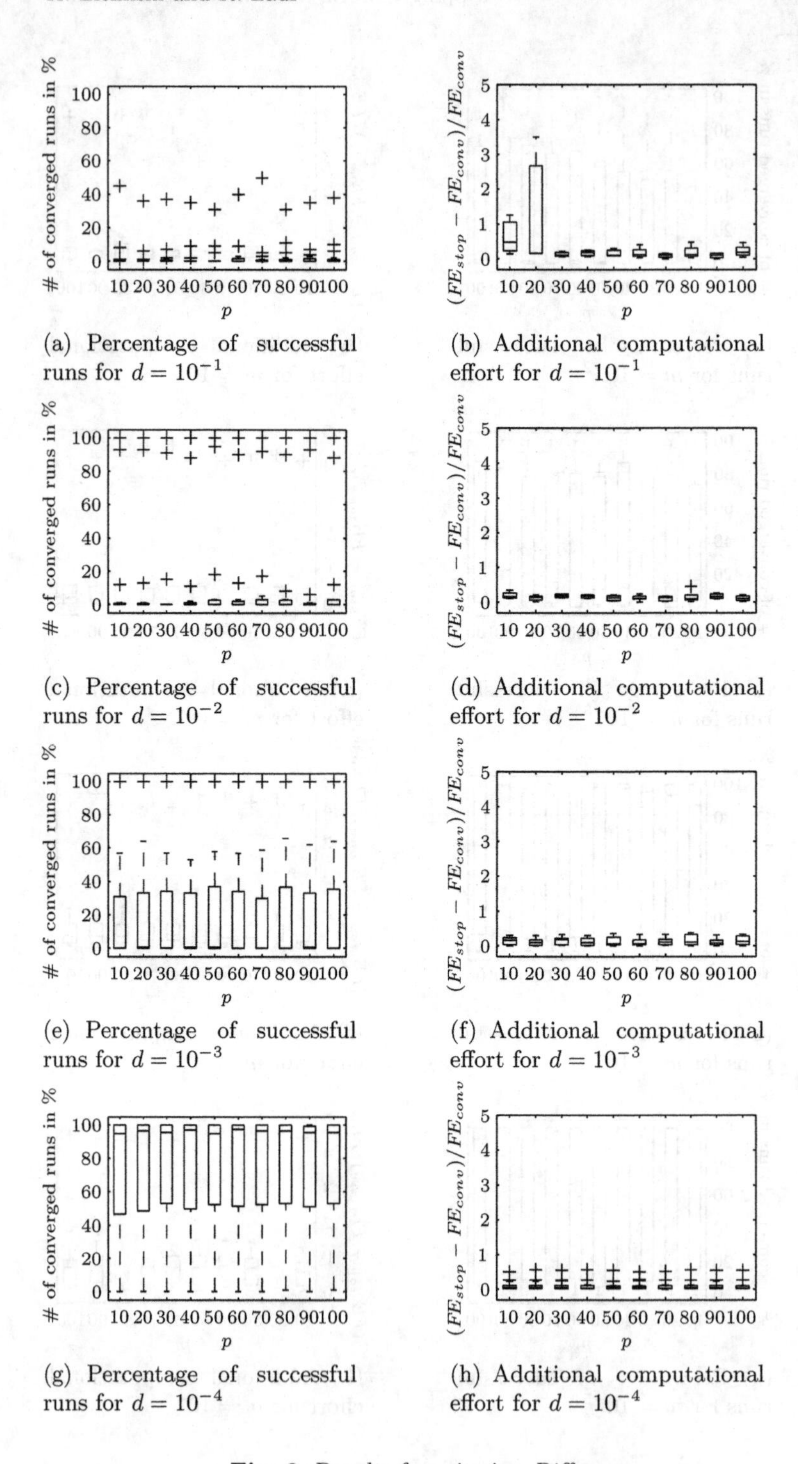

(a) Percentage of successful runs for $d = 10^{-1}$

(b) Additional computational effort for $d = 10^{-1}$

(c) Percentage of successful runs for $d = 10^{-2}$

(d) Additional computational effort for $d = 10^{-2}$

(e) Percentage of successful runs for $d = 10^{-3}$

(f) Additional computational effort for $d = 10^{-3}$

(g) Percentage of successful runs for $d = 10^{-4}$

(h) Additional computational effort for $d = 10^{-4}$

Fig. 8. Results for criterion *Diff*

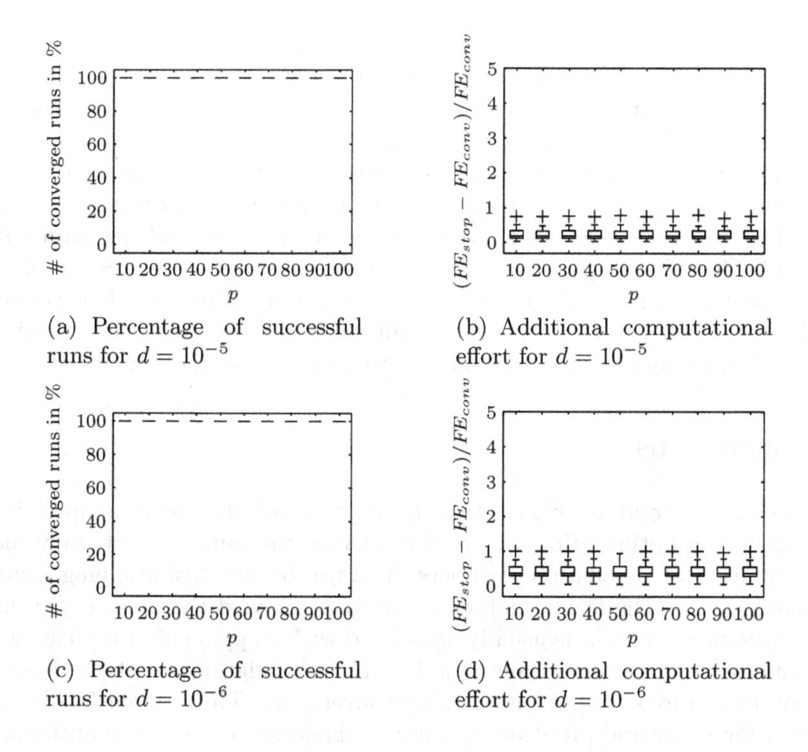

(a) Percentage of successful runs for $d = 10^{-5}$

(b) Additional computational effort for $d = 10^{-5}$

(c) Percentage of successful runs for $d = 10^{-6}$

(d) Additional computational effort for $d = 10^{-6}$

Fig. 9. Results for criterion *Diff* (continued)

5.7 Criterion *Diff*

The results of *Diff* are the most promising ones of this examination (see Fig. 8 and 9). Convergence rates of 100% have been achieved for all functions when the difference threshold is set to $d \leq 10^{-5}$ which is exactly one order of magnitude smaller than the demanded accuracy of $\epsilon = 10^{-4}$. For $\epsilon = 10^{-2}$ a similar result is obtained as convergence rates of 100% have been achieved for all functions with $d \leq 10^{-3}$ which is again one order of magnitude smaller than the desired accuracy. It can be concluded that choosing suitable settings of parameter d is an easy task because a connection to the demanded accuracy can be made.

For parameter p that denotes the percentage of the population that is demanded to be feasible, no dependence can be noticed, neither regarding convergence rate nor regarding the additional computational effort. Thus, parameter p can be omitted for criterion *Diff*. In this case, the number of parameters for *Diff* is reduced from two to one, contributing to its simplicity.

The additional computation effort is relatively low when compared to the results of other stopping criteria (see Fig. 8 and 9) which is also an advantage of *Diff*.

One limitation of *Diff* is that it yields bad results when an optimization problem contains an objective function with a flat surface, meaning that the same objective function value is yielded for a large subset of the search space

[20]. Fortunately, it can be argued that this is a special case that rarely occurs. Moreover, it can be discovered easily when the optimization run is monitored, so in that case another stopping criterion could be employed. Otherwise, the performance of *Diff* was also good in former work [21].

In summary, *Diff* has led to the best results of the stopping criteria that were examined here because all functions could be successfully terminated. Additionally, parameter p can be omitted completely as no dependence on it could be found. Choosing of parameter d is also simple because the results of this work indicate that it is sufficient to set it to one order of magnitude lower than the desired accuracy. Moreover, the results of *Diff* concerning the additional computational effort are good in contrast to other stopping criteria.

6 Conclusions

In this chapter stopping criteria have been presented that react adaptively to the state of an optimization run by considering the improvement, movement or distribution of population members. The use of adaptive stopping criteria was motivated by the fact that they could help to avoid the unnecessary high computational effort that is usually associated with stopping after a preassigned fixed number of function evaluations because of variations in the number of function evaluations that are needed for convergence. This approach is mainly intended for real-world problems because in this case normally the optimum is unknown.

The best results were yielded by criterion *Diff* that can be classified as a distribution-based criterion in objective space as it terminates an optimization run if the difference between best and worst objective function value in the current generation has fallen below a given threshold d. This work has shown that the setting of d is linked with the desired accuracy of the result, so choosing a suitable parameter setting for d is simple. The second parameter p that corresponds to the demanded feasible percentage of the population can be omitted completely as no dependence on it could be seen. The only limitation of *Diff* was revealed in former work where *Diff* failed for a function with a flat surface [20]. However, this property can be detected easily when observing an optimization run so this limitation is not grave.

The results for other stopping criteria were not as distinct as for *Diff*. Two distribution-based criteria in parameter space were examined that terminate an optimization run if the maximum distance of a specified subset of the population to the best population member (*MaxDistQuick*) or the standard deviation of a subset of population members (*StdDevQuick*) is below a user-defined threshold m, respectively. Most functions could be successfully terminated in reasonable time but the parameter settings that yielded these results varied for different functions. There are few functions for which none of the examined parameter settings were able to induce convergence rates of 100% (even with a decreased demanded accuracy of $\epsilon = 10^{-2}$) but it is assumed that higher settings of m might result in better performance for these functions.

Apart from distribution-based stopping criteria, also a movement-based criterion in parameter space *MovPar* was examined that induces termination of an optimization run if the movement of the individuals in parameter space falls below a given percentage t in a predefined number of generations g. For three functions bad results concerning the convergence rate were found regardless of parameter settings for an accuracy of $\epsilon = 10^{-4}$ while for the decreased accuracy of $\epsilon = 10^{-2}$ all functions could be successfully terminated after convergence has been obtained. It can be concluded that the criterion is basically able to stop optimization runs reliably if suitable parameter settings have been found.

Two implementations of improvement-based criteria yielded the worst results of this examination, where criterion *ImpBest* that observes the improvement of the best objective function value is yet worse than criterion *ImpAv* that monitors the improvement averaged over all individuals. For many functions no convergence rates of 100% could be found so these criteria cannot be considered as reliable, and furthermore choosing of parameter settings is not easy.

A third improvement-based criterion *NoAcc* was based on the number of generations g in which no trial vector has been accepted. Problems occurred because parameter g has to be different for varying functions in order to give good performance. Particularly, the missing scalability to values smaller than 1 may lead to high additional computational effort for certain optimization problems. If suitable settings for g can be found, *NoAcc* shows reliable performance.

It should be noted that although a large test set that contains a broad range of functions was used here, still optimization problems may exist for which the obtained conclusions do not hold. However, it is expected that at least similar behavior will be found in these cases.

Most of the stopping criteria that are described in this work can also be used for other evolutionary algorithms besides DE but it has to be noted that the performance may be different as it was shown for Particle Swarm Optimization in former work [20, 22]. It would be interesting to try the stopping criteria also for other optimization algorithms in future work.

The stopping criteria presented in this chapter are mainly designated for real-world problems with unknown optimum because for optimization problems with known optimum other good alternatives exist for terminating an optimization run. Only single-objective optimization was addressed yet but real-world problems often contain multiple objectives, thus future work must include the development of reliable stopping criteria for multi-objective optimization. Unfortunately, the situation is more difficult in multi-objective optimization because usually optimization goals are contradicting. Thus, not one single optimal point exists but several trade-off solutions which are usually called the Pareto-optimal front [17].

As a consequence, it is not easy to detect convergence even if the Pareto-optimal front is known. It was shown here that distribution-based criteria provide the best results for single-objective optimization, but for multi-objective problems with unknown Pareto-optimal front this concept will be generally not transferable because usually multiple Pareto-optimal solutions exist. Monitoring the movement of individuals may also lead to false conclusions because the individuals may still move along the Pareto-optimal front after the population has converged to it. If the improvement of individuals should be taken as basis for stopping criteria, the problem arises how to define improvement in the presence of several objectives.

Apart from the mechanisms presented in this work, there are some concepts inherent in multi-objective optimization which may also possibly be exploited for the definition of stopping conditions. For instance, a stopping criterion based on the observation of crowding distance was tested in [33]. A further possibility is to monitor the improvement of performance measures like hypervolume [17]. Because multi-objective optimization is a research topic that is currently discussed intensively in the evolutionary algorithms community and its importance can still be expected to grow in the following years, this is an interesting field for future work.

References

1. Storn, R., Price, K.: Differential Evolution - A Simple and Efficient Adaptive Scheme for Global Optimization over Continuous Spaces. Technical Report TR-95-012, International Computer Science Institute, Berkeley, CA (1995)
2. Mezura-Montes, E., Velázquez-Reyes, J., Coello Coello, C.A.: A Comparative Study of Differential Evolution Variants for Global Optimization. In: Proceedings of the Genetic and Evolutionary Computation Conference (2006)
3. Huang, V., Qin, A., Suganthan, P.: Self-adaptive Differential Evolution Algorithm for Constrained Real-Parameter Optimization. In: Proceedings of the IEEE Congress on Evolutionary Computation (2006)
4. Takahama, T., Sakai, S.: Constrained Optimization by the ϵ Constrained Differential Evolution with Gradient-Based Mutation and Feasible Elites. In: Proceedings of the IEEE Congress on Evolutionary Computation, pp. 308–315 (2006)
5. Tasgetiren, M.F., Suganthan, P.: A Multi-Populated Differential Evolution Algorithm for Solving Constrained Optimization Problem. In: Proceedings of the IEEE Congress on Evolutionary Computation, pp. 340–347 (2006)
6. Price, K.V., Storn, R.M., Lampinen, J.A.: Differential Evolution - A Practical Approach to Global Optimization. Springer, Heidelberg (2005)
7. Brest, J., Žumer, V., Maučec, M.S.: Self-Adaptive Differential Evolution Algorithm in Constrained Real-Parameter Optimization. In: Proceedings of the IEEE Congress on Evolutionary Computation, pp. 919–926 (2006)
8. Kukkonen, S., Lampinen, J.: Constrained Real-Parameter Optimization with Generalized Differential Evolution. In: Proceedings of the IEEE Congress on Evolutionary Computation, pp. 911–918 (2006)
9. Mezura-Montes, E., Velázquez-Reyes, J., Coello Coello, C.A.: Modified Differential Evolution for Constrained Optimization. In: Proceedings of the IEEE Congress on Evolutionary Computation, pp. 332–339 (2006)
10. Zielinski, K., Laur, R.: Constrained Single-Objective Optimization Using Differential Evolution. In: Proceedings of the IEEE Congress on Evolutionary Computation, Vancouver, BC, Canada, pp. 927–934 (2006)

11. Liang, J., Runarsson, T.P., Mezura-Montes, E., Clerc, M., Suganthan, P., Coello Coello, C.A., Deb, K.: Problem Definitions and Evaluation Criteria for the CEC 2006 Special Session on Constrained Real-Parameter Optimization. Technical report, Nanyang Technological University, Singapore (2006)
12. Onwubolu, G.C.: Optimizing CNC Drilling Machine Operations: Traveling Salesman Problem-Differential Evolution Approach. In: Onwubolu, G.C., Babu, B. (eds.) New Optimization Techniques in Engineering, pp. 537–566. Springer, Heidelberg (2004)
13. Lampinen, J., Storn, R.: Differential Evolution. In: Onwubolu, G.C., Babu, B. (eds.) New Optimization Techniques in Engineering, pp. 123–166. Springer, Heidelberg (2004)
14. Storn, R., Price, K.: Differential Evolution - A Simple and Efficient Heuristic for Global Optimization over Continuous Spaces. Journal of Global Optimization 11, 341–359 (1997)
15. Gämperle, R., Müller, S.D., Koumoutsakos, P.: A Parameter Study for Differential Evolution. In: Grmela, A., Mastorakis, N. (eds.) Advances in Intelligent Systems, Fuzzy Systems, Evolutionary Computation, pp. 293–298. WSEAS Press (2002)
16. Zielinski, K., Weitkemper, P., Laur, R., Kammeyer, K.D.: Parameter Study for Differential Evolution Using a Power Allocation Problem Including Interference Cancellation. In: Proceedings of the IEEE Congress on Evolutionary Computation, Vancouver, BC, Canada, pp. 6748–6755 (2006)
17. Deb, K.: Multi-Objective Optimization using Evolutionary Algorithms. Wiley, Chichester (2001)
18. Deb, K., Pratap, A., Agrawal, S., Meyarian, T.: A Fast and Elitist Multiobjective Genetic Algorithm: NSGA-II. IEEE Transactions on Evolutionary Computation 6(2), 182–197 (2002)
19. Onwubolu, G.C.: Differential Evolution for the Flow Shop Scheduling Problem. In: Onwubolu, G.C., Babu, B. (eds.) New Optimization Techniques in Engineering, pp. 585–611. Springer, Heidelberg (2004)
20. Zielinski, K., Peters, D., Laur, R.: Stopping Criteria for Single-Objective Optimization. In: Proceedings of the Third International Conference on Computational Intelligence, Robotics and Autonomous Systems, Singapore (2005)
21. Zielinski, K., Weitkemper, P., Laur, R., Kammeyer, K.D.: Examination of Stopping Criteria for Differential Evolution based on a Power Allocation Problem. In: Proceedings of the 10th International Conference on Optimization of Electrical and Electronic Equipment, Braşov, Romania, vol. 3, pp. 149–156 (2006)
22. Zielinski, K., Laur, R.: Stopping Criteria for a Constrained Single-Objective Particle Swarm Optimization Algorithm. Informatica 31(1), 51–59 (2007)
23. Schwefel, H.P.: Evolution and Optimum Seeking. John Wiley and Sons, Chichester (1995)
24. van den Bergh, F.: An Analysis of Particle Swarm Optimizers. PhD thesis, University of Pretoria (2001)
25. Syrjakow, M., Szczerbicka, H.: Combination of Direct Global and Local Optimization Methods. In: Proceedings of the IEEE International Conference on Evolutionary Computing (ICEC 1995), Perth, WA, Australia, pp. 326–333 (1995)
26. Espinoza, F.P.: A Self-Adaptive Hybrid Genetic Algorithm for Optimal Groundwater Remediation Design. PhD thesis, University of Illinois (2003)
27. Vasconcelos, J., Saldanha, R., Krähenbühl, L., Nicolas, A.: Genetic Algorithm Coupled with a Deterministic Method for Optimization in Electromagnetics. IEEE Transactions on Magnetics 33(2), 1860–1863 (1997)

28. Zaharie, D., Petcu, D.: Parallel Implementation of Multi-Population Differential Evolution. In: Proceedings of the 2nd Workshop on Concurrent Information Processing and Computing (2003)
29. Babu, B.V., Angira, R.: New Strategies of Differential Evolution for Optimization of Extraction Process. In: Proceedings of International Symposium & 56th Annual Session of IIChE (CHEMCON 2003), Bhubaneswar, India (2003)
30. Mezura-Montes, E., Coello Coello, C.A.: A Simple Multimembered Evolution Strategy to Solve Constrained Optimization Problems. IEEE Transactions on Evolutionary Computation 9(1), 1–17 (2005)
31. Mezura-Montes, E., Coello Coello, C.A.: What Makes a Constrained Problem Difficult to Solve by an Evolutionary Algorithm. Technical report, Centro de Investigación y de Estudios Avanzados del Instituto Politécnico Nacional, Mexico (2004)
32. Michalewicz, Z., Schoenauer, M.: Evolutionary Algorithms for Constrained Parameter Optimization Problems. Evolutionary Computation 4(1), 1–32 (1996)
33. Rudenko, O., Schoenauer, M.: A Steady Performance Stopping Criterion for Pareto-based Evolutionary Algorithms. In: Proceedings of the 6th International Multi-Objective Programming and Goal Programming Conference, Hammamet, Tunisia (2004)

Constrained Optimization by ε Constrained Differential Evolution with Dynamic ε-Level Control

Tetsuyuki Takahama[1] and Setsuko Sakai[2]

[1] Department of Intelligent Systems, Hiroshima City University, Asaminami-ku, Hiroshima 731-3194 Japan
takahama@its.hiroshima-cu.ac.jp

[2] Faculty of Commercial Sciences, Hiroshima Shudo University, Asaminami-ku, Hiroshima 731-3195 Japan
setuko@shudo-u.ac.jp

Summary. In this chapter, the improved ε constrained differential evolution (εDE) is proposed to solve constrained optimization problems with very small feasible region, such as problems with equality constraints, efficiently. The εDE is the combination of the ε constrained method and differential evolution. In general, it is very difficult to solve constrained problems with very small feasible region. To solve such problems, static control schema of allowable constraint violation is often used, where solutions are searched within enlarged region specified by the allowable violation and the region is reduced to the feasible region gradually. However, the proper control depends on the initial population and searching process. In this study, the dynamic control of allowable violation is proposed to solve problems with equality constraints efficiently. In the εDE, the amount of allowable violation can be specified by the ε-level. The effectiveness of the εDE with dynamic ε-level control is shown by comparing with the original εDE and well known optimization method on some nonlinear constrained problems with equality constraints.

1 Introduction

Constrained optimization problems, especially nonlinear optimization problems, where objective functions are minimized under given constraints, are very important and frequently appear in the real world. In this study, the following optimization problem (P) with inequality constraints, equality constraints, upper bound constraints and lower bound constraints will be discussed.

$$\begin{aligned}
\text{(P)}\ \text{minimize}\ \ & f(\boldsymbol{x}) \\
\text{subject to}\ \ & g_j(\boldsymbol{x}) \le 0,\ j = 1, \dots, q \\
& h_j(\boldsymbol{x}) = 0,\ j = q+1, \dots, m \\
& l_i \le x_i \le u_i,\ i = 1, \dots, n,
\end{aligned} \tag{1}$$

where $\boldsymbol{x} = (x_1, x_2, \cdots, x_n)$ is an n dimensional vector, $f(\boldsymbol{x})$ is an objective function, $g_j(\boldsymbol{x}) \le 0$ and $h_j(\boldsymbol{x}) = 0$ are q inequality constraints and $m - q$

U.K. Chakraborty (Ed.): Advances in Differential Evolution, SCI 143, pp. 139–154, 2008.
springerlink.com

equality constraints, respectively. Functions f, g_j and h_j are linear or nonlinear real-valued functions. Values u_i and l_i are the upper bound and the lower bound of x_i, respectively. Also, let the feasible space in which every point satisfies all constraints be denoted by $\mathcal{F}$ and the search space in which every point satisfies the upper and lower bound constraints be denoted by $\mathcal{S}$ ($\supset \mathcal{F}$).

There exist many studies on solving constrained optimization problems using evolutionary algorithms[1] and particle swarm optimization[2]. These studies can be classified into several categories according to the way the constraints are treated as follows:

(1) Constraints are only used to see whether a search point is feasible or not[3]. In this category, the searching process begins with one or more feasible points and continues to search for new points within the feasible region. When a new search point is generated and the point is not feasible, the point is repaired or discarded. In this category, generating initial feasible points is difficult and computationally demanding when the feasible region is very small. It is almost impossible to find initial feasible points in problems with equality constraints.

(2) The constraint violation, which is the sum of the violation of all constraint functions, is combined with the objective function. The penalty function method is in this category[4]. In the penalty function method, an extended objective function is defined by adding the constraint violation to the objective function as a penalty. The optimization of the objective function and the constraint violation is realized by the optimization of the extended objective function. The main difficulty of the penalty function method is the difficulty of selecting an appropriate value for the penalty coefficient that adjusts the strength of the penalty.

(3) The constraint violation and the objective function are used separately. In this category, both the constraint violation and the objective function are optimized by a lexicographic order in which the constraint violation precedes the objective function. Takahama and Sakai proposed the α constrained method[5, 6] and the ε constrained method[7], which adopt a lexicographic ordering with relaxation of the constraints. Deb[8] proposed a method in which the extended objective function that realizes the lexicographic ordering is used. Runarsson and Yao[9] proposed the stochastic ranking method in which the stochastic lexicographic order, which ignores the constraint violation with some probability, is used. These methods were successfully applied to various problems.

(4) The constraints and the objective function are optimized by multiobjective optimization methods. In this category, the constrained optimization problems are solved as the multiobjective optimization problems in which the objective function and the constraint functions are objectives to be optimized[10, 11, 12]. But in many cases, solving multiobjective optimization problems is a more difficult and expensive task than solving single objective optimization problems.

In this study, the improved ε constrained differential evolution (εDE), is proposed to solve constrained optimization problems with very small feasible region, such as problems with equality constraints, efficiently. The εDE is the combination of the ε constrained method and differential evolution. The ε constrained methods can convert algorithms for unconstrained problems to algorithms for constrained problems using the ε-level comparison, which compares the search points based on the constraint violation of them. The ε constrained method is in the promising category (3) and is proposed based on the α constrained method. The α constrained method was applied to Powell's direct search method in [5, 6], the nonlinear simplex method by Nelder and Mead in [13, 14, 15], a genetic algorithm (GA) using linear ranking selection in [16, 17] and particle swarm optimization (PSO) in [18]. The ε constrained method was applied to PSO in [7, 19, 20], GA in [21] and differential evolution (DE)[22, 23].

In general, it is very difficult to solve constrained problems with very small feasible region. To solve such problems, static control schema of allowable constraint violation is often used, where solutions are searched within enlarged region specified by the allowable violation and the region is reduced to the feasible region gradually. However, the proper control depends on the initial population and searching process. It is very difficult to decide the control beforehand. In this study, the dynamic control of allowable violation is proposed to solve problems with equality constraints efficiently. In the εDE, the amount of allowable violation can be specified by the ε-level. The effectiveness of the εDE with dynamic ε-level control is shown by comparing with the εDE with static control and well known optimization method on some nonlinear constrained problems with equality constraints.

The rest of this chapter is organized as follows: Section 2 describes the ε constrained method briefly. Section 3 describes the improved εDE by introducing dynamic control of the ε-level. Section 4 presents experimental results on various benchmark problems discussed in [9]. Comparisons with the results in [9] are included in this section. Finally, Section 5 concludes with a brief summary of this chapter and a few remarks.

2 The ε Constrained Method

2.1 Constraint Violation and ε-Level Comparison

In the ε constrained method, constraint violation $\phi(\boldsymbol{x})$ is defined. The constraint violation can be given by the maximum of all constraints or the sum of all constraints.

$$\phi(\boldsymbol{x}) = \max\{\max_j\{0, g_j(\boldsymbol{x})\}, \max_j |h_j(\boldsymbol{x})|\} \tag{2}$$

$$\phi(\boldsymbol{x}) = \sum_j ||max\{0, g_j(\boldsymbol{x})\}||^p + \sum_j ||h_j(\boldsymbol{x})||^p \tag{3}$$

where p is a positive number.

The *ε-level comparison* is defined as an order relation on the set of $(f(\boldsymbol{x}), \phi(\boldsymbol{x}))$. If the constraint violation of a point is greater than 0, the point is not feasible and its worth is low. The ε-level comparisons are defined by a lexicographic order in which $\phi(\boldsymbol{x})$ proceeds $f(\boldsymbol{x})$, because the feasibility of $\boldsymbol{x}$ is more important than the minimization of $f(\boldsymbol{x})$.

Let $f_1\,(f_2)$ and $\phi_1\,(\phi_2)$ be the function values and the constraint violation at a point $\boldsymbol{x}_1\,(\boldsymbol{x}_2)$, respectively. Then, for any ε satisfying $\varepsilon \geq 0$, ε-level comparison $<_\varepsilon$ and $\leq_\varepsilon$ between (f_1, ϕ_1) and (f_2, ϕ_2) is defined as follows:

$$(f_1, \phi_1) <_\varepsilon (f_2, \phi_2) \Leftrightarrow \begin{cases} f_1 < f_2, \text{ if } \phi_1, \phi_2 \leq \varepsilon \\ f_1 < f_2, \text{ if } \phi_1 = \phi_2 \\ \phi_1 < \phi_2, \text{ otherwise} \end{cases} \quad (4)$$

$$(f_1, \phi_1) \leq_\varepsilon (f_2, \phi_2) \Leftrightarrow \begin{cases} f_1 \leq f_2, \text{ if } \phi_1, \phi_2 \leq \varepsilon \\ f_1 \leq f_2, \text{ if } \phi_1 = \phi_2 \\ \phi_1 < \phi_2, \text{ otherwise} \end{cases} \quad (5)$$

In case of ε=∞, the ε-level comparison $<_\infty$ and $\leq_\infty$ are equivalent to the ordinal comparison $<$ and $\leq$ between function values. Also, in case of $\varepsilon = 0$, $<_0$ and $\leq_0$ are equivalent to the lexicographic order in which the constraint violation $\phi(\boldsymbol{x})$ precedes the function value $f(\boldsymbol{x})$.

2.2 The Properties of the ε Constrained Method

The ε constrained method converts a constrained optimization problem into an unconstrained one by replacing the order relation in direct search methods with the ε-level comparison. An optimization problem solved by the ε constrained method, that is, a problem in which the ordinary comparison is replaced with the ε-level comparison, $(\mathrm{P}_{\leq_\varepsilon})$, is defined as follows:

$$(\mathrm{P}_{\leq_\varepsilon}) \text{ minimize}_{\leq_\varepsilon} f(\boldsymbol{x}), \quad (6)$$

where minimize$_{\leq_\varepsilon}$ means the minimization based on the ε-level comparison $\leq_\varepsilon$. Also, a problem (P^ε) is defined that the constraints of (P), that is, $\phi(\boldsymbol{x}) = 0$, is relaxed and replaced with $\phi(\boldsymbol{x}) \leq \varepsilon$:

$$\begin{aligned} (\mathrm{P}^\varepsilon) \text{ minimize } & f(\boldsymbol{x}) \\ \text{subject to } & \phi(\boldsymbol{x}) \leq \varepsilon \end{aligned} \quad (7)$$

It is obvious that (P^0) is equivalent to (P).

For the three types of problems, (P^ε), $(\mathrm{P}_{\leq_\varepsilon})$ and (P), the following theorems are given based on the α constrained method[5, 6, 7].

Theorem 1. *If an optimal solution* (P^0) *exists, any optimal solution of* $(\mathrm{P}_{\leq_\varepsilon})$ *is an optimal solution of* (P^ε).

Theorem 2. *If an optimal solution of* (P) *exists, any optimal solution of* $(\mathrm{P}_{\leq_0})$ *is an optimal solution of* (P).

Theorem 3. *Let* $\{\varepsilon_n\}$ *be a strictly decreasing non-negative sequence and converge to 0. Let* $f(\boldsymbol{x})$ *and* $\phi(\boldsymbol{x})$ *be continuous functions of* $\boldsymbol{x}$. *Assume that an optimal solution* $\boldsymbol{x}^*$ *of* (P^0) *exists and an optimal solution* $\hat{\boldsymbol{x}}_n$ *of* $(\mathrm{P}_{\leq_{\varepsilon_n}})$ *exists for any* ε_n. *Then, any accumulation point to the sequence* $\{\hat{\boldsymbol{x}}_n\}$ *is an optimal solution of* (P^0).

Theorem 1 and 2 show that a constrained optimization problem can be transformed into an equivalent unconstrained optimization problem by using the ε-level comparison. So, if the ε-level comparison is incorporated into an existing unconstrained optimization method, constrained optimization problems can be solved. Theorem 3 shows that, in the ε constrained method, an optimal solution of (P^0) can be given by converging ε to 0 as well as by increasing the penalty coefficient to infinity in the penalty method.

3 The εDE

In this section, we first describe differential evolution. Then, we describe the εDE, which is the integration of the ε constrained method and DE. Static and dynamic control functions of relaxing equality constraints are also defined.

3.1 Differential Evolution

Differential evolution is an evolutionary algorithm proposed by Storn and Price [24, 25]. DE is a stochastic direct search method using population or multiple search points. DE has been successfully applied to the optimization problems including non-linear, non-differentiable, non-convex and multi-modal functions. It has been shown that DE is fast and robust to these functions.

The main feature of DE is that DE uses simple arithmetic operations to avoid the control of Gaussian mutation adopted in evolution strategy. In general, the mutation process must be adaptive to the step size of the Gaussian mutation, because the ideal step size depends on the gene or element that is mutated and the state of the evolution process. DE adopts the sum of a base vector and the scaled difference vectors as the mutation operation instead of Gaussian mutation. The base vector is an individual selected from the population. The difference vectors are formed by the differences between a pair of individuals randomly selected from the population. As the search space by the population contracts and expands over generations, the step size in each dimension, which is given by the difference vectors, adapts automatically.

There are some variants of DE that have been proposed, such as DE/best /1/bin and DE/rand/1/exp. The variants are classified using the notation DE/*base*/*num*/*cross*. "*base*" indicates the method of selecting a parent that will form the base vector. For example, DE/rand/*num*/*cross* selects the parent for the base vector at random from the population. DE/best/*num*/*cross* selects the best individual in the population. "*num*" indicates the number of difference vectors used to perturb the base vector. "*cross*" indicates the crossover mechanism used to create a child. For example, DE/*base*/*num*/bin shows that crossover is controlled by binomial crossover using constant crossover rate. DE/*base*/*num*/exp shows that crossover is controlled by a binomial crossover using exponentially decreasing the crossover rate.

In DE, initial individuals are randomly generated within the search space and form an initial population. Each individual contains n genes as decision variables or a decision vector. At each generation or iteration, all individuals are selected as parents. Each parent is processed as follows: The mutation process begins by choosing $1 + 2\ num$ individuals from the parents except for the parent in the processing. The first individual is a base vector. All subsequent individuals are paired to create num difference vectors. The difference vectors are scaled by the scaling factor F and added to the base vector. The resulting vector is then recombined or crossovered with the parent. The probability of recombination at an element is controlled by the crossover factor CR. This crossover process produces a trial vector. Finally, for survivor selection, the trial vector is accepted for the next generation if the trial vector is better than the parent.

3.2 The Algorithm of the εDE

The algorithm of the εDE based on DE/rand/1/exp variant, which is used in this study, is as follows:

Step0 Initialization. Initial N individuals $\boldsymbol{x}^i$ are generated as the initial search points, where there is an initial population $P(0) = \{\boldsymbol{x}^i, i = 1, 2, \cdots, N\}$. An initial ε-level is given by the ε-level control function $\varepsilon(0)$.

Step1 Termination condition. If the number of generations (iterations) exceeds the maximum generation $T_{\max}$, the algorithm is terminated.

Step2 Mutation. For each individual $\boldsymbol{x}^i$, three different individuals $\boldsymbol{x}^{p1}$, $\boldsymbol{x}^{p2}$ and $\boldsymbol{x}^{p3}$, each of which is also different from $\boldsymbol{x}^i$, are chosen from the population. A new vector $\boldsymbol{x}'$ is generated by the base vector $\boldsymbol{x}^{p1}$ and the difference vector $\boldsymbol{x}^{p2} - \boldsymbol{x}^{p3}$ as follows:

$$\boldsymbol{x}' = \boldsymbol{x}^{p1} + F(\boldsymbol{x}^{p2} - \boldsymbol{x}^{p3}) \tag{8}$$

where F is a scaling factor.

Step3 Crossover. The vector $\boldsymbol{x}'$ is crossovered with the parent $\boldsymbol{x}^i$. A crossover point j is chosen randomly from all dimensions $[1, n]$. The element at the j-th dimension of the trial vector $\boldsymbol{x}^{\text{new}}$ is inherited from the j-th element of the vector $\boldsymbol{x}'$. The elements of subsequent dimensions are inherited from $\boldsymbol{x}'$ with exponentially decreasing probability defined by a crossover factor CR. Otherwise, the elements are inherited from the parent $\boldsymbol{x}^i$. In real processing, Step2 and Step3 are integrated as one operation.

Step4 Survivor selection. The trial vector $\boldsymbol{x}^{\text{new}}$ is accepted for the next generation if the trial vector is better than the parent $\boldsymbol{x}^i$.

Step5 Controlling the ε-level. The ε-level is updated by the ε-level control function $\varepsilon(t)$.

Step6 Go back to Step1.

Fig. 1 shows the algorithm of the εDE.

```
εDE/rand/1/exp()
{
  P(0)=Generate N individuals {x^i} randomly;
  ε=ε(0);
  for(t=1; t ≤ T_max; t++) {
    for(i=1; i ≤ N; i++) {
      (p1,p2,p3)=select randomly from [1,N]
                  s.t. p1 ≠ p2 ≠ p3 ≠ i;
      x^new=x^i ∈ P(t-1);
      j=select randomly from [1,n];
      k=1;
      do {
        x_j^new=x_j^p1+F(x_j^p2 - x_j^p3);
        j=(j+1)%n;
        k++;
      } while(k ≤ n && u(0,1) < CR);
      if((f(x^new),φ(x^new)) <_ε (f(x^i),φ(x^i)))
        z^i=x^new;
      else
        z^i=x^i;
    }
    P(t)={z^i, i=1,2,···,N}
    ε=ε(t);
  }
}
```

Fig. 1. The algorithm of the ε constrained differential evolution with control of the ε-level, where $\varepsilon(t)$ is the ε-level control function, F is a scaling factor, CR is a crossover factor, and $u(0,1)$ is a uniform random number generator in $[0,1]$

3.3 Controlling the ε-Level

Usually, the ε-level does not need to be controlled. Many constrained problems can be solved based on the lexicographic order where the ε-level is constantly 0. However for problems with equality constraints, the ε-level should be controlled properly to obtain high quality solutions.

Static control

A simple static control of the ε-level proposed in [7] can be defined according to the equation (9). The initial ε-level $\varepsilon(0)$ is the constraint violation of the top θ-th individual in the initial search points. The ε-level is updated until the number of iterations t becomes the control generation T_c. After the number of iterations exceeds T_c, the ε-level is set to 0 to obtain solutions with minimum constraint violation.

$$\begin{aligned} \varepsilon_s(0) &= \phi(\boldsymbol{x}_\theta) \\ \varepsilon_s(t) &= \begin{cases} \varepsilon_s(0)(1-\frac{t}{T_c})^{cp}, & 0 < t < T_c, \\ 0, & t \geq T_c \end{cases} \end{aligned} \tag{9}$$

where $\boldsymbol{x}_\theta$ is the top θ-th individual. When θ is too small, although a feasible solution can be found rapidly, the stability to find an optimal solution becomes low. When θ is too large, although the optimal solution can be found, the efficiency to find feasible region and the optimal solution becomes low. By using $\theta = 0.2N$, the stability and the efficiency can be balanced in many problems.

Dynamic control

To improve the efficiency of the εDEwith static control, we propose dynamic control of the ε-level according to the equation (10). Modified generation t' ($t' \geq t$) is introduced to speed up the convergence of enlarged region into the feasible region. If the violation is reduced enough in search process, the generation t' is increased faster than usual generation t and the ε-level is decreased faster. If not, the generation t' is increased by 1 like usual generation t.

$$\begin{aligned} \varepsilon_d(0) &= \phi(\boldsymbol{x}_\theta) \\ \varepsilon_d(t) &= \begin{cases} \varepsilon_d(0)(1-\frac{t'}{T_c})^{cp}, & 0 < t' < T_c, \\ 0, & t' \geq T_c \end{cases} \end{aligned} \tag{10}$$

The modified generation t' is updated in each generation as follows.

$$t' = \begin{cases} 0, & t = 0, \\ t'+1, & \phi(\boldsymbol{x}_\eta) \geq \varepsilon_d(t) \\ t'+2, & \phi(\boldsymbol{x}_\eta) < \varepsilon_d(t) \text{ and } t'+2 \geq \varepsilon_s^{-1}(\phi(\boldsymbol{x}_\eta)) \\ \frac{1}{2}(t'+2) + \frac{1}{2}\varepsilon_s^{-1}(\phi(\boldsymbol{x}_\eta)), & \text{otherwise} \end{cases} \tag{11}$$

where $\boldsymbol{x}_\eta$ is the worst η-th individual. $\varepsilon_s^{-1}(\varepsilon)$ is the inverse function of $\varepsilon_s(t)$ that returns the generation of the ε-level being ε in $\varepsilon_s(t)$ and is defined as follows.

$$\varepsilon_s^{-1}(\varepsilon) = \left(1 - \sqrt[cp]{\frac{\varepsilon}{\varepsilon_d(0)}}\right) T_c \tag{12}$$

4 Solving Constrained Nonlinear Programming Problems with Equality Constraints

In this section, four benchmark problems that are mentioned in some studies [9, 26, 15] are optimized.

4.1 Test Problems and Experimental Conditions

Four test problems, which are nonlinear optimization problems with equality constraints, are shown as follows.

g03 [27]:

$$\begin{aligned} \text{maximize} \quad & f(\boldsymbol{x}) = (\sqrt{n})^n \prod_{i=1}^{n} x_i, \\ \text{subject to} \quad & h_1(\boldsymbol{x}) = \sum_{i=1}^{n} x_i^2 - 1 = 0, \\ & 0 \le x_i \le 1 \, (i = 1, \cdots, n), \quad n = 10 \end{aligned}$$

The optimal solution $x_i^* = \frac{1}{\sqrt{n}} \, (i=1,\cdots,n)$ and the optimal value $f(\boldsymbol{x}^*)=1$.

g05 [28]:

$$\begin{aligned} \text{minimize} \quad & f(\boldsymbol{x}) = 3x_1 + 0.000001x_1^3 + 2x_2 + \frac{0.000002}{3} x_2^3, \\ \text{subject to} \quad & g_1(\boldsymbol{x}) = x_3 - x_4 - 0.55 \le 0, \\ & g_2(\boldsymbol{x}) = -x_3 + x_4 - 0.55 \le 0, \\ & h_3(\boldsymbol{x}) = 1000 \sin(-x_3 - 0.25) + 1000 \sin(-x_4 - 0.25) \\ & \qquad +894.8 - x_1 = 0, \\ & h_4(\boldsymbol{x}) = 1000 \sin(x_3 - 0.25) + 1000 \sin(x_3 - x_4 - 0.25) \\ & \qquad +894.8 - x_2 = 0, \\ & h_5(\boldsymbol{x}) = 1000 \sin(x_4 - 0.25) + 1000 \sin(x_4 - x_3 - 0.25) \\ & \qquad +1294.8 = 0, \\ & 0 \le x_i \le 1200 \, (i = 1, 2), \quad -0.55 \le x_i \le 0.55 \, (i = 3, 4) \end{aligned}$$

The minimum value is unknown. The known best value is $f(\boldsymbol{x}) = 5126.4981$ [29].

g11 [29]:

$$\begin{aligned} \text{minimize} \quad & f(\boldsymbol{x}) = x_1^2 + (x_2 - 1)^2, \\ \text{subject to} \quad & h(\boldsymbol{x}) = x_2 - x_1^2 = 0, \\ & -1 \le x_i \le 1 \, (i = 1, 2) \end{aligned}$$

The optimal solution is $\boldsymbol{x}^* = \left(\pm\frac{1}{\sqrt{2}}, \frac{1}{2}\right)$ and the optimal value $f(\boldsymbol{x}^*) =$ 0.75.

g13 [28]:

$$\begin{aligned} \text{minimize} \quad & f(\boldsymbol{x}) = e^{x_1 x_2 x_3 x_4 x_5}, \\ \text{subject to} \quad & h_1(\boldsymbol{x}) = x_1^2 + x_2^2 + x_3^2 + x_4^2 + x_5^2 - 10 = 0, \\ & h_2(\boldsymbol{x}) = x_2 x_3 - 5 x_4 x_5 = 0, \\ & h_3(\boldsymbol{x}) = x_1^3 + x_2^3 + 1 = 0, \\ & -2.3 \le x_i \le 2.3 \, (i = 1, 2), \quad -3.2 \le x_i \le 3.2 \, (i = 3, 4, 5) \end{aligned}$$

The optimal solution is $\boldsymbol{x}^*$ = (-1.717143, 1.595709, 1.827247, -0.7636413, -0.763645) and the optimal value $f(\boldsymbol{x}^*) = 0.0539498$.

In many other methods, problems with equality constraints cannot be solved directly. Thus, the equality constraints are relaxed, that is, all equality constraints $h_j(\boldsymbol{x}) = 0, \; j = q+1, \cdots, m$ are replaced by inequalities:

Table 1. Results using static control and dynamic control; 25 independent runs

Problem	Control	Best	Mean	Median	Worst
↑g03	static	1.00050010	1.00050010	1.00050010	1.00050010
(1.0)	dynamic	1.00050010	1.00050010	1.00050010	1.00050010
g05	static	5126.49671	5126.49671	5126.49671	5126.49671
(5126.498)	dynamic	5126.49671	5126.49671	5126.49671	5126.49671
g11	static	0.74990000	0.74990000	0.74990000	0.74990000
(0.750)	dynamic	0.74990000	0.74990000	0.74990000	0.74990000
g13	static	0.05394151	0.05394151	0.05394151	0.05394151
(0.053950)	dynamic	0.05394151	0.05394151	0.05394151	0.05394151

$$|h_j(\boldsymbol{x})| \leq \delta,\ \delta > 0 \tag{13}$$

In the experiments, $\delta = 0.001$.

In εDE, the same settings are used for all problems. The parameters for the ε constrained method are defined as follows: The constraint violation ϕ is given by the sum of all constraints (p=1) in equation (3). The ε-level is controlled by equation (9) in static control and (10) in dynamic control where $T_c = 0.5T$. The εDE can solve problems with equality constraints directly. However, to compare with other method, $\delta = 0.001$ is tested. The parameters for DE are as follows: The number of agents $N = 40$, $F = 0.7$, $CR = 0.9$. The maximum number of generation $T = 5000$, and independent 25 runs are performed in each problem.

4.2 Experimental Results

Experimental results on the test problems are shown in Table 1, in which each value is the average of 25 runs. The column labeled "problem" shows the problem number. The optimal value in each problem is shown in parentheses under the problem number. The column labeled "control" shows the type of control for the ε-level, where the parameter for dynamic control is $\eta = 5$. Also, "best", "mean", "median", and "worst" are the best value, the average value, the median value, the worst value, respectively. Problem g03 is the maximization problem and is shown with an up arrow.

In all problems, both of εDE with static control and εDE with dynamic control found the same solutions in all runs. In all cases, the εDEs found smaller values than the optimal values. The solutions obtained by the εDEs were away about 0.001 from the feasible region because the constraints are relaxed with equation (13). Thus, it is thought that the εDEs' ability to search for feasible solutions is very high for the problems with the equality constraints. This result shows that both of static and dynamic control have equivalent ability to find optimal solutions.

To compare the efficiency of the static control and the dynamic control, another experiment is performed with changing the parameter $\eta = 0, 1, 2, 3, 4, 5, 6$.

Table 2. Results with changing the parameter η; 25 independent runs

Problem	Param	Success	Best	Worst	Mean	Sigma	Ratio
g03	static	25	87,031	92,438	90,034.2	1,347.6	1
	0	25	79,062	84,124	81,821.4	1,309.3	0.91
	1	25	63,647	72,966	69,184.8	2,281.1	0.77
	2	25	55,334	62,539	59,032.2	1,849.0	0.66
	3	25	47,261	54,213	51,170.0	1,861.2	0.57
	4	25	41,342	**48,950**	**44,467.4**	2,033.8	**0.49**
	5	25	**36,584**	69,402	45,551.7	6,118.2	0.51
	6	19	39,870	105,202	73,796.9	17,928.0	0.82
g05	static	25	96,688	98,290	97,572.0	362.0	1
	0	25	94,531	97,058	95,850.9	572.2	0.98
	1	25	91,776	94,321	93,099.6	732.8	0.95
	2	25	87,248	90,730	88,980.2	817.7	0.91
	3	25	81,998	85,982	84,225.6	1058.3	0.86
	4	25	76,026	81,286	78,620.8	1,299.8	0.81
	5	25	71,180	76,797	73,722.4	1,296.6	0.76
	6	25	**65,644**	**71,312**	**67,783.2**	1,315.4	**0.69**
g11	static	25	22,558	62,184	45,046.8	9,330.4	1
	0	25	7,044	47,736	34,894.0	8,811.0	0.77
	1	25	21,714	49,530	34,130.1	7,496.9	0.76
	2	25	12,545	41,469	30,411.5	6,805.9	0.68
	3	25	7,810	37,409	23,740.1	6,818.6	0.53
	4	25	17,403	33,025	26,380.2	4,100.5	0.59
	5	25	**4,676**	29,511	19,533.7	5,873.8	0.43
	6	25	5,918	**28,345**	**18,572.0**	6,849.0	**0.41**
g13	static	25	76,582	88,947	85,037.3	2,694.5	1
	0	25	75,854	84,101	80,048.8	2,178.8	0.94
	1	25	67,573	78,873	73,090.9	2,396.2	0.86
	2	25	57,850	70,317	63,716.6	2,902.4	0.75
	3	25	51,056	59,640	55,575.9	1,902.8	0.65
	4	25	42,432	51,819	47,945.0	2,401.3	0.56
	5	25	37,404	46,565	42,308.2	2,562.8	0.50
	6	25	**31,639**	**40,491**	**36,964.5**	2,216.6	**0.43**

Table 2 shows the number of function evaluations needed for satisfying the success condition of $f^{\text{best}} - f^* \leq 0.0001$ and $\boldsymbol{x}^{\text{best}}$ being feasible, where $\boldsymbol{x}^{\text{best}}$ and f^{best} are the best solution found in each run and its value, respectively. The column labeled "param" shows the parameter value of η for dynamic control, and the result of static control is also shown. The column labeled "success" shows the number of runs in which solutions satisfying success condition are found. Also, "best", "worst", "mean" and "sigma" are the best value, the worst value, the average value and the standard deviation of the number of function evaluations for satisfying success condition. The ratio of the number of function evaluations

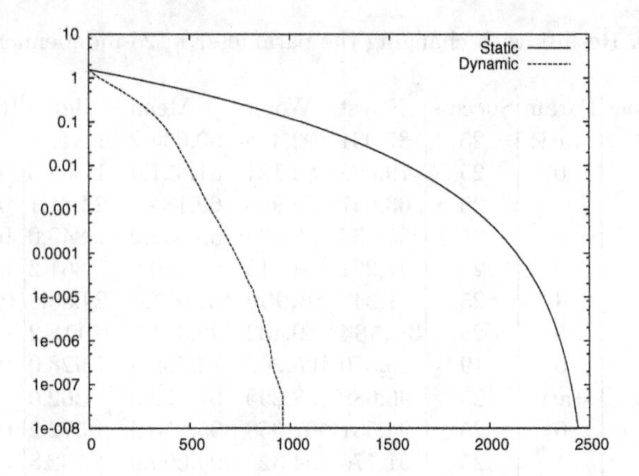

Fig. 2. The control of the ε-level in problem g03

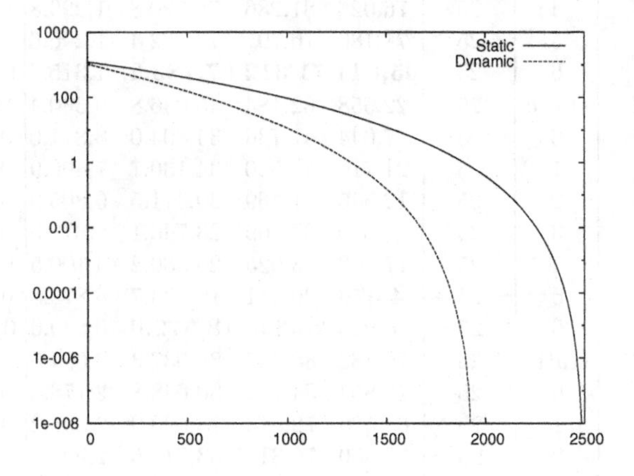

Fig. 3. The control of the ε-level in problem g05

between the static control and dynamic control is shown in the column labeled "ratio". The better cases are highlighted using boldface.

When the parameter η is in $[0, 5]$, solutions satisfying success condition are found in all runs. If η is large, the convergence speed of enlarged region into feasible region becomes high and feasible solutions and the optimal solution can be found faster. When the parameter η is 5, the number of function evaluations is reduced about 50% in g03, g11 and g13 and it is reduced about 24% in g05. So, the efficiency of the εDE with dynamic control is very higher than that of the εDE with static control. However, when η is 6, success condition cannot be attained in 6 runs out of 25 runs for g03. It is thought that the decrease of the ε-level is too fast, enlarged region is reduced too fast, and search process cannot find the optimal solution. Thus, the value of η should be selected properly to find optimal solutions, and the proper value of parameter η is 5.

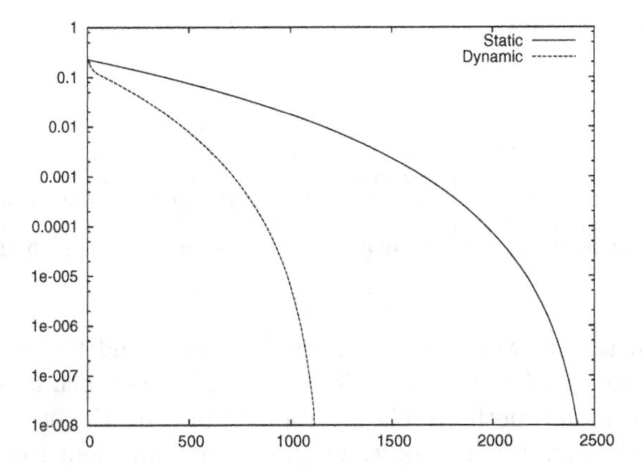

Fig. 4. The control of the ε-level in problem `g11`

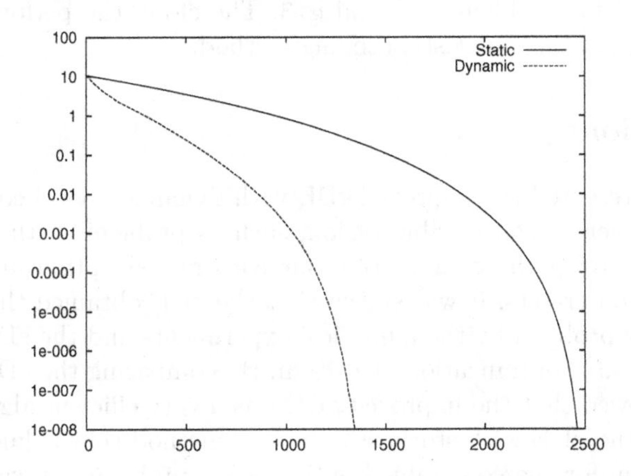

Fig. 5. The control of the ε-level in problem `g13`

Figures 2, 3, 4 and 5 show the change of the ε-level over generations for static control and dynamic control with $\eta = 5$ in problems `g03`, `g05`, `g11` and `g13`, respectively. Apparently, the convergence speed of the ε-level in dynamic control with $\eta = 5$ is higher than that in static control. These graphs show the effectiveness of the dynamic control.

4.3 Comparison with the Stochastic Ranking Method

To show the effectiveness of the εDE with dynamic control, the solutions found by this method are compared to those found by Runarsson and Yao's stochastic ranking method[9]. In [9], the maximum number of evaluations in each run was $1750 \times 200 = 350,000$ and all equality constraints were relaxed using $\delta = 10^{-4}$. Table 3 shows the comparison of the two methods. The better cases are highlighted using boldface. The results of the εDE were taken from Table 1 where

Table 3. Comparison between our (indicated by εDE) and Runarsson and Yao's (indicated by RY[9]) algorithms

f	Optimal	Best Result		Median Result		Mean Result		Worst Result	
		εDE	RY	εDE	RY	εDE	RY	εDE	RY
↑g03	1.000	**1.001**	1.000	**1.001**	1.000	**1.001**	1.000	**1.001**	1.000
g05	5126.498	5126.497	5126.497	**5126.497**	5127.372	**5126.497**	5128.881	**5126.497**	5142.472
g11	0.750	0.750	0.750	0.750	0.750	0.750	0.750	0.750	0.750
g13	0.053950	**0.053942**	0.053957	**0.053942**	0.057006	**0.053942**	0.067543	**0.53942**	0.438803

all equality constraints were relaxed using $\delta = 10^{-4}$ and the maximum number of evaluations was $5000 \times 40 = 200,000$ which was much less than that of the stochastic ranking method. The solutions found by the stochastic ranking method were very high quality solutions that were equivalent to the known optimal solutions. Nevertheless, the εDE found better solutions for problems g03 and g13. Also, the stability of the εDE was better than that of the stochastic ranking method for problems g05 and g13. Therefore, the performance of the εDE is better than the stochastic ranking method.

5 Conclusions

This chapter presented the improved εDE with dynamic ε-level control to solve problems with very small feasible region, such as problems with equality constraints. By applying the εDE to the four constrained optimization problems with equality constraints, it was shown that the εDE obtained the optimal solution for every problem by the numerical experiments and the εDE was a high precision and stable optimization algorithm. By comparing the εDE with static control, we showed that the improved εDE was a very efficient algorithm. Also, by comparing the εDE with stochastic ranking method that is known as an efficient algorithm for the constrained optimization problems, it was shown that the εDE was a very stable and good algorithm.

In the future, we will explore ways to prevent the ε-level from converging too fast. Also, we will apply the εDE to various application fields.

Acknowledgements

This research is supported in part by Grant-in-Aid for Scientific Research (C) (No. 16500083, 17510139) of Japan society for the promotion of science and Hiroshima City University Grant for Special Academic Research (General Studies) 7111.

References

1. Michalewicz, Z.: A survey of constraint handling techniques in evolutionary computation methods. In: Proceedings of the 4th Annual Conference on Evolutionary Programming, pp. 135–155. MIT Press, Cambridge (1995)

2. Coath, G., Halgamuge, S.K.: A comparison of constraint-handling methods for the application of particle swarm optimization to constrained nonlinear optimization problems. In: Proc. of IEEE Congress on Evolutionary Computation, Canberra, Australia, pp. 2419–2425 (2003)
3. Hu, X., Eberhart, R.C.: Solving constrained nonlinear optimization problems with particle swarm optimization. In: Proc. of the Sixth World Multiconference on Systemics, Cybernetics and Informatics, Orlando, Florida (2002)
4. Parsopoulos, K.E., Vrahatis, M.N.: Particle swarm optimization method for constrained optimization problems. In: Sincak, P., Vascak, J., et al. (eds.) Intelligent Technologies — Theory and Application: New Trends in Intelligent Technologies. Frontiers in Artificial Intelligence and Applications, vol. 76, pp. 214–220. IOS Press, Amsterdam (2002)
5. Takahama, T., Sakai, S.: Tuning fuzzy control rules by α constrained method which solves constrained nonlinear optimization problems. The Transactions of the Institute of Electronics, Information and Communication Engineers J82-A(5), 658–668 (1999) (in Japanese)
6. Takahama, T., Sakai, S.: Tuning fuzzy control rules by the α constrained method which solves constrained nonlinear optimization problems. Electronics and Communications in Japan, Part3: Fundamental Electronic Science 83(9), 1–12 (2000)
7. Takahama, T., Sakai, S.: Constrained optimization by ϵ constrained particle swarm optimizer with ϵ-level control. In: Proc. of the 4th IEEE International Workshop on Soft Computing as Transdisciplinary Science and Technology (WSTST 2005), May 2005, pp. 1019–1029 (2005)
8. Deb, K.: An efficient constraint handling method for genetic algorithms. Computer Methods in Applied Mechanics and Engineering 186(2/4), 311–338 (2000)
9. Runarsson, T.P., Yao, X.: Stochastic ranking for constrained evolutionary optimization. IEEE Transactions on Evolutionary Computation 4(3), 284–294 (2000)
10. Camponogara, E., Talukdar, S.N.: A genetic algorithm for constrained and multiobjective optimization. In: Alander, J.T. (ed.) 3rd Nordic Workshop on Genetic Algorithms and Their Applications (3NWGA), August 1997, pp. 49–62. University of Vaasa, Vaasa, Finland (1997)
11. Surry, P.D., Radcliffe, N.J.: The COMOGA method: Constrained optimisation by multiobjective genetic algorithms. Control and Cybernetics 26(3), 391–412 (1997)
12. Ray, T., Liew, K.M., Saini, P.: An intelligent information sharing strategy within a swarm for unconstrained and constrained optimization problems. Soft Computing – A Fusion of Foundations, Methodologies and Applications 6(1), 38–44 (2002)
13. Takahama, T., Sakai, S.: Learning fuzzy control rules by α-constrained simplex method. The Transactions of the Institute of Electronics, Information and Communication Engineers J83-D-I(7), 770–779 (2000) (in Japanese)
14. Takahama, T., Sakai, S.: Learning fuzzy control rules by α-constrained simplex method. Systems and Computers in Japan 34(6), 80–90 (2003)
15. Takahama, T., Sakai, S.: Constrained optimization by applying the α constrained method to the nonlinear simplex method with mutations. IEEE Transactions on Evolutionary Computation 9(5), 437–451 (2005)
16. Takahama, T., Sakai, S.: Constrained optimization by α constrained genetic algorithm (αGA). The Transactions of the Institute of Electronics, Information and Communication Engineers J86-D-I(4), 198–207 (2003) (in Japanese)
17. Takahama, T., Sakai, S.: Constrained optimization by α constrained genetic algorithm (αGA). Systems and Computers in Japan 35(5), 11–22 (2004)

18. Takahama, T., Sakai, S.: Constrained optimization by combining the α constrained method with particle swarm optimization. In: Proc. of Joint 2nd International Conference on Soft Computing and Intelligent Systems and 5th International Symposium on Advanced Intelligent Systems (2004)
19. Takahama, T., Sakai, S., Iwane, N.: Constrained optimization by the ε constrained hybrid algorithm of particle swarm optimization and genetic algorithm. In: Zhang, S., Jarvis, R. (eds.) AI 2005. LNCS (LNAI), vol. 3809, pp. 389–400. Springer, Heidelberg (2005)
20. Takahama, T., Sakai, S.: Solving constrained optimization problems by the ε constrained particle swarm optimizer with adaptive velocity limit control. In: Proc. of the 2nd IEEE International Conference on Cybernetics & Intelligent Systems, June 2006, pp. 683–689 (2006)
21. Takahama, T., Sakai, S.: Constrained optimization by the ε constrained genetic algorithm. IPSJ Journal 47(6), 1861–1871 (2006)
22. Takahama, T., Sakai, S., Iwane, N.: Solving nonlinear constrained optimization problems by the ε constrained differential evolution. In: Proc. of the 2006 IEEE Conference on Systems, Man, and Cybernetics, October 2006, pp. 2322–2327 (2006)
23. Takahama, T., Sakai, S.: Constrained optimization by the ε constrained differential evolution with gradient-based mutation and feasible elites. In: Proc. of 2006 IEEE Congress on Evolutionary Computation, July 2006, pp. 308–315 (2006)
24. Storn, R., Price, K.: Minimizing the real functions of the ICEC 1996 contest by differential evolution. In: Proc. of the International Conference on Evolutionary Computation, pp. 842–844 (1996)
25. Storn, R., Price, K.: Differential evolution – a simple and efficient heuristic for global optimization over continuous spaces. Journal of Global Optimization 11, 341–359 (1997)
26. Mezura-Montes, E., Coello, C.A.C.: A simple multimembered evolution strategy to solve constrained optimization problems. IEEE Trans. on Evolutionary Computation 9(1), 1–17 (2005)
27. Mchalewicz, Z., Nazhiyath, G., Michalewicz, M.: A note on usefullness of geometricalcrossover of numerical optimization problems. In: Fogel, L.J., Angeline, P.J., Bäck, T. (eds.) Proc. 5th Annual Conference on Evolutionary Programming, pp. 305–312. MIT Press, Cambridge (1996)
28. Hock, W., Schittkowski, K.: Test examples for nonlinear programming codes. Lecture Notes in Economics and Mathematical Systems. Springer, Heidelberg (1981)
29. Koziel, S., Michalewicz, Z.: Evolutionary algorithms, homomorphous mappings, and constrained parameter optimization. Evolutionary Computation 7(1), 19–44 (1999)

Opposition-Based Differential Evolution

Shahryar Rahnamayan, Hamid R. Tizhoosh, and Magdy M.A. Salama

Faculty of Engineering, University of Waterloo, Waterloo, Canada

Summary. Although the concept of the opposition has an old history in other fields and sciences, this is the first time that it contributes to enhance an optimizer. This chapter presents a novel scheme to make the differential evolution (DE) algorithm faster. The proposed opposition-based DE (ODE) employs opposition-based optimization (OBO) for population initialization and also for generation jumping. In this work, opposite numbers have been utilized to improve the convergence rate of the classical DE. A test suite with 15 benchmark functions is employed for experimental verification. The contribution of the opposite numbers is empirically verified. Additionally, two time varying models for control parameter adjustment of ODE are investigated. Details of the ODE algorithm, the test set, and the comparison strategy are provided.

1 Introduction

The footprints of the opposition concept can be observed in many areas around us. This concept has sometimes been called by different names. Opposite particles in physics, antonyms in languages, complement of an event in probability, antithetic variables in simulation, opposite proverbs in culture, absolute or relative complement in set theory, subject and object in philosophy of science, good and evil in animism, opposition parties in politics, theses and antitheses in dialectic, opposition day in parliaments, and dualism in religions and philosophies are just some examples to mention (Table 1 contains more examples and corresponding details).

The Yin-Yang symbol in the ancient Chinese philosophy is probably the oldest opposition concept which was expressed by human kind (Figure 1). Black and white represent yin and yang, respectively. This symbol reflects the twisted duality of all things in nature, namely, receptive vs. creative, feminine vs. masculine, dark vs. bright, and finally passive force vs. active force. Greek's classical elements to explain patterns in nature (Figure 2) also mention and make use of the opposition concept, namely, fire (hot and dry) vs. water (cold and wet), earth (cold and dry) vs. air (hot and wet).

It seems that without using the opposition concept explaining of different entities around us is hard and maybe even impossible. In order to explain an entity or a situation we sometimes explain its opposite instead. In fact, opposition often offers a balance between the entities. For instance, the east, west, south, and north can not be defined

U.K. Chakraborty (Ed.): Advances in Differential Evolution, SCI 143, pp. 155–171, 2008.
springerlink.com

alone. The same is valid for cold and hot and many other examples. Extreme opposites constitute our upper and lower boundaries. Imagination of the infinity is vague but when we consider the limited, then it becomes more imaginable because its opposite is definable. We finish the introduction with Rumi's quote:

> *Therefore, the foundation of the creation was (based) upon opposites. Necessarily, we are battling because of loss and gain.* Rumi (1207 – 1273) in "Masnawi"

In the following section, we explain how opposition concept can be utilized in optimization. Firstly, the opposite point in one- and D-dimensional spaces are defined, and secondly, the opposition-based optimization is described.

Table 1. Footprints of opposition in variant fields

Example	**Field**	**Description**
Opposite Particles/Elements	Physics	Such as magnetic poles (N and S), opposite polarities (+ and −), electron-proton in an atom, action-reaction forces in Newton's third law, and so on.
Antonyms	Language	A word that means the opposite of another word (e.g., hot/cold, fast/slow, top/down, left/right, day/night, on/off).
Antithetic Variables	Simulation	Antithetic (negatively correlated) pair of random variables used for variance reduction.
Opposite Proverbs	Culture	Two proverbs with the opposite advice or meaning (e.g., The pen is mightier than the sword. Actions speak louder than words.); proverb or its opposite pair offers an applicable solution based on specific situation or condition.
Complements	Set theory	a) Relative complement ($B - A = \{x \in B \mid x \notin A\}$), b) Absolute complement ($A^c = U - A$, where U is the universal set).
Opposition	Politics	A political party or organized group opposed to the government.
Inverter	Digital design	Output of the inverter gate is one if input is zero and vice versa.
Opposition Day	Legislation	A day in the parliament in which small opposition parties are allowed to propose the subject for debate (e.g., Canada's parliament has 25 opposition days).
Dualism	Philosophy and Religion	Two fundamental principles/concepts, often in opposition to each other; such as "Yin" and "Yang" in Chinese philosophy and Taoist religion (Figure 1), "subject" and "object" in philosophy of science, "good" and "evil" in animism.
Dialectic	Philosophy	An exchange of "theses" and "antitheses" resulting in a "synthesis" (e.g. in Hinduism, these three elements are creation (Brahma), maintenance of order (Vishnu) and destruction or disorder (Shiva)).
Classical Elements	Archetype	A set of archetypal classical elements to explain patterns in nature (e.g., the Greek classical elements are Fire (hot and dry), Earth (cold and dry), Air (hot and wet), and Water (cold and wet), Figure 2).
if-then-else	Algorithm	if *condition* then *statements* [else *elsestatements*], the *else statements* are executed when the opposite of the *condition* happens.
Complement of an Event	Probability	$P(A') = 1 - P(A)$.
Revolution	socio-political	A significant Socio-political change in a short period of time.

Fig. 1. Yin-Yang symbol or Taijitu in ancient Chinese philosophy. Black and white are representing yin (receptive, feminine, dark, passive force) and yang (creative, masculine, bright, active force), respectively.

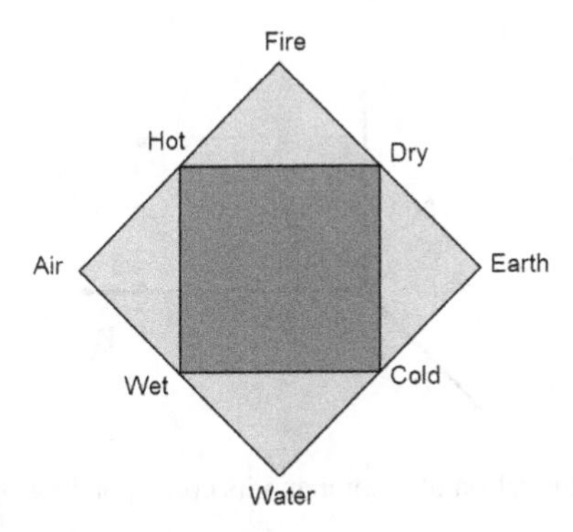

Fig. 2. Greek classical elements to explain patterns in nature are Fire (hot and dry), Earth (cold and dry), Air (hot and wet), and Water (cold and wet).

2 Opposition-Based Optimization

Generally speaking, evolutionary optimization methods start with some candidate solutions (initial population) and try to improve them toward some optimal solution(s). The process of searching terminates when some predefined criteria are satisfied. In the absence of a priori information about the solution, we usually start with a *random guesse*. The computation time, among others, is related to the distance of these initial guesses from the optimal solution. We can improve our chance of starting with a closer (fitter) solution by simultaneously checking *the opposite solution*. By doing this, the fitter one (guess or opposite guess) can be chosen as an initial solution. In fact, according to probability theory, 50% of the time a guess is farther from the solution than its opposite. So, starting with the closer of the two guesses (as judged by its fitness) has the potential to accelerate convergence. The same approach can be applied not only to initial solutions but also continuously to each solution in the current population. However, before concentrating on opposition-based optimization, we need to define the concept of opposite numbers [1]:

Definition (opposite number) - Let x be a real number in an interval $[a, b]$ ($x \in [a, b]$); the opposite of x, denoted by $\breve{x}$, is defined by

$$\breve{x} = a + b - x. \tag{1}$$

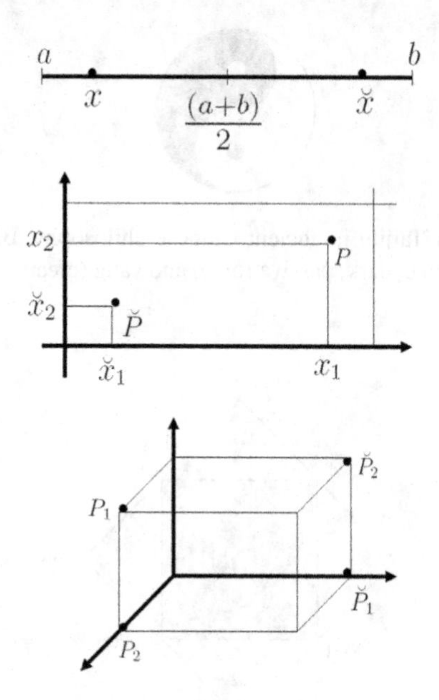

Fig. 3. Illustration of a point and its corresponding opposite

Figure 3 (top) illustrates x and its opposite $\breve{x}$ in interval $[a, b]$. As seen, x and $\breve{x}$ are located in equal distance from the interval center ($|(a+b)/2 - x| = |\breve{x} - (a+b)/2|$) and from the interval boundaries ($|x - a| = |b - \breve{x}|$) as well.

This definition can be extended to higher dimensions [1].

Definition (opposite point) - Let $P(x_1, x_2, ..., x_D)$ be a point in D-dimensional space, where $x_1, x_2, ..., x_D$ are real numbers and $x_i \in [a_i, b_i]$, $i = 1, 2, ..., D$. The opposite point of P is defined by $\breve{P}(\breve{x}_1, \breve{x}_2, ..., \breve{x}_D)$ where

$$\breve{x}_i = a_i + b_i - x_i. \tag{2}$$

Figure 3 illustrates a sample point and its corresponding opposite point in one, two, and three dimensional spaces.

Now, after the definition of the opposite points we are ready to define *Opposition-Based Optimization (OBO).*

Opposition-Based Optimization (OBO) - Let $P(x_1, x_2, ..., x_D)$, a point in a D-dimensional space with $x_i \in [a_i, b_i]$ ($i = 1, 2, 3, ..., D$), be a candidate solution. Assume $f(x)$ is a fitness function which is used to measure candidate's optimality. According to the opposite point definition, $\breve{P}(\breve{x}_1, \breve{x}_2, ..., \breve{x}_D)$ is the opposite of $P(x_1, x_2, ..., x_D)$. Now, if $f(\breve{P}) \geq f(P)$, then point P can be replaced with $\breve{P}$; otherwise we continue with P. Hence, the point and its opposite point are evaluated simultaneously to continue with the fitter one [1].

3 Opposition-Based Differential Evolution

Similar to all population-based optimization algorithms, two main steps are distinguishable for DE, namely population initialization and producing new generations by evolutionary operations such as mutation, crossover, and selection. We will enhance these two steps using the opposition-based optimization concept. The classical DE is chosen as a parent algorithm and the proposed opposition-based schemes are embedded in DE to accelerate its convergence speed. Corresponding pseudo-code and flowchart for the proposed approach (ODE) are given in Algorithm 1 and Figure 4, respectively. Newly added/extended code segments (which are shown by grey blocks in Figure 4) will be explained in the following subsections.

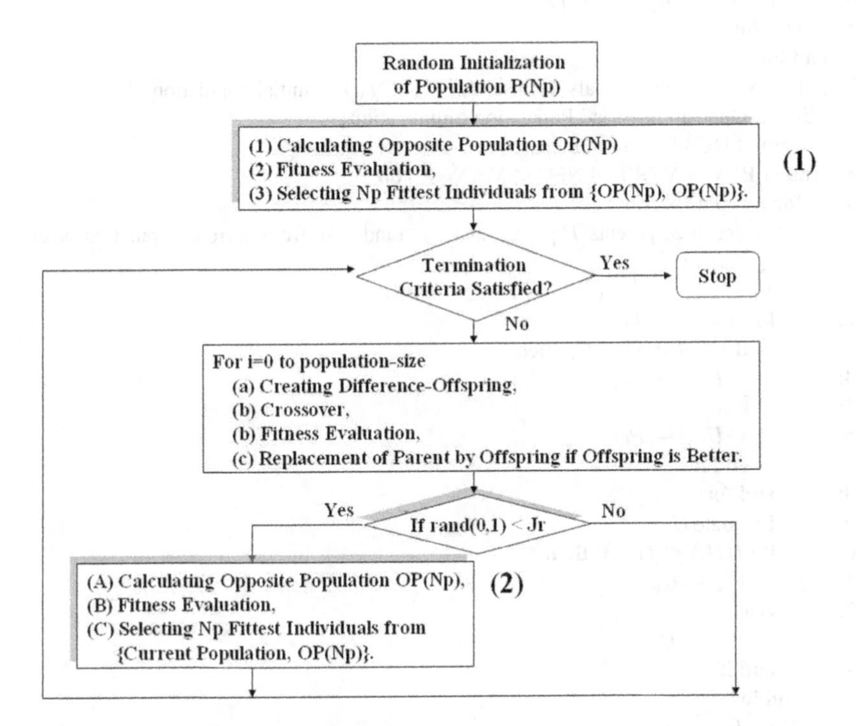

Fig. 4. New blocks are illustrated by gray boxes. Block (1): Opposition-based initialization, Block (2): Opposition-based generation jumping (J_r: jumping rate, $rand(0,1)$: uniformly generated random number, N_p: population size)

3.1 Opposition-Based Population Initialization

In absence of a priori knowledge, random number generation is generally the only choice to create an initial population. But as mentioned before, by utilizing OBO we can obtain fitter starting candidates even when there is no a priori knowledge about the solution(s). Block (1) from Figure 4 shows the implementation of opposition-based population initialization. Following steps explain that procedure:

Algorithm 1. Pseudo-code for Opposition-Based Differential Evolution (ODE). P_0: Initial population, OP_0: Opposite of initial population, P: Current population, OP: Opposite of current population, D: Problem dimension, $[a_j, b_j]$: Range of the j-th variable, J_r: Jumping rate, $\min_j^p/\max_j^p$: Minimum/maximum value of the j-th variable in the current population. Steps **2-7** and **27-34** are implementations of opposition-based population initialization and opposition-based generation jumping, respectively.

```
1: Generate uniformly distributed random population P_0
   {Begin of Opposition-Based Population Initialization}
2: for i = 0 to N_p do
3:    for j = 0 to D do
4:       OP_{0i,j} ← a_j + b_j − P_{0i,j}
5:    end for
6: end for
7: Select N_p fittest individuals from set the {P_0, OP_0} as initial population P_0
   {End of Opposition-Based Population Initialization}
   {Begin of DE's Evolution Steps}
8: while ( BFV > VTR and NFC < MAX_NFC ) do
9:    for i = 0 to N_p do
10:      Select three parents P_{i1}, P_{i2}, and P_{i3} randomly from current population where i ≠
         i1 ≠ i2 ≠ i3
11:      V_i ← P_{i1} + F × (P_{i2} − P_{i3})
12:      for j = 0 to D do
13:         if rand(0, 1) < C_r then
14:            U_{i,j} ← V_{i,j}
15:         else
16:            U_{i,j} ← P_{i,j}
17:         end if
18:      end for
19:      Evaluate U_i
20:      if (f(U_i) ≤ f(P_i)) then
21:         P'_i ← U_i
22:      else
23:         P'_i ← P_i
24:      end if
25:   end for
26:   P ← P'
      {End of DE's Evolution Steps}
      {Begin of Opposition-Based Generation Jumping}
27:   if rand(0, 1) < J_r then
28:      for i = 0 to N_p do
29:         for j = 0 to D do
30:            OP_{i,j} ← MIN_j^p + MAX_j^p − P_{i,j}
31:         end for
32:      end for
33:      Select N_p fittest individuals from set the {P, OP} as current population P
34:   end if
      {End of Opposition-Based Generation Jumping}
35: end while
```

step 1. Initialize the population $\mathrm{P}(N_p)$ randomly,
step 2. Calculate opposite population by

$$\mathrm{OP}_{i,j} = \mathrm{a}_j + \mathrm{b}_j - \mathrm{P}_{i,j}, \tag{3}$$

$$i = 1, 2, ..., N_p \,;\, j = 1, 2, ..., D.$$

where $\mathrm{P}_{i,j}$ and $\mathrm{OP}_{i,j}$ denote the jth variable of the ith population and the opposite-population vector, respectively.
step 3. Select the N_p fittest individuals from the set $\{\mathrm{P} \cup \mathrm{OP}\}$ as the initial population.

According to the above procedure, $2N_p$ function evaluations are required instead of N_p for the regular random population initialization. But, by the opposition-based initialization, the parent algorithm can start with the fitter initial individuals instead.

3.2 Opposition-Based Generation Jumping

By applying a similar approach to the current population, the evolutionary process can be forced to jump to a fitter generation. Based on a jumping rate J_r (i.e. jumping probability), after generating new populations by mutation, crossover, and selection, the opposite population is calculated and the N_p fittest individuals are selected from the union of the current population and the opposite population. As a difference to opposition-based initialization, it should be noted here that in order to calculate the opposite population for generation jumping, the opposite of each variable is calculated dynamically. That is, the maximum and minimum values of each variable in the *current population* ($[\mathrm{MIN}_j^p, \mathrm{MAX}_j^p]$) are used to calculate opposite points instead of using variables' predefined interval boundaries ($[a_j, b_j]$):

$$\mathrm{OP}_{i,j} = \mathrm{MIN}_j^p + \mathrm{MAX}_j^p - \mathrm{P}_{i,j}, i = 1, 2, ..., N_p; j = 1, 2, ..., D. \tag{4}$$

By staying within variables' static boundaries, it is possible to jump outside of the already shrunken search space and loose the knowledge of the current reduced space (converged population). Hence, we calculate opposite points by using variables' current interval in the population ($[\mathrm{MIN}_j^p, \mathrm{MAX}_j^p]$) which is, as the search does progress, increasingly smaller than the corresponding initial range $[a_j, b_j]$. Block (2) from Figure 4 indicates the implementation of opposition-based generation jumping.

A pictorial example is presented in Figure 5 to exhibit opposition-based generation jumping procedure in 2D space. The 'S' indicates location of the solution. Dark and light circles present the points and the opposite points, respectively. As seen, in the resulted population (shown by the current P), the average distance of the selected candidates (which contains some original points and the opposite of some others) from the solution is smaller than it was for population (P) and opposite population (OP), individually.

Generally speaking, in order to utilize the advantages of the opposition-based optimization to accelerate population-based algorithms, many schemes can be suggested and investigated. But, it seems that considering the following features during the design of these schemes are crucial:

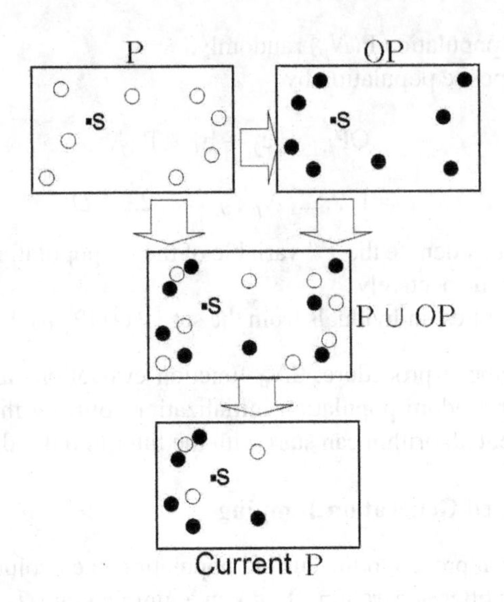

Fig. 5. A pictorial example to show the opposition-based generation jumping in 2D space ($N_p = 8$)

Generality - Proposing general schemes makes it easy to use OBO on a wider range of population-based optimization methods. Tailored schemes would be obviously more rigid for generalization. Manipulating the internal operators of the optimizer leads to lower generality, although, the customized schemes can result a higher performance.

Simplicity - This feature is always desirable in all science and engineering design problems. Simplicity supports a higher understandability and makes any design easy to implement and modify. Also, in practical environments, the simple schemes are widely appreciated.

Problem Independency - Proposed schemes have to be universal and capable to solve a wider range of optimization problems. By equipping the parent optimizer with the opposition-based schemes, it should not be specialized to solve a group of specific problems (e.g., unimodal). This case is experimentally verifiable by applying the algorithm to solve various global optimization problems. In other words, the proposed schemes should not reduce the universality of the parent optimizer to solve different problems.

Effectiveness - Considering opposite points needs more function calls and should be controlled smartly to prevent loosing the benefits through extra computations. Overall, the extra function calls should be reasonable and bring a benefit to the optimization process. The benefit can be faster convergence, higher robustness, or higher solution quality. Furthermore, improving one of these features should not affect the other benefits. During the experimental verification of the proposed algorithm, different measures are employed to investigate each criterion individually.

4 Experimental Verifications

4.1 Comparison of DE and ODE

A set of 15 well-known benchmark functions, which contains 7 unimodal and 8 multimodal functions, has been selected for performance verification of the ODE. The definition of the benchmark functions is given in Table 2.

Table 2. List of employed benchmark functions. S denotes the search space.

Name	*Definition*	*S*
1st De Jong	$f_1(X) = \sum_{i=1}^{D} {x_i}^2$	$[-5.12, 5.12]^D$
Axis Parallel Hyper-Ellipsoid	$f_2(X) = \sum_{i=1}^{D} i{x_i}^2$	$[-5.12, 5.12]^D$
Schwefel's Problem 1.2	$f_3(X) = \sum_{i=1}^{D} \left(\sum_{j=1}^{i} x_j \right)^2$	$[-65, 65]^D$
Rastrigin's Function	$f_4(X) = 10D + \sum_{i=1}^{D} (x_i^2 - 10\cos(2\pi x_i))$	$[-5.12, 5.12]^D$
Griewangk's Function	$f_5(X) = \sum_{i=1}^{D} \frac{x_i^2}{4000} - \prod_{i=1}^{D} \cos\left(\frac{x_i}{\sqrt{i}}\right) + 1$	$[-600, 600]^D$
Sum of Different Power	$f_6(X) = \sum_{i=1}^{D} \lvert x_i \rvert^{(i+1)}$	$[-1, 1]^D$
Ackley's Problem	$f_7(X) = -20\exp\left(-0.2\sqrt{\frac{\sum_{i=1}^{D} x_i^2}{D}}\right) - \exp\left(\frac{\sum_{i=1}^{D} \cos(2\pi x_i)}{D}\right) + 20 + e$	$[-32, 32]^D$
Levy Function	$f_8(X) = \sin^2(3\pi x_1) + \sum_{i=1}^{D-1} (x_i - 1)^2(1 + \sin^2(3\pi x_{i+1})) + (x_D - 1)(1 + \sin^2(2\pi x_D))$	$[-10, 10]^D$
Michalewicz Function	$f_9(X) = -\sum_{i=1}^{D} \sin(x_i)(sin(ix_i^2/\pi))^{2m}, (m = 10)$	$[0, \pi]^D$
Zakharov Function	$f_{10}(X) = \sum_{i=1}^{D} x_i^2 + \left(\sum_{i=1}^{D} 0.5ix_i\right)^2 + \left(\sum_{i=1}^{D} 0.5ix_i\right)^4$	$[-5, 10]^D$
Schwefel's Problem 2.22	$f_{11}(X) = \sum_{i=1}^{D} \lvert x_i \rvert + \prod_{i=1}^{D} \lvert x_i \rvert$	$[-10, 10]^D$
Step Function	$f_{12}(X) = \sum_{i=1}^{D} (\lfloor x_i + 0.5 \rfloor)^2$	$[-100, 100]^D$
Alpine Function	$f_{13}(X) = \sum_{i=1}^{D} \lvert x_i \sin(x_i) + 0.1x_i \rvert$	$[-10, 10]^D$
Exponential Problem	$f_{14}(X) = \exp\left(-0.5\sum_{i=1}^{D} x_i^2\right)$	$[-1, 1]^D$
Salomon Problem	$f_{15}(X) = 1 - \cos(2\pi \lVert x \rVert) + 0.1 \lVert x \rVert$, where $\lVert x \rVert = \sqrt{\sum_{i=1}^{D} x_i^2}$	$[-100, 100]^D$

We compare the convergence speed of DE and ODE by measuring the *number of function calls* (NFC) which is the most commonly used metric in literature [16, 11, 2, 3, 4, 5, 6]. A smaller NFC means higher convergence speed. The termination criterion is to find a value smaller than the value-to-reach (VTR) before reaching the maximum number of function calls MAX_{NFC}. In order to minimize the effect of the stochastic nature of the algorithms on the metric, the reported number of function calls (NFC) for each function is the average over 50 trials. The number of times, for which the algorithm successfully reaches the VTR for each test function is measured as the *success rate* SR:

$$\text{SR} = \frac{\text{number of times reached VTR}}{\text{total number of trials}}. \tag{5}$$

In order to combine these two measures (NFC and SR), a new measure, called *success performance*, has been introduced as follows [6]:

$$\text{SP} = \frac{\text{mean (NFC for successful runs)}}{\text{SR}}. \tag{6}$$

By this definition, the two following algorithms have equal performances (SP=100):

Algorithm A: mean (NFC for successful runs)=50 *and SR*=0.5,
Algorithm B: mean (NFC for successful runs)=100 *and SR*=1.

SP gives an equal importance weight for NFC and SR, but dependent to the different applications each of them can be more important than other one. Some times success rate is more crucial factor than convergence speed and vice versa. For our experiments, gathering results for unsuccessful case is more time consuming because the algorithm should meet the maximum number of function calls (MAX_{NFC}) for termination. Parameter settings for all conducted experiments are presented in Table 3.

Table 3. Parameter settings

Parameter name	*Setting*	*Reference*
population size (N_p)	100	[7, 9, 8]
differential amplification factor (F)	0.5	[10, 11, 13, 12, 7]
crossover probability constant (C_r)	0.9	[10, 11, 13, 12, 7]
jumping rate constant (J_r)	0.3	[4, 14, ?, 15]
maximum number of function calls (MAX_{NFC})	10^6	[4, 14, 15]
value to reach (VTR)	10^{-8}	[6]
mutation strategy	$DE/rand/1/bin$	[10, 16, 18, 7, 17]

In order to maintain a reliable and fair comparison (a) the parameter settings are kept the same for all conducted experiments, unless we mention new settings, (b) for all experiments, the reported values are the average of the results for 50 independent runs, and most importantly (c) extra fitness evaluations required for the opposite points (both in population initialization and also generation jumping phases) are counted as well.

Results for DE and ODE to solve test problems are given in Table 4 (the results in the last column will be discussed in section 4.2). As seen, ODE outperforms DE on 14 benchmark functions with respect to the success performance. Some sample performance comparison graphs are presented in Figure 6. With the same control parameter settings for both algorithms and fixing the jumping rate for the ODE ($J_r = 0.3$), their success rates are comparable while ODE shows better convergence speed than DE. The jumping rate is an important control parameter which, if optimally set, can achieve even better results. Detailed discussion about this parameter is given in [14, ?].

4.2 Contribution of Opposite Points

In this section, we want to verify that the achieved acceleration rate is really due to utilizing opposite points. For this purpose, all parts of the proposed algorithm are kept untouched and instead of using opposite points for the population initialization and the generation jumping, uniformly generated random points are employed. In order to have a fair competition for this case, exactly like what we did for opposite points, the current

Table 4. Comparison of DE, ODE, and RDE. The best result for each case is highlighted in boldface. Results for RDE has been explained in section 4.2 (corresponding results for replacing the opposite points with random points).

		DE			ODE			RDE		
F	*D*	*NFC*	*SR*	*SP*	*NFC*	*SR*	*SP*	*NFC*	*SR*	*SP*
f_1	30	87748	1	87748	47716	1	**47716**	115096	1	115096
f_2	30	96488	1	96488	53304	1	**53304**	126780	1	126780
f_3	20	177880	1	177880	168680	1	**168680**	231152	1	231152
f_4	10	328844	1	328844	70389	0.76	**92617**	501875	0.96	522786
f_5	30	113428	1	113428	69342	0.96	**72231**	149744	1	149744
f_6	30	25140	1	25140	8328	1	**8328**	29096	1	29096
f_7	30	169152	1	169152	98296	1	**98296**	222784	1	222784
f_8	30	101460	1	101460	70408	1	**70408**	138308	1	138308
f_9	10	191340	0.76	**251763**	213330	0.56	380946	306900	0.60	511500
f_{10}	30	385192	1	385192	369104	1	**369104**	498200	1	498200
f_{11}	30	187300	1	187300	155636	1	**155636**	244396	1	244396
f_{12}	30	41588	1	41588	23124	1	**23124**	54316	1	54316
f_{13}	30	411164	1	411164	337532	1	**337532**	927230	0.24	3863458
f_{14}	10	19528	1	19528	15704	1	**15704**	23156	1	23156
f_{15}	10	37824	1	37824	24260	1	**24260**	46800	1	46800
	SR_{ave}		0.98			0.95			0.92	

interval (dynamic interval, $[\mathrm{MIN}_j^p, \mathrm{MAX}_j^p]$) of the variables are used to generate new random points in the generation jumping phase. So, line 4 from Algorithm 1 should be changed to:

$$RP_{0i,j} \longleftarrow a_j + (b_j - a_j) \times rand(0,1),$$

where $rand(0,1)$ generates a uniformly distributed random number on the interval $(0,1)$. This change is for the initialization part, so the predefined boundaries of variables ($[a_j, b_j]$) have been used to generate new random numbers. In fact, instead of generating N_p, $2N_p$ random individuals are generated. In the same manner, line 30 should be replaced with

$$RP_{i,j} \longleftarrow \mathrm{MIN}_j^p + (\mathrm{MAX}_j^p - \mathrm{MIN}_j^p) \times rand(0,1).$$

As mentioned before, for generation jumping, the current boundaries of the variables ($[\mathrm{MAX}_j^p, \mathrm{MIN}_j^p]$) are used to generate random numbers. And finally, in order to have the same selection method, lines 7 and 33 in Algorithm 1 are substituted with

Select N_p fittest individuals from set the $\{P, RP\}$ as current population P;

After these modifications, the random version of ODE (called RDE) is introduced. Now, we are ready to apply this algorithm to solve our test problems. All control parameters are kept the same to have a fair comparison. Results for the current algorithm are presented in Table 4. As seen, RDE can not outperform DE or ODE on any of benchmark function with respect to the success performance (even, its average success rate is less than others). This clearly demonstrates that the achieved improvements are due to usage of opposite points, and that the same level of improvement cannot be achieved via additional random sampling.

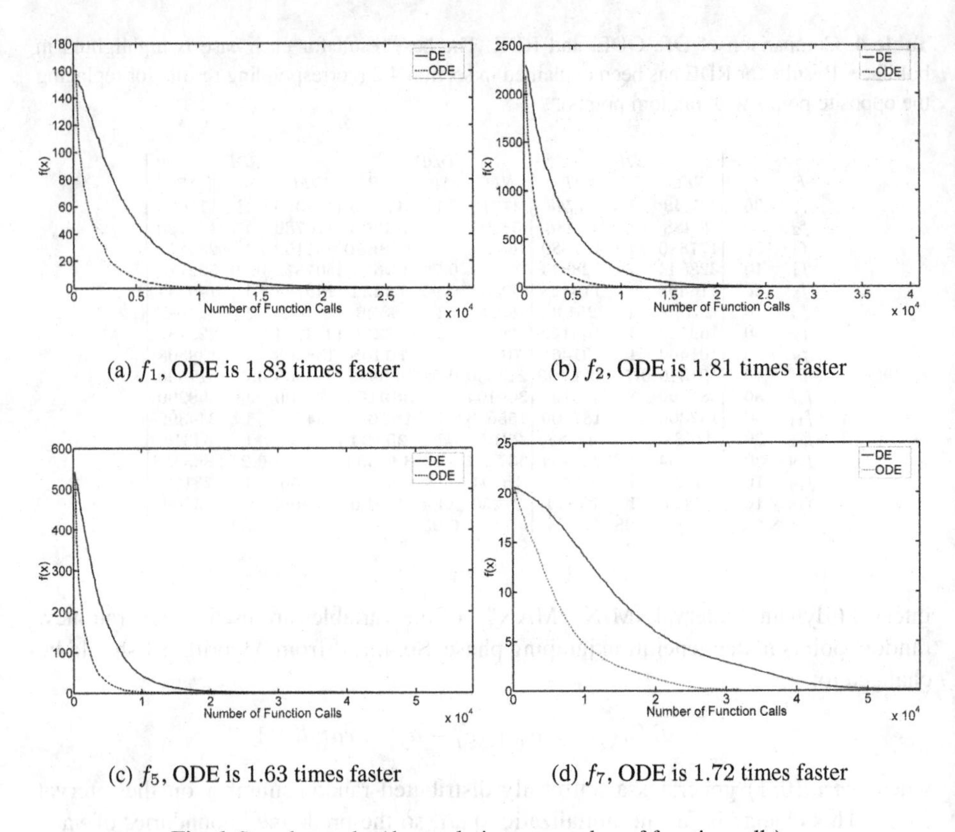

(a) f_1, ODE is 1.83 times faster

(b) f_2, ODE is 1.81 times faster

(c) f_5, ODE is 1.63 times faster

(d) f_7, ODE is 1.72 times faster

Fig. 6. Sample graphs (best solution vs. number of function calls)

5 ODE with Variable Jumping Rate

In this section, a time varying jumping rate (TVJR) model for opposition-based differential evolution (ODE) has been investigated. According to this model, the jumping rate changes during the evolution based on the number of function evaluations. The same test suite has been employed to compare performance of DE and ODE with variable jumping rate settings.

Generally speaking, parameter control in evolutionary algorithms (EAs) can be performed in following three ways [19]: deterministic, adaptive, and self-adaptive. The first one uses a predefined rule to modify the parameter value without gaining any feedback from the evolution process while the second one changes the parameter value based on the information which receives from the search process. The third one utilizes the same evolutionary approach not only to solve the problem but also to optimize own control parameters by encoding some strategic parameters inside the population.

The proposed idea in this section is similar to Das *et al.* work [20]. They utilized time varying approach for setting of the scale factor F in differential evolution

(DE), which can be considered as a deterministic approach according to the mentioned categorization.

5.1 Investigated Jumping Rate Models

For ODE a constant value for jumping rates was utilized. Here, two types of varying jumping rate are investigated (linearly increasing and decreasing functions). Three proposed settings for J_r are as follows:

- J_r (constant)$= J_{r_{ave}}$,
- $J_r(\text{TVJR1}) = (J_{r_{max}} - J_{r_{min}}) \times \left(\frac{\text{MAX}_{\text{NFC}} - \text{NFC}}{\text{MAX}_{\text{NFC}}}\right)$,
- $J_r(\text{TVJR2}) = (J_{r_{max}} - J_{r_{min}}) - (J_{r_{max}} - J_{r_{min}}) \times \left(\frac{\text{MAX}_{\text{NFC}} - \text{NFC}}{\text{MAX}_{\text{NFC}}}\right)$,

where $J_{r_{ave}}$, $J_{r_{max}}$, and $J_{r_{min}}$ are the average, maximum, and minimum jumping rates, respectively. MAX_{NFC} and NFC are the maximum number of function calls and the current number of function calls, respectively.

In order to support a fair comparison between these three different jumping rate settings, the average jumping rate should be the same for all of them. Obviously we should have $J_{r_{ave}} = \frac{(J_{r_{max}} + J_{r_{min}})}{2}$. Following values for these parameters are selected: $J_{r_{ave}} = 0.3$ and $J_{r_{min}} = 0$ (no jumping), so $J_{r_{max}} = 0.6$. Figure 7 shows the corresponding diagrams (jumping rate vs. NFCs) for three following settings:

- $J_r(\text{constant}) = 0.3$,
- $J_r(\text{TVJR1}) = 0.6 \times \left(\frac{\text{MAX}_{\text{NFC}} - \text{NFC}}{\text{MAX}_{\text{NFC}}}\right)$,
- $J_r(\text{TVJR2}) = 0.6 - 0.6 \times \left(\frac{\text{MAX}_{\text{NFC}} - \text{NFC}}{\text{MAX}_{\text{NFC}}}\right)$.

$J_r(\text{TVJR1})$ represents higher jumping rate during exploration and lower jumping rate during exploitation (fine-tuning); $J_r(\text{TVJR2})$ performs exactly in reverse manner. By these settings, we can investigate effects of generation jumping during optimization process.

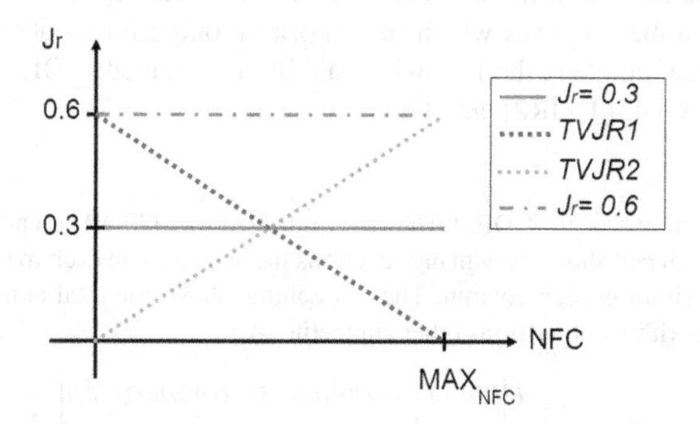

Fig. 7. Jumping rate vs. NFCs for $J_r(\text{ODE}) = 0.3$, $J_r(\text{TVJR1}) = 0.6 \times \left(\frac{\text{MAX}_{\text{NFC}} - \text{NFC}}{\text{MAX}_{\text{NFC}}}\right)$, and $J_r(\text{TVJR2}) = 0.6 - 0.6 \times \left(\frac{\text{MAX}_{\text{NFC}} - \text{NFC}}{\text{MAX}_{\text{NFC}}}\right)$

5.2 Empirical Results

The benchmark test set and all parameter settings are the same as before. The only difference is the maximum number of function calls, which is 2×10^5 for $f_1, f_2, f_3, f_6, f_8, f_{15}, f_{21}$; 5×10^5 for $f_5, f_{18}, f_{19}, f_{31}$; and 5×10^4 for $f_7, f_{23}, f_{41}, f_{56}$. Results of applying DE, ODE ($J_r = 0.3$), ODE (TVJR1), and ODE (TVJR2) to solve the test problems are given in Table 5. As seen, ODE (TVJR1) delivers best success performance (SP) for 13 benchmark functions, while this number for DE, ODE ($J_r = 0.3$), and ODE (TVJR2) is 0, 1, and 1, respectively.

Table 5. Comparison of DE, ODE ($J_r = 0.3$), ODE (TVJR1), and ODE (TVJR2). D: Dimension, NFC: Number of function calls (average over 50 trials), SR: Success rate, SP: Success performance.

		DE			*ODE* ($J_r = 0.3$)			*ODE (TVJR1)*			*ODE (TVJR2)*		
F	*D*	*NFC*	*SR*	*SP*	*NFC*	*SR*	*SP*	*NFC*	*SR*	*SP*	*NFC*	*SR*	*SP*
f_1	30	87748	1	87748	47716	1	47716	42300	1	**42300**	66305	1	66305
f_2	30	96488	1	96488	53304	1	53304	45720	1	**45720**	72990	1	72990
f_3	20	177880	1	177880	168680	1	168680	159775	1	**159775**	175460	1	175460
f_4	10	328844	1	328844	65056	0.64	101650	59063	0.80	**73829**	136070	1	136070
f_5	30	113428	1	113428	64920	0.75	86560	63594	0.90	**70660**	86235	1	86235
f_6	30	25140	1	25140	8328	1	8328	6080	1	**6080**	14175	1	14175
f_7	30	169152	1	169152	98296	1	98296	88355	1	**88355**	117095	1	117095
f_8	30	101460	1	101460	70408	1	70408	65247	0.95	**68681**	82245	1	82245
f_9	10	215260	0.56	384393	168470	0.76	**221671**	188440	0.65	289908	379660	0.60	632767
f_{10}	30	385192	1	385192	369104	1	369104	389955	1	389955	360595	1	**360595**
f_{11}	30	187300	1	187300	155636	1	155636	146795	1	**146795**	167685	1	167685
f_{12}	30	41588	1	41588	23124	1	23124	20290	1	**20290**	29165	1	29165
f_{13}	30	411164	1	411164	337532	1	337532	326350	1	**326350**	377425	1	377425
f_{14}	10	19528	1	19528	15704	1	15704	14270	1	**14270**	17735	1	17735
f_{15}	10	37824	1	37824	24260	1	24260	21400	1	**21400**	28710	1	28710
	SR_{ave}		0.97			0.94			0.95			0.97	

Pairwise comparison of these algorithms is presented in Table 6. The given number in each cell indicates on how many functions the algorithm in each row outperforms the corresponding algorithm in each column. The last column of the table shows the total numbers (number of cases which the algorithm outperforms other competitors); by comparing these numbers the following ranking is obtained: ODE (TVJR1) (best), ODE ($J_r = 0.3$), ODE (TVJR2), and DE.

Table 6. Pairwise comparison of DE, ODE ($J_r = 0.3$), ODE (TVJR1), and ODE (TVJR2). Given number in each cell shows how many functions the algorithm in each row outperforms the corresponding algorithm in each column. The last column shows the total numbers (number of cases which the algorithm outperforms other competitors).

	DE	*ODE* ($J_r = 0.3$)	*ODE (TVJR1)*	*ODE (TVJR2)*	Total
DE	-	0	1	1	**2**
ODE ($J_r = 0.3$)	15	-	2	12	**29**
ODE (TVJR1)	14	13	-	14	**41**
ODE (TVJR2)	14	3	1	-	**18**

The average success rate (shown in the last row of the Table 5) for DE and ODE (TVJR2) is marginally better than other two competitors.

6 Conclusion

In this chapter, the concept of opposition-based optimization (OBO) has been employed to accelerate differential evolution. The OBO was utilized to introduce opposition-based population initialization and opposition-based generation jumping. By embedding these two steps within DE, opposition-based differential evolution (ODE) was proposed. The experimental results clearly confirmed that ODE is faster than the classical DE. Our conclusion can be summarized as follows:

- By replacing opposite points with uniformly generated random points in the same variables' range, the resulted algorithm (RDE) performs slower than the parent algorithm (DE). Therefore, the contribution of opposite points to the acceleration process was confirmed and was not reproducible by additional random sampling.
- According to our comprehensive experiments (not included in this chapter), the range $[0.1, 0.4]$ is recommended for an unknown optimization problem. Most of the functions presented a reliable acceleration improvement and almost a smooth behavior in this interval. Although, the optimal jumping rate can be somewhere out of this range, higher jumping rates are generally not recommended.
- As an advantageous of an opposite versus random point, purely random resampling or selection of solutions from a given population has the higher chance of visiting or even revisiting unproductive regions of the search space compared to the opposite points [25] .
- The benefits of opposition-based optimization is not the same for different problems. This is because of using fix settings for the parameters instead of the optimal ones and/or the different characteristics of each problem (e.g., modality, dimension, surface features, separability of the variables and so on). Similar to all optimization approaches, ODE does not present a consistent behavior over different problems. However, in overall and over the employed benchmark test suite, ODE performed better than classical DE.
- The proposed opposition-based schemes are general enough to be applied on other population-based algorithms. The opposition-based schemes work at the population level and leave the evolutionary part of the algorithms untouched. This generality gives higher flexibility to these schemes to be embedded inside other population-based algorithms for further investigation.
- The optimal control parameters are problem-oriented such that developing a self-adaptive/ adaptive algorithm is a valuable attempt. Many studies confirm that for population-based algorithms the optimal parameters are problem-oriented. Running limited trials to determine desirable parameters is a common approach (if not a practical way for time consuming objective functions). For this reason, the self-adaptive/adaptive control parameter setting would be a valuable improvement for ODE.
- The time varying jumping rate for opposition-based differential evolution was proposed and two behaviorally reverse versions (linearly decreasing and increasing

functions) were compared with the constant setting [15]. The results show that the linearly decreasing jumping rate performs better than constant setting and also than linearly increasing policy. This means generation jumping in the exploration time is more desirable than during exploitation. Because during the fine-tuning, we are faced with shrunken search space and the jumping of the individuals may not be advantageous (because the point and the opposite-point are very close together). There is no exact border between exploration and exploitation stage. Hence, the gradual behavior for the decreasing and increasing functions are proposed.

- The proposed time varying jumping rate functions utilize the maximum number of function calls (MAX_{NFC}) which may not be exactly known for the black-box optimization problems; this can be regarded as a disadvantage for this method. Adaptive setting of the jumping rate can be a desirable solution.
- Results are promising, however, the opposition-based optimization is still in its infancy. Results confirm that the opposition concept has the potential to play desire and positive role in optimization. But, it is important to mention that the current study constitutes the first step of this newly opened direction. Like many other new concepts, opposition-based optimization needs further studies to disclose its exact benefits, weaknesses, and limitations. In fact, the main claim is not defeating DE or any of its numerous versions but to introduce a new notion into optimization via metaheuristics; this is the notion of opposition.

The opposition-based optimization is simple to implement and open to be used in many different ways for different purposes. This study is a starting point in this direction to confirm the potentials and motivate other researchers in optimization and machine learning fields to engage with the opposition concept.

References

1. Tizhoosh, H.R.: Opposition-Based Learning: A New Scheme for Machine Intelligence. In: Int. Conf. on Computational Intelligence for Modelling Control and Automation (CIMCA 2005), Vienna, Austria, vol. I, pp. 695–701 (2005)
2. Andre, J., Siarry, P., Dognon, T.: An Improvement of the Standard Genetic Algorithm Fighting Premature Convergence in Continuous Optimization. Advance in Engineering Software 32, 49–60 (2001)
3. Hrstka, O., Kučerová, A.: Improvement of Real Coded Genetic Algorithm Based on Differential Operators Preventing Premature Convergence. Advance in Engineering Software 35, 237–246 (2004)
4. Rahnamayan, S., Tizhoosh, H.R., Salama, M.M.A.: Opposition-Based Differential Evolution Algorithms. In: IEEE Congress on Evolutionary Computation (CEC 2006), IEEE World Congress on Computational Intelligence, Vancouver, Canada, pp. 7363–7370 (2006)
5. Rahnamayan, S., Tizhoosh, H.R., Salama, M.M.A.: Opposition-Based Differential Evolution for Optimization of Noisy Problems. In: IEEE Congress on Evolutionary Computation (CEC 2006), IEEE World Congress on Computational Intelligence, Vancouver, Canada, pp. 6756–6763 (2006)
6. Suganthan, P.N., Hansen, N., Liang, J.J., Deb, K., Chen, Y.-P., Auger, A., Tiwari, S.: Problem Definitions and Evaluation Criteria for the CEC 2005 Special Session on Real-Parameter Optimization. Technical Report, Nanyang Technological University, Singapore And KanGAL Report Number 2005005 (Kanpur Genetic Algorithms Laboratory, IIT Kanpur) (May 2005)

7. Brest, J., Greiner, S., Bošković, B., Mernik, M., Žumer, V.: Self-Adapting Control Parameters in Differential Evolution: A Comparative Study on Numerical Benchmark Problems. Journal of IEEE Transactions on Evolutionary Computation 10(6), 646–657 (2006)
8. Lee, C.Y., Yao, X.: Evolutionary programming using mutations based on the Lévy probability distribution. IEEE Transactions on Evolutionary Computation 8(1), 1–13 (2004)
9. Yao, X., Liu, Y., Lin, G.: Evolutionary programming made faster. IEEE Transactions on Evolutionary Computation 3(2), 82–102 (1999)
10. Storn, R., Price, K.: Differential Evolution- A Simple and Efficient Heuristic for Global Optimization over Continuous Spaces. Journal of Global Optimization 11, 341–359 (1997)
11. Vesterstroem, J., Thomsen, R.: A Comparative Study of Differential Evolution, Particle Swarm Optimization, and Evolutionary Algorithms on Numerical Benchmark Problems. In: Proceedings of the Congress on Evolutionary Computation (CEC 2004), vol. 2, pp. 1980–1987. IEEE Publications, Los Alamitos (2004)
12. Liu, J., Lampinen, J.: A fuzzy adaptive differential evolution algorithm. Soft Computing - A Fusion of Foundations, Methodologies and Applications 9(6), 448–462 (2005)
13. Ali, M.M., Törn, A.: Population set-based global optimization algorithms: Some modifications and numerical studies. Comput. Oper. Res. 31(10), 1703–1725 (2004)
14. Rahnamayan, S., Tizhoosh, H.R., Salama, M.M.A.: Opposition-Based Differential Evolution (ODE). Journal of IEEE Transactions on Evolutionary Computation 12(1), 64–79 (2008)
15. Rahnamayan, S., Tizhoosh, H.R., Salama, M.M.A.: Opposition-Based Differential Evolution (ODE) With Variable Jumping Rate. In: IEEE Symposium on Foundations of Computational Intelligence, Honolulu, Hawaii, USA, April 2007, pp. 81–88 (2007)
16. Price, K., Storn, R.M., Lampinen, J.A.: Differential Evolution: A Practical Approach to Global Optimization (Natural Computing Series), 1st edn. Springer, Heidelberg (2005)
17. Sun, J., Zhang, Q., Tsang, E.P.K.: DE/EDA: A new evolutionary algorithm for global optimization. Information Sciences 169, 249–262 (2005)
18. Onwubolu, G.C., Babu, B.V.: New Optimization Techniques in Engineering. Springer, Berlin (2004)
19. Eiben, A.E., Hinterding, R.: Paramater Control in Evolutionary Algorithms. IEEE Transactions on Evolutionary Computation 3(2), 124–141 (1999)
20. Das, S., Konar, A., Chakraborty, U.K.: Two Improved Differential Evolution Schemes for Faster Global Search. In: Proceedings of the 2005 conference on Genetic and evolutionary computation, Washington, USA, pp. 991–998 (2005)
21. Tizhoosh, H.R.: Reinforcement Learning Based on Actions and Opposite Actions. In: ICGST International Conference on Artificial Intelligence and Machine Learning (AIML 2005), Cairo, Egypt (2005)
22. Tizhoosh, H.R.: Opposition-Based Reinforcement Learning. Journal of Advanced Computational Intelligence and Intelligent Informatics 10(3) (2006)
23. Shokri, M., Tizhoosh, H.R., Kamel, M.: Opposition-Based Q(λ) Algorithm. In: 2006 IEEE World Congress on Computational Intelligence (IJCNN 2006), Vancouver, Canada, pp. 646–653 (2006)
24. Ventresca, M., Tizhoosh, H.R.: Improving the Convergence of Backpropagation by Opposite Transfer Functions. In: 2006 IEEE World Congress on Computational Intelligence (IJCNN 2006), Vancouver, Canada, pp. 9527–9534 (2006)
25. Rahnamayan, S., Tizhoosh, H.R., Salama, M.M.A.: Opposition versus Randomness in Soft Computing Techniques. Elsevier Journal on Applied Soft Computing 8, 906–918 (2008)
26. Brest, J., Greiner, S., Bošković, B., Mernik, M., Žumer, V.: Self-Adapting Control Parameters in Differential Evolution: A Comparative Study on Numerical Benchmark Problems. IEEE Transactions on Evolutionary Computation 10(6), 646–657 (2006)
27. Rahnamayan, S.: Opposition-Based Differential Evolution, PhD Thesis, Departement of Systems Design Engineering, University of Waterloo, Waterloo, Canada (May 2007)

Multi-objective Optimization Using Differential Evolution: A Survey of the State-of-the-Art

Efrén Mezura-Montes[1], Margarita Reyes-Sierra[2], and Carlos A. Coello Coello[3,*]

[1] Laboratorio Nacional de Informática Avanzada (LANIA A.C.)
Rébsamen 80, Centro, Xalapa, Veracruz, 91000 Mexico
emezura@lania.mx

[2] Universidad Veracruzana
Facultad de Estadística e Informática
Avenida Xalapa s/n esquina Ávila-Camacho, Xalapa, Veracruz, 91020 Mexico
reyes_sierra@hotmail.com

[3] CINVESTAV-IPN (Evolutionary Computation Group)
Departamento de Computación
Av. IPN No. 2508, Col. San Pedro Zacatenco, México D.F. 07360, Mexico
ccoello@cs.cinvestav.mx

Summary. Differential Evolution is currently one of the most popular heuristics to solve single-objective optimization problems in continuous search spaces. Due to this success, its use has been extended to other types of problems, such as multi-objective optimization. In this chapter, we present a survey of algorithms based on differential evolution which have been used to solve multi-objective optimization problems. Their main features are described and, based precisely on them, we propose a taxonomy of approaches. Some theoretical work found in the specialized literature is also provided. To conclude, based on our findings, we suggest some topics that we consider to be promising paths for future research in this area.

1 Introduction

Nowadays, evolutionary algorithms (EAs) are considered a very effective alternative to solve complex search problems, including either global (i.e., single-objective) [39] or multi-objective optimization problems [10]. The first attempts to use EAs to solve multi-objective problems relied mainly on genetic algorithms (GAs) [16] and evolution strategies (ES) [6]. A comprehensive review of these approaches can be found in [10].

In 1995, Storn and Price proposed the most recent evolutionary algorithm called Differential Evolution (DE) [45] to solve real-parameter optimization problems. DE uses a simple mutation operator based on differences between pairs of solutions (called vectors) with the aim of finding a search direction based on the distribution of solutions in the current population. DE also utilizes a steady-state-like replacement mechanism, where the newly generated offspring (called

* Corresponding author.

U.K. Chakraborty (Ed.): Advances in Differential Evolution, SCI 143, pp. 173–196, 2008.
springerlink.com

trial vector) competes only against its corresponding parent (old object vector) and replaces it if the offspring has a higher fitness value. In the remainder of this chapter, we will use trial vector and offspring as synonymous. The same applies for parent and old object vector.

DE shares some characteristics with previous EAs and also has some differences. The similarities are the following: DE is a population-based approach, recombination and mutation are the variation operators used to generate new solutions and a replacement mechanism provides capabilities to maintain a fixed size in the population.

However, unlike GAs, where binary encoding can be used, solutions in DE are coded with real values as in ES. But, DE does not use a fixed distribution (as the Gaussian distribution adopted in ES) to control the behavior of the mutation operator; instead, the current distribution of the solutions in the search space determines the stepsize and the search direction for each individual. This last feature seems to be one of its main advantages.

Due to the multicriteria nature of most real-world problems, multi-objective optimization problems are very common, particularly in engineering applications. As the name indicates, multi-objective optimization problems involve multiple objectives, which should be optimized simultaneously and that often are in conflict with each other. This results in a group of alternative solutions which must be considered equivalent in the absence of information concerning the relevance of the others.

Since Evolutionary Algorithms (EAs) deal with a group of candidate solutions, it seems natural to use them in multi-objective optimization problems to find a group of optimal solutions. Indeed, EAs have proved very efficient in solving multi-objective optimization problems [10, 11].

With the rise of new bio-inspired heuristics for numerical optimization, like Particle Swarm Optimization (PSO) [27] and also DE, it is important to analyze how they are adapted to solve different types of problems, like, in our case, multi-objective optimization problems. This work focuses on a review of the state-of-the-art in multi-objective optimization with DE as a search engine.

This chapter is organized as follows: In Section 2, DE is explained in detail and its main variants are presented. Section 3, provides the statement of the multi-objective optimization problems and also some related definitions. In Section 4 some multi-objective issues included in evolutionary multi-objective optimization are addressed. After that, in Section 5 we show our proposed taxonomy of DE-based approaches for multi-objective optimization. Some theoretical results regarding DE for multi-objective optimization are summarized in Section 6. Finally, Section 7 includes our conclusions and future paths of research.

2 Differential Evolution Variants

There are some variants of the DE algorithm. They vary on (1) the type of the criterion to select one of the individuals to be used in the mutation operator

```
1   Begin
2       G=0
3       Create a random initial population x_{i,G} ∀i, i = 1,...,NP
4       Evaluate f(x_{i,G}) ∀i, i = 1,...,NP
5       For G=1 to MAX_GEN Do
6           For i=1 to NP Do
7 ⇒             Select randomly r1 ≠ r2 ≠ r3 :
8 ⇒             j_rand = randint(1, D)
9 ⇒             For j=1 to D Do
10 ⇒                If (rand_j[0,1) < CR or j = j_rand) Then
11 ⇒                    u_{i,j,G+1} = x_{r3,j,G} + F(x_{r1,j,G} − x_{r2,j,G})
12 ⇒                Else
13 ⇒                    u_{i,j,G+1} = x_{i,j,G}
14 ⇒                End If
15 ⇒            End For
16              If (f(u_{i,G+1}) ≤ f(x_{i,G})) Then
17                  x_{i,G+1} = u_{i,G+1}
18              Else
19                  x_{i,G+1} = x_{i,G}
20              End If
21          End For
22          G = G + 1
23      End For
24  End
```

Fig. 1. "DE/rand/1/bin" algorithm. randint(min,max) is a function that returns an integer number between min and max. rand$[0, 1)$ is a function that returns a real number between 0 and 1. Both are based on a uniform probability distribution. "NP", "MAX_GEN", "CR" and "F" are user-defined parameters. "D" is the dimensionality of the problem. Steps pointed with arrows change from variant to variant.

(called donor vector), (2) the number of differences computed also in the mutation operator and, finally, (3) in the recombination operator chosen.

The most popular variant is called *"DE/rand/1/bin"*, where "DE" refers to the name of the algorithm, the word "rand" indicates that the donor vector selected to compute the mutation values is chosen at random, "1" is the number of pairs of solutions chosen (most of the time chosen at random) to calculate the mutation differential and finally "bin" means that a binomial recombination is used. The corresponding algorithm of this variant is presented in Figure 1.

Besides typical parameters used in EAs (number of individuals and number of generations), two parameters are adopted in DE: "CR" and "F". "CR" controls the influence of the parent in the generation of the offspring. Higher values mean less influence of the parent in the features of its offspring. "F" scales the influence of the set of pairs of solutions selected to calculate the mutation value (one pair in the case of the algorithm in Figure 1).

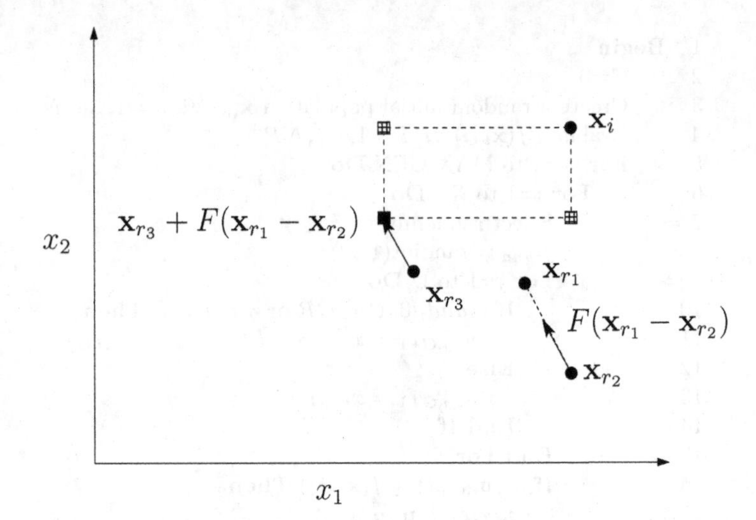

Fig. 2. DE/rand/1/bin recombination and mutation operators example. $\mathbf{x}_i$ is the current parent, $\mathbf{x}_{r_3}$ is the donor individual chosen at random (but it can be the best solution in the population in other variants), $\mathbf{x}_{r_1}$ and $\mathbf{x}_{r_2}$ are the individuals chosen at random to calculate the scaled difference between them and to define a search direction. The black square represents the mutation vector which can be the location of the only offspring generated after performing recombination. Additionally, the filled squares are the other two possible locations for the only offspring after recombination.

In Figure 2 the effect of the DE mutation and recombination operator in its most popular variant (*DE/rand/1/bin*) is explained. $\mathbf{x}_{r_3}$ is the donor solution which can be chosen either at random or it can be the best solution in the population. $\mathbf{x}_{r_1}$ and $\mathbf{x}_{r_2}$ are the pair of solutions chosen always at random and used to compute the difference between them in order to define a search direction. This difference is scaled with the "F" parameter. After that, it is added to $\mathbf{x}_{r_3}$ to define the location of the "mutation vector" (black square in Figure 2). This "mutation vector" is combined with the original parent with a binomial (discrete) recombination and the location of the mutation vector plus the two filled squares in the figure represent the possible positions of the offspring generated. Finally, this offspring will compete against its parent (based on fitness) and the best one will remain in the population for the next generation.

As it was mentioned before, the difference among the different DE variants are mainly on the way the donor solution (from the set chosen to compute the "mutation vector") is selected, the number of pairs of randomly chosen solutions and the type of recombination operator adopted. Among the main variations we distinguish the following:

- Variants with discrete recombination operator (either binomial or exponential):
  - *DE/rand/1/bin*
  - *DE/rand/1/exp*

  - *DE/best/1/bin*
  - *DE/best/1/exp*

  The "rand" variants select the donor solution ($\mathbf{x}_{r_3}$) and the pair of solutions to calculate the mutation differential ($\mathbf{x}_{r_1}$ and $\mathbf{x}_{r_2}$) at random. In contrast, the "best" variants use the best solution in the population as the donor solution and the pair of solutions are chosen at random.
- Variants with arithmetic recombination:
  - *DE/current-to-rand/1*
  - *DE/current-to-best/1*

  The only difference between them is that the first one selects the donor solution ($\mathbf{x}_{r_3}$) and the pair of solutions to calculate the differential mutation ($\mathbf{x}_{r_1}$ and $\mathbf{x}_{r_2}$) at random. The second one uses the best solution in the population as the donor solution and, again, the pair of solutions to calculate the differential mutation are chosen randomly.
- Variants with combined arithmetic-discrete recombination:
  - *DE/current-to-rand/1/bin*

The implementation details of each DE variant are summarized in Table 1.

Table 1. DE basic variants. j_r is a random integer number generated between $[0, n]$, where n is the number of variables of the problem. $U_j(0, 1)$ is a real number generated at random between 0 an 1. Both numbers are generated using a uniform distribution. p is the number of pairs of solutions used to compute the differences in the mutation operator. $\mathbf{u}_i$ is the offspring (or trial vector), $\mathbf{x}_{r_3}$ is the donor solution chosen at random, $\mathbf{x}_{best}$ is the best solution in the population as donor solution, $\mathbf{x}_i$ is the current parent (old object vector) and $\mathbf{x}_{r_1^p}$ and $\mathbf{x}_{r_2^p}$ are the "pth" pair to compute the mutation differential.

Variant
rand/p/bin:
$u_{i,j} = \begin{cases} x_{r_3,j} + F \cdot \sum_{k=1}^{p}(x_{r_1^p,j} - x_{r_2^p,j}) & \text{if } U_j(0,1) < CR \text{ or } j = j_r \\ x_{i,j} & \text{otherwise} \end{cases}$
rand/p/exp:
$u_{i,j} = \begin{cases} x_{r_3,j} + F \cdot \sum_{k=1}^{p}(x_{r_1^p,j} - x_{r_2^p,j}) & \text{from } U_j(0,1) < CR \text{ or } j = j_r \\ x_{i,j} & \text{otherwise} \end{cases}$
best/p/bin:
$u_{i,j} = \begin{cases} x_{best,j} + F \cdot \sum_{k=1}^{p}(x_{r_1^p,j} - x_{r_2^p,j}) & \text{if } U_j(0,1) < CR \text{ or } j = j_r \\ x_{i,j} & \text{otherwise} \end{cases}$
best/p/exp:
$u_{i,j} = \begin{cases} x_{best,j} + F \cdot \sum_{k=1}^{p}(x_{r_1^p,j} - x_{r_2^p,j}) & \text{from } U_j(0,1) < CR \text{ or } j = j_r \\ x_{i,j} & \text{otherwise} \end{cases}$
current-to-rand/p:
$\mathbf{u}_i = \mathbf{x}_i + K \cdot (\mathbf{x}_{r_3} - \mathbf{x}_i) + F \cdot \sum_{k=1}^{p}(\mathbf{x}_{r_1^p} - \mathbf{x}_{r_2^p})$
current-to-best/p:
$\mathbf{u}_i = \mathbf{x}_i + K \cdot (\mathbf{x}_{best} - \mathbf{x}_i) + F \cdot \sum_{k=1}^{p}(\mathbf{x}_{r_1^p} - \mathbf{x}_{r_2^p})$
current-to-rand/p/bin:
$u_{i,j} = \begin{cases} x_{i,j} + K \cdot (x_{r_3,j} - x_{i,j}) + F \cdot \sum_{k=1}^{p}(x_{r_1^p,j} - x_{r_2^p,j}) & \text{if } U_j(0,1) < CR \text{ or } j = j_r \\ x_{i,j} & \text{otherwise} \end{cases}$

3 Multi-objective Optimization

We are interested in solving problems of the type[1]:

$$\text{minimize } \mathbf{f}(\mathbf{x}) := [f_1(\mathbf{x}), f_2(\mathbf{x}), \ldots, f_k(\mathbf{x})] \tag{1}$$

subject to:

$$g_i(\mathbf{x}) \leq 0 \quad i = 1, 2, \ldots, m \tag{2}$$

$$h_i(\mathbf{x}) = 0 \quad i = 1, 2, \ldots, p \tag{3}$$

where $\mathbf{x} = [x_1, x_2, \ldots, x_n]^T$ is the vector of decision variables, $f_i : \mathbb{R}^n \rightarrow \mathbb{R}$, $i = 1, ..., k$ are the objective functions and $g_i, h_j : \mathbb{R}^n \rightarrow \mathbb{R}$, $i = 1, ..., m$, $j = 1, ..., p$ are the constraint functions of the problem.

To describe the concept of optimality in which we are interested, we will introduce next a few definitions.

Definition 1. Given two vectors $\mathbf{x}, \mathbf{y} \in \mathbb{R}^k$, we say that $\mathbf{x} \leq \mathbf{y}$ if $x_i \leq y_i$ for $i = 1, ..., k$, and that **x dominates y** (denoted by $\mathbf{x} \prec \mathbf{y}$) if $\mathbf{x} \leq \mathbf{y}$ and $\mathbf{x} \neq \mathbf{y}$.

Figure 3 shows a particular case of the **dominance relation** in the presence of two objective functions.

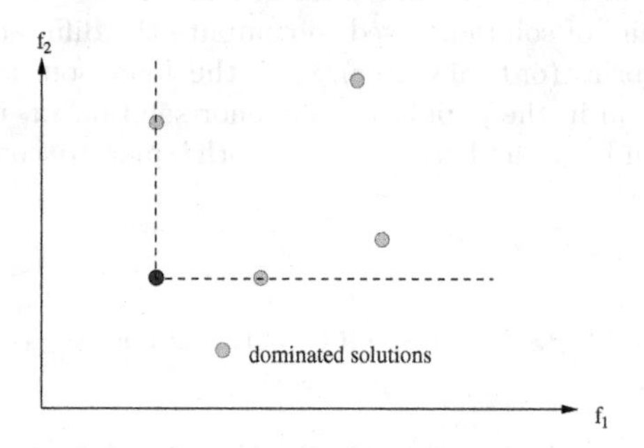

Fig. 3. Dominance relation in a bi-objective space

Definition 2. We say that a vector of decision variables $\mathbf{x} \in \mathcal{X} \subset \mathbb{R}^n$ is **nondominated** with respect to $\mathcal{X}$, if there does not exist another $\mathbf{x}' \in \mathcal{X}$ such that $\mathbf{f}(\mathbf{x}') \prec \mathbf{f}(\mathbf{x})$.

Definition 3. We say that a vector of decision variables $\mathbf{x}^* \in \mathcal{F} \subset \mathbb{R}^n$ ($\mathcal{F}$ is the feasible region) is **Pareto-optimal** if it is nondominated with respect to $\mathcal{F}$.

Definition 4. The **Pareto Optimal Set** $\mathcal{P}^*$ is defined by:

$$\mathcal{P}^* = \{\mathbf{x} \in \mathcal{F} | \mathbf{x} \text{ is Pareto-optimal}\}$$

[1] Without loss of generality, we will assume only minimization problems.

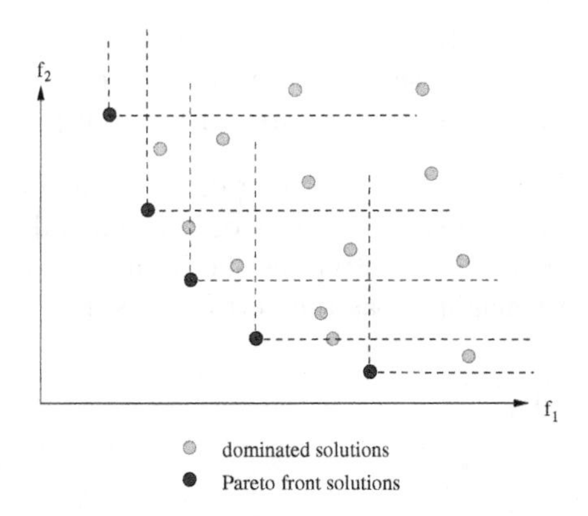

Fig. 4. The Pareto front of a set of solutions in a two objective space

Definition 5. The **Pareto Front** $\mathcal{PF}^*$ is defined by:

$$\mathcal{PF}^* = \{\mathbf{f}(\mathbf{x}) \in \mathbb{R}^k | \mathbf{x} \in \mathcal{P}^*\}$$

Figure 4 shows a particular case of the **Pareto front** in the presence of two objective functions.

We thus wish to determine the Pareto optimal set from the set $\mathcal{F}$ of all the decision variable vectors that satisfy (2) and (3). Note however that in practice, not all the Pareto optimal set is normally desirable (e.g., it may not be desirable to have different solutions that map to the same values in objective function space) or achievable.

4 Differential Evolution for Multi-objective Problems

In order to apply the DE strategy for solving multi-objective optimization problems, the original scheme has to be modified since the solution set of a problem with multiple objectives does not consist of a single solution (as in global optimization). Instead, in multi-objective optimization, we aim to find a set of different solutions (the so-called Pareto optimal set), as mentioned in Section 3.

Two are the main aspects that have been considered by researchers who have extended the DE approach to multi-objective optimization:

1. How to promote diversity into the population?
2. How to select and/or retain the best individuals? That is, how to perform *elitism*?

We briefly discuss these two design aspects in the following Sections.

4.1 Promoting Diversity

Promoting diversity may be done through the selection process by means of mechanisms based on some *quality* measures that indicate the closeness of the individuals within the population. In order to help understanding the specific approaches that are going to be described later on, we present here two of the most important density measures used in the area of multi-objective optimization:

- **Crowding distance** [14]. The crowding distance factor gives us an idea of how crowded are the closest neighbors of a given individual, in objective function space. This measure estimates the perimeter of the cuboid formed by using the nearest neighbors as the vertices. See Figure 5.

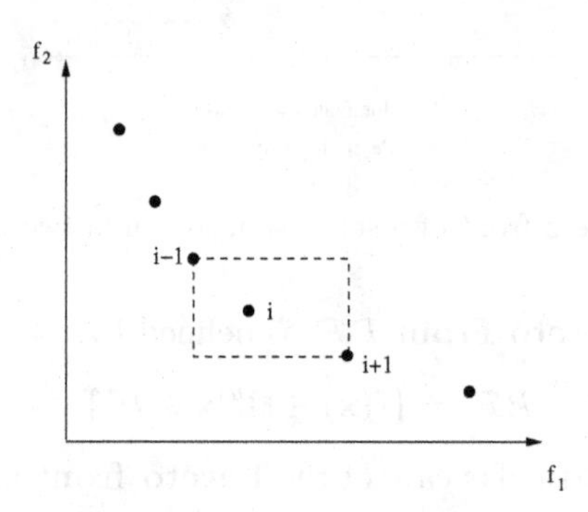

Fig. 5. The crowding distance factor for an example with two objective functions. Individuals with a larger value of this factor are preferred.

- **Fitness sharing** [17, 13]: When an individual is sharing resources with others, its fitness is degraded in proportion to the number and closeness to individual that surround it within a certain perimeter. A neighborhood of an individual is defined in terms of a parameter called σ_{share} that indicates the radius of the neighborhood. Such neighborhoods are called *niches.* See Figure 6.

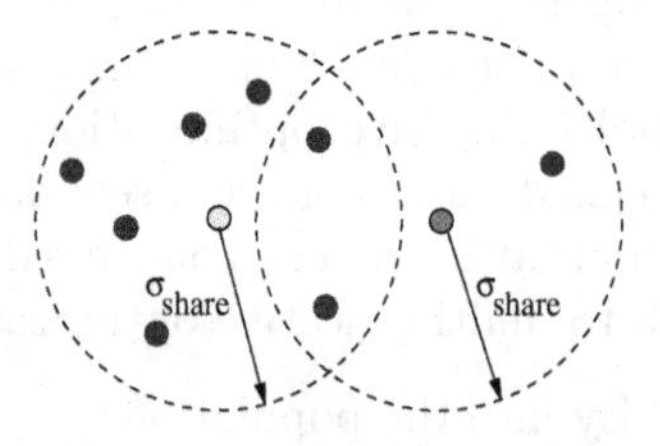

Fig. 6. For each individual, a niche is defined. Individuals whose niche is less crowded are preferred.

4.2 Performing Elitism

In evolutionary multi-objective optimization, elitism is usually implemented through an external archive, also called secondary population, which stores the nondominated individuals found along the search. Such an archive will allow the entrance of a solution only if: (a) it is nondominated with respect to the contents of the archive or (b) it dominates any of the solutions within the archive (in this case, the dominated solutions have to be deleted from the archive).

Besides, elitism can also be implemented through the use of $(\mu + \lambda)$-selection (also called *plus* selection), by which, at each generation, parents and children are compared in order to select the best of them to conform the next population.

One of the most popular mechanisms used to select the best individuals from the combined population of parents and children is the so-called *nondominated sorting* approach. This approach is based on the Pareto ranking mechanism firstly proposed by Goldberg in 1989 [16]. The nondominated sorting mechanism ranks the individuals of the population in different levels in the following way. All nondominated individuals are classified into one category with rank 1 (level 1), then, this group of individuals with rank 1 is ignored and the process is repeated. This time, the nondominated individuals will have rank 2 (level 2). The process continues until all individuals are classified. Individuals with lower rank are always preferred for selection. Figure 7 shows the ranking process of the nondominated sorting approach.

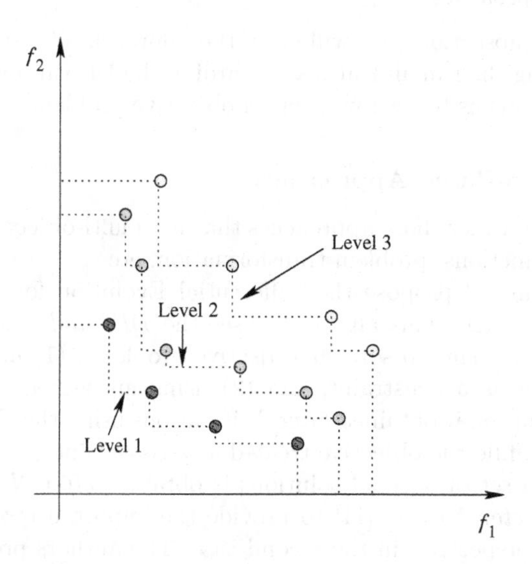

Fig. 7. Ranking process of the nondominated sorting approach

According to Goldberg [16], to maintain appropriate diversity, the nondominated sorting procedure should be used in conjunction with niching techniques as, for example, the fitness sharing mechanism previously mentioned.

The first multi-objective evolutionary algorithm (MOEA) which used the nondominated sorting approach proposed by Goldberg was the Nondominated Sorting Genetic Algorithm (NSGA) proposed by Srinivas and Deb in 1994 [54]. The NSGA algorithm combined the nondominated sorting approach with fitness sharing in its corresponding fitness assignment process. Later, the improved version of the NSGA, called NSGA-II [14], incorporated the nondominated sorting approach in order to perform a $(\mu + \lambda)$-selection. In the NSGA-II, the population for the next generation is obtained by first introducing all individuals with rank 1 (level 1), then all individuals with rank 2 (level 2) and the process continues until the population is complete. However, when there is not enough space to include all individuals with rank i (level i), a crowding distance mechanism is applied in order to select the best individuals from such level.

5 Taxonomy

In this section, we propose a classification of approaches, based on common features to adapt DE for multi-objective optimization. The proposed classes are enumerated as follows:

1. Non-Pareto-based approaches.
2. Pareto-based approaches.
   a) Using Pareto dominance
   b) Using Pareto ranking.
3. Combined approaches.

In the following Subsections, we will show the approaches located in each category, by describing their main features, regarding the DE variant used and their companion mechanisms to deal with multi-objective problems.

5.1 Non-Pareto-Based Approaches

In this class, we consider those approaches that use multi-objective concepts like combination of functions, problem transformation, etc.

Babu and Jehan [5] propose the Differential Evolution for Multi-Objective Optimization approach. This algorithm uses the *DE/rand/1/bin* variant with two different mechanisms to solve bi-objective problems: (1) incorporating one objective function as a constraint, and (2) using an aggregating function. A single optimal solution is obtained after N iterations using the *Penalty Function Method* [11] to handle the objective treated as a constraint in the first case. On the other hand, a set of optimal solutions is obtained after N iterations using the *Weighting Factor Method* [11] to provide the importance of each objective from the user's perspective, in the second case. The authors present results for two bi-objective problems and compare them with respect to a simple GA. The authors indicate that the DE algorithm provides the exact optimum with a lower number of evaluations than the GA.

Li and Zhang [36] propose a multi-objective differential evolution algorithm based on decomposition (MODE/D) for continuous multi-objective optimization problems with variable linkages. The authors use the weighted Tchebycheff approach to decompose a multi-objective optimization problem into several scalar optimization subproblems. The differential evolution operator based on the *DE/rand/1/bin* variant is used for generating new trail solutions, and a neighborhood relationship among all the subproblems generated is defined, such that they all have similar optimal solutions. For validating their approach, the authors adopt test problems with variable linkages [41] and propose variants of some of the Zitzler-Deb-Thiele (ZDT) test problems [59]. Results are compared with respect to the NSGA-II [14], the Nondominated Sorting Differential Evolution (NSDE) [25] and GD3 [30]. The authors report that MODE/D clearly outperformed the other approaches with respect to which it was compared.

5.2 Pareto-Based Approaches

In this group we classify those methods that use Pareto concepts to deal with multiple objectives. We divided this class into two subclasses, because we detected two different ways to apply them: (1) as a criterion to select the best solution in the DE selection mechanism and (2) as a ranking procedure.

Pareto dominance

In this subclass, we describe those approaches where Pareto Dominance was used as a criterion to select the best solution between the old population vector and the trial vector. One of the features that distinguishes each approach is the decision made when both solutions are nondominated between each other. Furthermore, other authors use Pareto Dominance as a filter to get only nondominated solutions.

Apparently, Chang et al. [9] constitutes the first reported attempt to extend differential evolution for multi-objective problems. In this paper, the authors use *DE/rand/1/bin* with an external archive (called "Pareto optimal set" by the authors) to store the nondominated solutions obtained during the search. The approach also incorporates fitness sharing to maintain diversity. An interesting aspect of this approach is that the selection mechanism of the differential evolution algorithm is modified in order to enforce that the members of the new generation are nondominated not only regarding their objective values but also regarding a set of distance metric values (one assigned to each objective) which ensure the new solutions are at certain minimum distance from the previously found nondominated solutions. This approach is adopted to fine-tune the fuzzy automatic train operation (ATO) for a typical mass transit system, in which three objectives are considered: (1) punctuality (least deviation from scheduled arrival time), (2) least energy consumption and (3) maximum passenger comfort. This application is discussed in further detail in [8].

Abbass et al. [4, 3, 51] propose the Pareto Differential Evolution (also abbreviated as PDE) algorithm. The authors use an special case of the *DE/-current_to_rand/1/bin* variant with $K = 0$ (see last row in Table 1), because the old population vector (the parent) is used in the calculation of the trial vector (combined with the difference vector) and also in the discrete recombination. The algorithm works as follows. The initial population is initialized using a Gaussian distribution with mean 0.5 and standard deviation 0.15. Only the nondominated solutions are retained in the population for recombination (all dominated solutions are removed). Three parents are randomly selected (one as the main parent and also trial solution) and a child is generated with them. The offspring is placed in the population only if it dominates the main parent; otherwise, a new selection process takes place. This process continues until the population is completed. If the number of nondominated solutions exceeds a certain threshold (50 was adopted in [4]), a distance metric is adopted to remove parents which are too close from each other (this can be seen as a niching procedure in which this distance metric is the niche radius). In this approach, the step-length parameter F is generated from a Gaussian distribution $N(0,1)$ and the boundary constraints are preserved either by reversing the sign if the variable is ≤ 0 or by repetitively subtracting 1 if it is ≥ 0, until the variable is within the allowable boundaries. This algorithm also incorporates a mutation operator which is applied with certain probability (after the crossover operator), by adding to each variable a small random perturbation. PDE is compared with respect to SPEA [60] in [3] (without mutation) and also with respect to many other approaches (including PAES [28], the NSGA [54] and the NPGA [22]) in [4] (including mutation).

In [2], a new version of PDE is introduced. This version is called Self-adaptive Pareto Differential Evolution (SPDE) algorithm, because it self-adapts its crossover and its mutation rates.

In [1], Abbass proposes an approach called Memetic Pareto Artificial Neural Networks (MPANN). This approach consists of a version of Pareto Differential Evolution (PDE) [3] enhanced with the Back-Propagation (BP) local search algorithm, in order to speed up convergence. MPANN is used to evolve neural networks in which an attempt is made to obtain a trade-off between the architecture and generalization ability of the network. So, two objectives are minimized: (1) error and (2) the number of hidden units. MPANN is validated using two benchmark data sets: the Australian credit card assessment problem and the diabetes problem (both were taken from the UCI Machine Learning Repository [40]). Results are compared with respect to 23 algorithms, which include decision trees, rule-based methods, neural networks and statistical algorithms. MPANN was able to outperform the traditional backpropagation approach and obtained results competitive against the other 23 algorithms with respect to which it was compared.

Kukkonen and Lampinen extended *DE/rand/1/bin* to solve multi-objective optimization problems in their approach called Generalized Differential Evolution (GDE). In fact, GDE is able to solve global and multi-objective optimization problems (either constrained or unconstrained). The first version of their approach [32] modified the original DE selection operation by introducing Pareto Dominance as a selection criterion between the old population member and the trial vector. Also, Pareto dominance in the constraint space is considered to handle the constraints of the problem.

To promote a better distribution of the nondominated solutions, a second version of the approach, called GDE2 [29] was introduced. In this version, a crowding distance measure was used to select the best solution when the old population vector and the trial vector are feasible and nondominated with respect to each other, in such a way that the vector located in the less crowded region will be part of the population of the next generation. The authors acknowledge that GDE2 was sensitive to its initial parameters and that the modified selection mechanism slows down convergence.

Santana-Quintero and Coello Coello [50] propose the ϵ-MyDE. This approach keeps two populations: the main population (which is used to select the parents) and a secondary (external) population, in which the concept of ϵ-dominance [35] is adopted to retain the nondominated solutions found and to distribute them in an uniform way. The concept of ϵ-dominance does not allow two solutions with a difference less than ϵ_i in the i-th objective to be nondominated with respect to each other, thereby allowing a good spread of solutions. ϵ-MyDE uses real numbers representation, and incorporates a constraint-handling mechanism that allows infeasible solutions to intervene during recombination. *DE/rand/1/bin* variant is used to evolve the main population. However, after a user-defined number of generations, the three random solutions used in the mutation operator are selected from the secondary population in such a way that they are close among them in the objective function space. If none of the solutions satisfies this condition, a random solution from this secondary population is chosen. Finally, to improve exploration capabilities, a uniform mutation operator is added.

Portilla Flores [44] proposes a multi-objective version of differential evolution, which is used for concurrent design of pinion–rack continuously variable transmission (CVT). This mechatronic design problem is formulated as a dynamic multi-objective optimization problem in which two objectives are considered: (1) maximize the mechanical CVT efficiency, and (2) minimize the controller energy. The *DE/rand/1/bin* variant is used in this approach and Pareto dominance between a parent and its offspring works like the selection criterion. Also, this technique incorporates a secondary population to retain the nondominated solutions found during the evolutionary process. Finally, it uses the feasibility rules from [38] to handle the constraints of the problem.

However, the approach does not include an explicit mechanism to maintain diversity (although a set of diverse solutions is actually generated). An interesting aspect of this work is that results are compared with respect to a mathematical programming technique: the goal attainment method. The comparison of results indicated that, as expected, the goal attainment method was very sensitive to its initial search point. Also, in several runs, it was not able to converge to a feasible solution. In contrast, the differential evolution algorithm was able to converge to feasible solutions in all the runs performed. However, the solutions generated by the goal attainment method were nondominated with respect to the solutions produced by differential evolution. Additionally, the CPU time required by differential evolution was about twice the time required by the goal attainment method.

Pareto Ranking

This subsection includes those approaches where a Pareto ranking procedure was added to them. The aim is to perform a $(\mu + \lambda)$-selection after the set of trial vectors have been generated from the current population.

In [37], Madavan proposes the Pareto-Based Differential Evolution (PBDE) approach. In this algorithm, Differential Evolution is extended to multi-objective optimization by incorporating Pareto-based mechanisms proposed previously by Deb et al. [12, 14]. It is interesting to note that this approach uses the same special case of the *DE/current_to_rand/1/bin* variant used by Abbass [4, 3, 51] where $K = 0$. The PBDE algorithm modifies the selection procedure in the basic DE algorithm by incorporating the key elements of the NSGA-II algorithm: the nondominated sorting and ranking selection procedure. In this way, once the new candidate vectors are obtained using DE operators (where the basic crossover operator is applied using the trial vector as the main parent), the new population is combined with the existing parents population and then the best members of the combined population (parents plus offspring) are chosen. As in the NSGA-II algorithm, the population for the next generation is filled by taking the individuals from the best nondominated rank down and discarding individuals with the same rank based on the diversity measure (crowding distance). This algorithm is not compared with respect to any other approach and is tested on 10 different unconstrained problems performing 250,000 evaluations. The authors indicate that the approach has difficulties to converge to the true Pareto front in two problems (Kursawe's test function [31] and ZDT4 [59]).

Xue et al. [56, 55] propose the Multi-Objective Differential Evolution (MODE) approach. This algorithm uses a variant of DE created by the authors in which the best individual is incorporated to create the offspring. This variant has some similarities with the traditional *DE/best/1/bin*. A Pareto-based approach is introduced to implement the selection of the best individual: if the trial solution is dominated, a set of nondominated individuals can be identified, and the "best" turns out to be any individual (randomly picked) from this set. On the other hand, if the trial solution is nondominated, it will be the "best" solution itself. The formula used by Xue et al. to create the offspring is the following:

$$p'_i = \gamma \cdot p_{best} + (1 - \gamma)p_i + F \cdot \sum_{k=1}^{K} (p_{i_{ak}} - p_{i_{bk}})$$

where p_{best} is the best individual selected, $\gamma \in [0, 1]$ represents the greediness of the operator, and K is the number of perturbation vectors (they use $K = 2$). It is worth noting that the previous formula is applied with certain mutation probability (p_m). Also, the authors adopt $(\mu+\lambda)$-selection, Pareto ranking (according to Goldberg [16]) and crowding distance [14] in order to produce and maintain well-distributed solutions. Actually, the authors incorporate a new parameter, called σ_{crowd}, which is used to penalize the fitness of the individuals, based on the crowding distance values, in order to improve the $(\mu + \lambda)$-selection approach. MODE is used to solve five high dimensionality unconstrained problems with 250,000 evaluations and the results are compared only to those obtained by SPEA [61].

Iorio and Li [25] propose the Nondominated Sorting Differential Evolution (NSDE). This approach is a simple modification of the NSGA-II [14]. The only difference between this approach and the NSGA-II is in the method for generating new individuals. The NSGA-II uses a real-coded crossover and mutation operator, but in the NSDE, these operators are replaced with the operators of Differential Evolution. New candidates are generated using the *DE/current-to-rand/1* variant, which is known to be rotationally invariant. A number of experiments are conducted on a uni-modal rotated problem from the literature. NSDE is used to solve a uni-modal rotated problem with a certain degree of rotation on each plane. The results of the NSDE outperformed those produced by the NSGA-II, and thus, it is shown that Differential Evolution can provide rotationally invariant behavior on a multi-objective optimization problem. In further work, Iorio and Li [26] propose three new versions of NSDE that incorporate directional information, by selecting parents for the generation of new individuals according to measures of both convergence and spread. For convergence, the authors modify the selection process (of the main parent) of NSDE in order to calculate differential vectors that point towards regions where better ranked individuals are located. For spread, the authors modify NSDE so that it favors the selection process (of the supporting parents) from different regions of decision variable space, but with the same rank. The modified approach is called NSDE-DCS (DCS stands for "directional convergence and spread") and is compared with respect to the NSGA-II, the original NSDE [25], NSDE-DC (NSDE only with the directional convergence mechanism), and NSDE-DS (NSDE only with the directional spread mechanism). Results indicate that all the NSDE versions outperform the NSGA-II, but NSDE-DS practically provides the same results as NSDE-DCS. This is a very interesting outcome that indicates that improving spread may, in some cases, also improve convergence.

Robič and Filipič [48] propose an approach called Differential Evolution for Multi-Objective Optimization (DEMO). They used the *DE/rand/1/bin* variant. DEMO modifies the mechanism followed to decide when a new vector replaces the parent: if the new vector dominates the parent, the new vector replaces the parent; if the parent dominates the new vector, the new vector is discarded; otherwise, the new vector is added in the population. In this way, the population can be extended and the newly created vectors take part immediately in the creation of the subsequent vectors. After the creation process of new vectors has finished, DEMO applies a nondominated sorting mechanism (combined with the use of the crowding distance measure) in order to truncate the population and maintain a fixed number of vectors at each iteration. This enables a fast convergence towards the true Pareto front, while the use of nondominated sorting and crowding distance (derived from the NSGA-II [14]) of the extended population promotes the uniform spread of solutions. Robič and Filipič also propose two additional versions of DEMO in which the newly created vector is not compared against the parent, but against the most similar individual in either the decision variable space or the objective space. The three DEMO variants are compared in five high-dimensionality unconstrained problems outperforming in some problems to the NSGA-II, PDE [2], PAES [28], SPEA [61] and MODE [56]. However, the authors didn't find any variant of DEMO to be significantly better than another, so they recommend to use the original version of DEMO (which compares the new vector against the parent), since it is the most efficient one (computationally speaking).

To deal with the shortcomings of GDE2 (described in the previous group) regarding slow convergence, Kukkonen and Lampinen proposed an improved version called GDE3 [30] (a combination of the earlier GDE versions and the Pareto-Based Differential Evolution algorithm [37]). This version added a growing population size and nondominated sorting (as in the NSGA-II [14]) to improve the distribution of solutions in the final Pareto front and to decrease the sensitivity of the approach to its initial parameters. In GDE3, when the old population vector and the trial vector are feasible and nondominated with respect to each other, both of them are maintained. Hence, the population size will grow. To maintain a fixed population size for the next generation, nondominated sorting is performed after each generation to prune the population size. GDE3 is compared with respect to the NSGA-II in several test functions, including some from the Deb-Thiele-Laumanns-Zitzler (DTLZ) test suite [15].

5.3 Combined Approaches

Finally, this class considers those approaches where a set of schemes have been mixed in the DE-based multi-objective algorithm. There are approaches which consider either Pareto concepts and also population-based concepts in the same approach, or techniques where, besides global search, local search is considered.

In [42], Parsopoulos et al. introduce a parallel multi-population DE called the Vector Evaluated Differential Evolution (VEDE) approach, for multi-objective optimization. VEDE is inspired by the Vector Evaluated Genetic Algorithm (VEGA) [52] approach. A number M of sub-populations are considered in a ring topology. Each population is evaluated using one of the objective functions of the problem, and there is an exchange of information among the populations through the migration of the best individuals. In this way, only the versions of DE that use the best individual to create new vectors can take full advantage of this information exchange. Also, the algorithm incorporates a domination selection procedure to enhance its performance by favoring non-dominated individuals in the population. The selection mechanism introduced by Parsopoulos et al. is similar to that used by Abbass et al. [4], in which the new vector is introduced in the population if it dominates the main parent. Finally, VEDE uses an external archive for the maintenance of the Pareto optimal set. VEDE is validated using four bi-objective unconstrained problems and is compared with respect to VEGA. Furthermore, VEDE was tested on three versions, using different DE variants: *DE/best/1/bin*, *DE/best/2/bin* and *DE/current_to_best/1*. The authors indicate that the proposed approach outperformed VEGA in all cases, however, among the three DE variants, none of the them was clearly superior to the other two.

Santana-Quintero's approach (ϵ-MyDE) was further hybridized with rough sets to give raise to a new approach called DEMORS (Differential Evolution for Multiobjective Optimization with Rough Sets) [20]. DEMORS operates in two phases. During the first phase, an improved version of ϵ-MyDE is applied for 2000 fitness function evaluations. The main improvement on ϵ-MyDE is the incorporation of the so-called Pareto-adaptive ϵ-grid [21] for the secondary population. The concept of Pareto-adaptive ϵ-dominance eliminates several of the drawbacks of ϵ-dominance [35]. During the second phase, a local search procedure based on rough sets theory [43] is applied for 1000 fitness function evaluations, in order to improve the solutions produced at the previous phase. The idea is to combine the high convergence rate of differential evolution with the high local search capabilities of rough sets. DEMORS is able to converge to the true Pareto front (or very close to it) in test problems with up to 30 decision variables, while only performing 3000 fitness function evaluations. Results are compared with respect to the NSGA-II.

Landa-Becerra and Coello Coello [34] propose the use of the ε-constraint technique [18] hybridized with a single-objective evolutionary optimizer: the cultured differential evolution [33]. The variant used in this approach is *DE/rand/1/bin*, however, the influence of the knowledge of the problem during the process, allows to change the variant to *DE/best/1/bin*. In fact, some modifications to the original *DE/rand/1/bin* are used (e.g. using the absolute value of the differences, adding another scaling factor besides "F" and using historical values of the best solution during the evolutionary process). The ε-constraint method transforms a multi-objective optimization problem into several single-objective optimization problems (each of these optimizations leads to a single Pareto optimal point). This method has been normally disregarded in the evolutionary multi-objective optimization literature due to its high computational cost [53, 46]. However, the authors argue that, if care is placed in the single-objective optimizer, this sort of hybrid can generate the true Pareto front of very difficult multi-objective optimization problems at a reasonable computational cost. Such a hypothesis is validated by solving DTLZ8 and DTLZ9 from the benchmark proposed in [15] together with several other test problems from the benchmark proposed in [23, 24]. All of these test functions are considered very hard to solve by current MOEAs, and this is illustrated by showing the results obtained by the NSGA-II in them. In most cases, even when performing a very high number of fitness function evaluations, the NSGA-II is unable to reach the true Pareto front. In contrast, the hybrid algorithm proposed in this paper is able to converge to the true Pareto front (or very close to it) of all the problems.

6 Convergence Properties of Multi-Objective Differential Evolution

Some theoretical studies about multi-objective extensions of differential evolution have been done recently. In [56, 57], Xue et al. perform a mathematical modeling and convergence analysis of a *continuous* Multi-Objective Differential Evolution (C-MODE) algorithm. The convergence properties of C-MODE are studied in a similar manner to the work presented by Hanne in [19], where convergence has been defined as follows:

Definition 6. A MOEA is said to **converge to the entire set of Pareto optimal solutions $\mathcal{P}^*$ with probability one** if

$$d(\mathcal{P}^*,\mathcal{P}^t) \rightarrow 0 \text{ with probability one as } t \rightarrow \infty,$$

where $d(\mathcal{P}^*,\mathcal{P}^t)$ is a distance function between $\mathcal{P}^*$ and $\mathcal{P}^t$, and it is defined as:

$$d(\mathcal{P}^*,\mathcal{P}^t) = | \mathcal{P}^* \cup \mathcal{P}^t | - | \mathcal{P}^* \cap \mathcal{P}^t |$$

The approach of Xue et al. employs underlying geometric structures (*cones*) based primarily on convex sets, to prove the convergence of the population of the C-MODE to the Pareto optimal set with probability one. Readers are directed to the associated references for a detailed description and associated theorem proof details.

On the other hand, Xue et al. study the C-MODE operators and their effects on the convergence properties of the algorithm, under the Gaussian initial population assumption. They show that the limiting properties of C-MODE depend on the factor $(2KF^2 + (1-\gamma)^2)$, where K, F and γ are the parameters associated to the approach. If this factor is greater than 1, the population variance matrix explodes, and C-MODE successfully identifies the optimal solution set; otherwise, the population variance matrix vanishes.

Xue et al. confirm the mathematical results developed by simulation results obtained by applying C-MODE to numerical examples with different parameter settings. Also, they conduct simulation results on complicated continuous benchmark functions and show that the C-MODE performs better when the parameters are set to meet the obtained conditions. In this way, the results obtained by Xue et al. can also be used to guide the parameter setting of the C-MODE when applied in real world applications.

In [56, 58], Xue et al. extend their theoretical work by modeling a *discrete* version of MODE, D-MODE, in the framework of Markov processes and develop the corresponding convergence properties. They study the Markov model for the D-MODE with finite population size. Two situations are considered: one with a population large enough to contain all the Pareto optimal solutions while the other is the opposite. In the second situation, an external archive is needed to store all visited Pareto optimal solutions. In both cases, Xue et al. prove the convergence with probability one of D-MODE to the set of Pareto optimal solutions in a similar manner to the work presented by Rudolph in [49].

7 Conclusions and Future Research Paths

In this chapter, we have presented a survey of Differential Evolution approaches modified to solve multi-objective optimization problems. We found that the techniques can be categorized in three classes: (1) Non-Pareto-based, (2) Pareto-based and (3) combined approaches. In fact, Pareto-based approaches were divided into two sub-classes: Using Pareto dominance and Using Pareto ranking. Combined approaches, as the name indicates, combines different schemes (e.g. global and local search, Pareto dominance and ranking) into one single approach.

In other heuristics to solve multi-objective optimization problems, such as particle swarm optimization (PSO) [47], key features of the heuristic itself have been adapted as to get some benefit in the way the problem is being solved (e.g. leader selection for creating new solutions). In contrast, from our findings, we observed that in the case of DE, the selection of the individuals used for the generation of new solutions has not been modified in most cases, with the exception of a few proposals [26]. We also found that the most popular schemes added to DE for multi-objective optimization were Pareto dominance for the selection mechanism between parent and offspring and Pareto ranking after the whole set of offspring have been generated.

Based on the aforementioned findings, we enumerate the topics we consider as promising paths for future research:

- **Diversity:** DE has shown a high convergence rate, like other metaheuristics such as PSO [27], but with a higher degree of robustness. However, DE present problems to actually reach the true Pareto front (it gets trapped in local optimum fronts). Furthermore, DE has some problems to spread solutions along the obtained front. This seems to indicate that multi-objective DE-based approaches require alternative (i.e., more effective) diversity maintenance schemes.
- **Variants:** Most approaches included in this survey use the most popular variant (*DE/rand/1/bin*) [9, 5, 32, 29, 30, 48, 50, 44, 20, 34, 36]. Despite the fact that other authors have used others variants like *DE/current_to_best/1* [42], *DE/current_to_rand/1* [26], special cases of *DE/current_to_rand/1/bin* [4, 1, 3, 37] and new variants [56, 55], it is not clear which variant is more suited for multi-objective optimization (i.e., which type of mutation and recombination operator is able to bias the search in a better way as to reach the true Pareto front in a more effective manner).
- **DE mutation operator:** In DE for global optimization, it is common to assume that the vectors that will be used to calculate the differences when computing the trial vector are chosen at random. However, as it was shown by Iorio and Li [26], in multi-objective optimization, some additional criteria might be taken into account for the selection of the pairs of solutions to participate in the mutation operator.
- **Parameter adaptation:** Online and self-adaptation attempts are still scarce in multi-objective differential evolution. Novel schemes to adapt key parameters like "CR", "F" or even the number of differences for the mutation operator are promising topics for future research.
- **Alternative encodings:** DE was proposed for continuous search spaces. Thus, one topic of interest is to develop alternative encodings that allow the use of differential evolution in problems requiring alternative encodings (e.g., combinatorial optimization problems). The use of encodings such as the random keys [7] or other proposals may be alternatives worth exploring in such cases.
- **Theory:** Studies about convergence of different DE variants, and runtime analysis, among other topics, will improve the current DE theory.

- **Applications:** Another path of research is the application of previously proposed DE-based approaches to the solution of real-world multi-objective optimization problems. Interesting behaviors may be found when applying DE in real-world multi-objective search spaces.

Acknowledgments

The first author acknowledges support from CONACyT through project number 52048-Y. The third author acknowledges support from CONACyT through project number 45683-Y.

References

1. Abbass, H.A.: A Memetic Pareto Evolutionary Approach to Artificial Neural Networks. In: Stumptner, M., Corbett, D.R., Brooks, M. (eds.) Canadian AI 2001. LNCS (LNAI), vol. 2256, pp. 1–12. Springer, Heidelberg (2001)
2. Abbass, H.A.: The Self-Adaptive Pareto Differential Evolution Algorithm. In: Congress on Evolutionary Computation (CEC 2002), Piscataway, New Jersey, May 2002, vol. 1, pp. 831–836. IEEE Service Center (2002)
3. Abbass, H.A., Sarker, R.: The Pareto Differential Evolution Algorithm. International Journal on Artificial Intelligence Tools 11(4), 531–552 (2002)
4. Abbass, H.A., Sarker, R., Newton, C.: PDE: A Pareto-frontier Differential Evolution Approach for Multi-objective Optimization Problems. In: Proceedings of the Congress on Evolutionary Computation 2001 (CEC 2001), Piscataway, New Jersey, May 2001, vol. 2, pp. 971–978. IEEE Service Center (2001)
5. Babu, B.V., Jehan, M.M.L.: Differential Evolution for Multi-Objective Optimization. In: Proceedings of the 2003 Congress on Evolutionary Computation (CEC 2003), Canberra, Australia, December 2003, vol. 4, pp. 2696–2703. IEEE Press, Los Alamitos (2003)
6. Bäck, T.: Evolutionary Algorithms in Theory and Practice. Oxford University Press, New York (1996)
7. Bean, J.C.: Genetics and random keys for sequencing and optimization. ORSA Journal on Computing 6(2), 154–160 (1994)
8. Chang, C.S., Xu, D.Y.: Differential Evolution Based Tuning of Fuzzy Automatic Train Operation for Mass Rapid Transit System. IEE Proceedings of Electric Power Applications 147(3), 206–212 (2000)
9. Chang, C.S., Xu, D.Y., Quek, H.B.: Pareto-optimal set based multiobjective tuning of fuzzy automatic train operation for mass transit system. IEE Proceedings on Electric Power Applications 146(5), 577–583 (1999)
10. Coello, C.A.C., Lamont, G.B., Van Veldhuizen, D.A.: Evolutionary Algorithms for Solving Multi-Objective Problems, 2nd edn. Springer, New York (2007)
11. Deb, K.: Multi-Objective Optimization using Evolutionary Algorithms. John Wiley & Sons, Chichester (2001)
12. Deb, K., Agrawal, S., Pratab, A., Meyarivan, T.: A Fast Elitist Non-Dominated Sorting Genetic Algorithm for Multi-Objective Optimization: NSGA-II. In: Deb, K., Rudolph, G., Lutton, E., Merelo, J.J., Schoenauer, M., Schwefel, H.-P., Yao, X. (eds.) PPSN 2000. LNCS, vol. 1917, pp. 849–858. Springer, Heidelberg (2000)

13. Deb, K., Goldberg, D.E.: An Investigation of Niche and Species Formation in Genetic Function Optimization. In: David Schaffer, J. (ed.) Proceedings of the Third International Conference on Genetic Algorithms, San Mateo, California, June 1989, pp. 42–50. George Mason University, Morgan Kaufmann Publishers (1989)
14. Deb, K., Pratap, A., Agarwal, S., Meyarivan, T.: A Fast and Elitist Multiobjective Genetic Algorithm: NSGA–II. IEEE Transactions on Evolutionary Computation 6(2), 182–197 (2002)
15. Deb, K., Thiele, L., Laumanns, M., Zitzler, E.: Scalable Test Problems for Evolutionary Multiobjective Optimization. In: Abraham, A., Jain, L., Goldberg, R. (eds.) Evolutionary Multiobjective Optimization. Theoretical Advances and Applications, pp. 105–145. Springer, USA (2005)
16. Goldberg, D.E.: Genetic Algorithms in Search, Optimization and Machine Learning. Addison-Wesley Publishing Company, Reading (1989)
17. Goldberg, D.E., Richardson, J.: Genetic algorithms with sharing for multimodal function optimization. In: Proceedings of the Second International Conference on Genetic Algorithms, pp. 41–49. Lawrence Erlbaum Associates, Mahwah (1987)
18. Haimes, Y.Y., Lasdon, L.S., Wismer, D.A.: On a Bicriterion Formulation of the Problems of Integrated System Identification and System Optimization. IEEE Transactions on Systems, Man, and Cybernetics 1(3), 296–297 (1971)
19. Hanne, T.: On the convergence of multiobjective evolutionary algorithms. European Journal of Operational Research 117(3), 553–564 (1999)
20. Hernández-Díaz, A.G., Santana-Quintero, L.V., Coello, C.C., Caballero, R., Molina, J.: A New Proposal for Multi-Objective Optimization using Differential Evolution and Rough Sets Theory. In: Keijzer, M., et al. (eds.) 2006 Genetic and Evolutionary Computation Conference (GECCO 2006), Seattle, Washington, USA, July 2006, vol. 1, pp. 675–682. ACM Press, New York (2006)
21. Hernández-Díaz, A.G., Santana-Quintero, L.V., Coello, C.A.C., Molina, J.: Pareto-adaptive ϵ-dominance. Evolutionary Computation 15(4), 493–517 (2007)
22. Horn, J., Nafpliotis, N., Goldberg, D.E.: A Niched Pareto Genetic Algorithm for Multiobjective Optimization. In: Proceedings of the First IEEE Conference on Evolutionary Computation, IEEE World Congress on Computational Intelligence, Piscataway, New Jersey, June 1994, vol. 1, pp. 82–87. IEEE Service Center (1994)
23. Huband, S., Barone, L., While, L., Hingston, P.: A Scalable Multi-objective Test Problem Toolkit. In: Coello Coello, C.A., Hernández Aguirre, A., Zitzler, E. (eds.) EMO 2005. LNCS, vol. 3410, pp. 280–295. Springer, Heidelberg (2005)
24. Huband, S., Hingston, P., Barone, L., While, L.: A Review of Multiobjective Test Problems and a Scalable Test Problem Toolkit. IEEE Transactions on Evolutionary Computation 10(5), 477–506 (2006)
25. Iorio, A.W., Li, X.: Solving rotated multi-objective optimization problems using differential evolution. In: Webb, G.I., Yu, X. (eds.) AI 2004. LNCS (LNAI), vol. 3339, pp. 861–872. Springer, Heidelberg (2004)
26. Iorio, A.W., Li, X.: Incorporating Directional Information within a Differential Evolution Algorithm for Multi-objective Optimization. In: Keijzer, M., et al. (eds.) 2006 Genetic and Evolutionary Computation Conference (GECCO 2006), Seattle, Washington, USA, July 2006, vol. 1, pp. 691–697. ACM Press, New York (2006)
27. Kennedy, J., Eberhart, R.C.: Swarm Intelligence. Morgan Kaufmann Publishers, San Francisco, California (2001)
28. Knowles, J.D., Corne, D.W.: Approximating the Nondominated Front Using the Pareto Archived Evolution Strategy. Evolutionary Computation 8(2), 149–172 (2000)

29. Kukkonen, S., Lampinen, J.: An Extension of Generalized Differential Evolution for Multi-objective Optimization with Constraints. In: Yao, X., Burke, E.K., Lozano, J.A., Smith, J., Merelo-Guervós, J.J., Bullinaria, J.A., Rowe, J.E., Tiňo, P., Kabán, A., Schwefel, H.-P. (eds.) PPSN 2004. LNCS, vol. 3242, pp. 752–761. Springer, Heidelberg (2004)
30. Kukkonen, S., Lampinen, J.: GDE3: The third Evolution Step of Generalized Differential Evolution. In: 2005 IEEE Congress on Evolutionary Computation (CEC 2005), Edinburgh, Scotland, September 2005, vol. 1, pp. 443–450. IEEE Service Center (2005)
31. Kursawe, F.: A Variant of Evolution Strategies for Vector Optimization. In: Schwefel, H.-P., Männer, R. (eds.) PPSN 1990. LNCS, vol. 496, pp. 193–197. Springer, Heidelberg (1991)
32. Lampinen, J.: De's selection rule for multiobjective optimization. Technical report, Lappeenranta University of Technology, Department of Information Technology (2001)
33. Becerra, R.L., Coello, C.A.C.: Cultured differential evolution for constrained optimization. Computer Methods in Applied Mechanics and Engineering 195(33-36), 4303–4322 (2006)
34. Becerra, R.L., Coello, C.A.C.: Solving Hard Multiobjective Optimization Problems Using ε-Constraint with Cultured Differential Evolution. In: Runarsson, T.P., Beyer, H.-G., Burke, E.K., Merelo-Guervós, J.J., Whitley, L.D., Yao, X. (eds.) PPSN 2006. LNCS, vol. 4193, pp. 543–552. Springer, Heidelberg (2006)
35. Laumanns, M., Thiele, L., Deb, K., Zitzler, E.: Combining Convergence and Diversity in Evolutionary Multi-objective Optimization. Evolutionary Computation 10(3), 263–282 (2002)
36. Li, H., Zhang, Q.: A Multiobjective Differential Evolution Based on Decomposition for Multiobjective Optimization with Variable Linkages. In: Runarsson, T.P., Beyer, H.-G., Burke, E.K., Merelo-Guervós, J.J., Whitley, L.D., Yao, X. (eds.) PPSN 2006. LNCS, vol. 4193, pp. 583–592. Springer, Heidelberg (2006)
37. Madavan, N.K.: Multiobjective Optimization Using a Pareto Differential Evolution Approach. In: Congress on Evolutionary Computation (CEC 2002), Piscataway, New Jersey, May 2002, vol. 2, pp. 1145–1150. IEEE Service Center (2002)
38. Mezura-Montes, E., Coello, C.A.C.: A Simple Multimembered Evolution Strategy to Solve Constrained Optimization Problems. IEEE Transactions on Evolutionary Computation 9(1), 1–17 (2005)
39. Michalewicz, Z.: Genetic Algorithms + Data Structures = Evolution Programs, 3rd edn. Springer, Heidelberg (1996)
40. Newman, D.J., Hettich, S., Blake, C.L., Merz, C.J.: UCI repository of machine learning databases (1998), `http://www.ics.uci.edu/~mlearn/MLRepository.html`
41. Okabe, T., Jin, Y., Olhofer, M., Sendhoff, B.: On Test Functions for Evolutionary Multi-objective Optimization. In: Yao, X., Burke, E.K., Lozano, J.A., Smith, J., Merelo-Guervós, J.J., Bullinaria, J.A., Rowe, J.E., Tiňo, P., Kabán, A., Schwefel, H.-P. (eds.) PPSN 2004. LNCS, vol. 3242, pp. 792–802. Springer, Heidelberg (2004)
42. Parsopoulos, K.E., Taoulis, D.K., Pavlidis, N.G., Plagianakos, V.P., Vrahatis, M.N.: Vector Evaluated Differential Evolution for Multiobjective Optimization. In: 2004 Congress on Evolutionary Computation (CEC 2004), Portland, Oregon, USA, June 2004, vol. 1, pp. 204–211. IEEE Service Center (2004)
43. Pawlak, Z.: Rough sets. International Journal of Computer and Information Sciences 11(1), 341–356 (1982)

44. Portilla Flores, E.A.: Integración Simultánea de Aspectos Estructurales y Dinámicos para el Diseño Óptimo de un Sistema de Transmisión de Variación Continua. PhD thesis, Departamento de Ingeniería Eléctrica, Sección de Mecatrónica, CINVESTAV-IPN, México, D.F., México (June 2006) (in Spanish)
45. Price, K.V.: An Introduction to Differential Evolution. In: Corne, D., Dorigo, M., Glover, F. (eds.) New Ideas in Optimization, pp. 79–108. McGraw-Hill, London (1999)
46. Ranji Ranjithan, S., Kishan Chetan, S., Dakshima, H.K.: Constraint Method-Based Evolutionary Algorithm (CMEA) for Multiobjective Optimization. In: Zitzler, E., Deb, K., Thiele, L., Coello Coello, C.A., Corne, D.W. (eds.) EMO 2001. LNCS, vol. 1993, pp. 299–313. Springer, Heidelberg (2001)
47. Reyes-Sierra, M., Coello, C.A.C.: Multi-Objective Particle Swarm Optimizers: A Survey of the State-of-the-Art. International Journal of Computational Intelligence Research 2(3), 287–308 (2006)
48. Robič, T., Filipič, B.: DEMO: Differential Evolution for Multiobjective Optimization. In: Coello Coello, C.A., Hernández Aguirre, A., Zitzler, E. (eds.) EMO 2005. LNCS, vol. 3410, pp. 520–533. Springer, Heidelberg (2005)
49. Rudolph, G.: Some Theoretical Properties of Evolutionary Algorithms under Partially Ordered Fitness Values. In: Fabian, Cs., Intorsureanu, I. (eds.) Proceedings of the Evolutionary Algorithms Workshop (EAW 2001), Bucharest, Romania, January 2001, pp. 9–22 (2001)
50. Santana-Quintero, L.V., Coello, C.A.C.: An Algorithm Based on Differential Evolution for Multi-Objective Problems. International Journal of Computational Intelligence Research 1(2), 151–169 (2005)
51. Sarker, R., Abbass, H., Newton, C.: Solving Two Multi-objective Optimization Problems using Evolutionary Algorithm. In: Mohammadian, M., Sarker, R., Yao, X. (eds.) Computational Intelligence in Control. Idea Group Publishing, USA (2002)
52. David Schaffer, J.: Multiple Objective Optimization with Vector Evaluated Genetic Algorithms. In: Genetic Algorithms and their Applications: Proceedings of the First International Conference on Genetic Algorithms, Hillsdale, New Jersey, pp. 93–100. Lawrence Erlbaum, Mahwah (1985)
53. Srigiriraju, K.C.: Noninferior Surface Tracing Evolutionary Algorithm (NSTEA) for Multi Objective Optimization. Master's thesis, North Carolina State University, Raleigh, North Carolina (August 2000)
54. Srinivas, N., Deb, K.: Multiobjective Optimization Using Nondominated Sorting in Genetic Algorithms. Evolutionary Computation 2(3), 221–248 (1994)
55. Xue, F.: Multi-Objective Differential Evolution: Theory and Applications. PhD thesis, Rensselaer Polytechnic Institute, Troy, New York (September 2004)
56. Xue, F., Sanderson, A.C., Graves, R.J.: Pareto-based Multi-Objective Differential Evolution. In: Proceedings of the 2003 Congress on Evolutionary Computation (CEC 2003), Canberra, Australia, December 2003, vol. 2, pp. 862–869. IEEE Press, Los Alamitos (2003)
57. Xue, F., Sanderson, A.C., Graves, R.J.: Modeling and convergence analysis of a continuous multi-objective differential evolution algorithm. In: 2005 IEEE Congress on Evolutionary Computation (CEC 2005), Edinburgh, Scotland, September 2005, vol. 1, pp. 228–235. IEEE Service Center (2005)
58. Xue, F., Sanderson, A.C., Graves, R.J.: Multi-objective differential evolution - algorithm, convergence analysis, and applications. In: 2005 IEEE Congress on Evolutionary Computation (CEC 2005), Edinburgh, Scotland, September 2005, vol. 1, pp. 743–750. IEEE Service Center (2005)

59. Zitzler, E., Deb, K., Thiele, L.: Comparison of Multiobjective Evolutionary Algorithms: Empirical Results. Evolutionary Computation 8(2), 173–195 (2000)
60. Zitzler, E., Teich, J., Bhattacharyya, S.S.: Evolutionary Algorithm Based Exploration of Software Schedules for Digital Signal Processors. In: Banzhaf, W., Daida, J., Eiben, A.E., Garzon, M.H., Honavar, V., Jakiela, M., Smith, R.E. (eds.) Proceedings of the Genetic and Evolutionary Computation Conference (GECCO 1999), July 1999, vol. 2, pp. 1762–1769. Morgan Kaufmann, San Francisco (1999)
61. Zitzler, E., Thiele, L.: Multiobjective Evolutionary Algorithms: A Comparative Case Study and the Strength Pareto Approach. IEEE Transactions on Evolutionary Computation 3(4), 257–271 (1999)

A Review of Major Application Areas of Differential Evolution

V.P. Plagianakos[1], D.K. Tasoulis[2], and M.N. Vrahatis[1]

[1] Computational Intelligence Laboratory, Department of Mathematics, University of Patras Artificial Intelligence Research Center–UPAIRC, University of Patras, GR–26110 Patras, Greece
{vpp,vrahatis}@math.upatras.gr

[2] Institute for Mathematical Sciences, Imperial College London, South Kensington, London SW7 2PG, United Kingdom
d.tasoulis@imperial.ac.uk

Summary. In this chapter we present an overview of the major applications areas of differential evolution. In particular we pronounce the strengths of DE algorithms in tackling many difficult problems from diverse scientific areas, including single and multiobjective function optimization, neural network training, clustering, and real life DNA microarray classification. To improve the speed and performance of the algorithm we employ distributed computing architectures and demonstrate how parallel, multi–population DE architectures can be utilised in single and multi-objective optimization. Using data mining we present a methodology that allows the simultaneous discovery of multiple local and global minimizers of an objective function. At a next step we present applications of DE in real life problems including the training of integer weight neural networks and the selection of genes of DNA microarrays in order to boost predictive accuracy of classification models. The chapter concludes with a discussion on promising future extensions of the algorithm, and presents novel mutation operators, that are the result of a genetic programming procedure, as very interesting future research direction.

1 Introduction

Evolutionary Algorithms (EAs) refer to problem solving optimization algorithms which employ computational models of evolutionary processes. A variety of evolutionary algorithms have been proposed. The major ones include: Genetic Algorithms [21, 24], Evolutionary Programming [17, 19], Evolution Strategies [39], Genetic Programming [29], Particle Swarm Optimization [27], and the Differential Evolution algorithm [46]. All these algorithms share the common conceptual base of simulating the evolution of a population of individuals using a predefined set of operators. Commonly two kinds of operators are used: the *selection* and the *search* operators. The most widely used search operators are *mutation* and *recombination*.

The selection operator enforces the natural selection and the survival of the fittest on the population of individuals. The recombination and the mutation operators stochastically perturb the individuals providing efficient exploration of the search space. This

U.K. Chakraborty (Ed.): Advances in Differential Evolution, SCI 143, pp. 197–238, 2008.
springerlink.com

perturbation is primarily controlled by the user defined recombination and mutation rates. Although simplistic from a biologist's point of view, these algorithms are sufficiently complex to provide robust and powerful search mechanisms and have shown their strength in solving hard optimization problems.

In this chapter we focus on the Differential Evolution algorithm and present some of its applications, including single and multiobjective function optimization, neural network training, clustering, and real life DNA microarray problems. Furthermore, we discuss promising future extensions of the algorithm by the incorporation of genetically programmed mutation operators.

2 The Differential Evolution Algorithm

More than ten years ago, Storn and Price [46] have presented a novel optimization method, called Differential Evolution (DE), which has been designed to handle nondifferentiable, nonlinear and multimodal objective functions. To fulfill this requirement, DE has been designed as a stochastic parallel direct search method, which utilizes concepts borrowed from the broad class of evolutionary algorithms, but requires few easily chosen control parameters. Early experimental results have shown that DE has good convergence properties and outperforms other well known evolutionary algorithms [45, 46]. In the following paragraphs we outline the workings of DE.

2.1 The Workings of DE

In each population, new individuals (vectors) are generated by the combination of randomly chosen vectors. This operation in our context can be referred as *mutation*. The outcoming vectors are then mixed with another predetermined vector – the *target* vector – and this operation can be called *recombination*. This operation yields the so–called *trial* vector. The trial vector is accepted for the next generation if and only if it reduces the value of the objective function f. This operation can be referred as *selection*. A high-level description of the above mentioned operators (for one generation) is given below:

Step 1. Do for each Vector
Step 2. MutantVector := MUTATION(Vector)
Step 3. TrialVector := RECOMBINATION(MutantVector)
Step 4. **If** f(TrialVector) $\leqslant$ f(Vector)
Step 5. Vector := TrialVector
Step 6. **EndIf**
Step 7. EndDo

We now briefly review the two basic DE variation operators. The first DE operator we consider is mutation. Specifically, for each individual x_g^i, $i = 1, 2, \ldots, NP$, where g denotes the current generation and *NP* the number of individuals in the population, a new individual v_{g+1}^i (mutant vector) is generated according to one of the following equations:

$$v_{g+1}^i = x_g^{\text{best}} + \mu(x_g^{r_1} - x_g^{r_2}), \quad (1)$$
$$v_{g+1}^i = x_g^{r_1} + \mu(x_g^{r_2} - x_g^{r_3}), \quad (2)$$
$$v_{g+1}^i = x_g^i + \mu(x_g^{\text{best}} - x_g^i) + \mu(x_g^{r_1} - x_g^{r_2}), \quad (3)$$
$$v_{g+1}^i = x_g^{\text{best}} + \mu(x_g^{r_1} - x_g^{r_2}) + \mu(x_g^{r_3} - x_g^{r_4}), \quad (4)$$
$$v_{g+1}^i = x_g^{r_1} + \mu(x_g^{r_2} - x_g^{r_3}) + \mu(x_g^{r_4} - x_g^{r_5}), \quad (5)$$

where x_g^{best} is the best member of the previous generation; $\mu > 0$ is a real parameter, called *mutation constant*, which controls the amplification of the difference between two individuals so as to avoid the stagnation of the search process; and $r_1, r_2, r_3, r_4, r_5 \in \{1, 2, \ldots, i-1, i+1, \ldots, NP\}$, are random integers mutually different and not equal to the running index i.

Trying to rationalize the above equations, we observe that Equation (2) is similar to the crossover operator used by some Genetic Algorithms and Equation (1) derives from it, when the best member of the previous generation is employed. Equations (3), (4) and (5) are modifications obtained by the combination of Equations (1) and (2). It is clear that more such relations can be generated using the above ones as building blocks. For example, the recently proposed trigonometric mutation operator [14] performs mutations with probability τ_μ according to the following equation:

$$v_{g+1}^i = (x_g^{r_1} + x_g^{r_2} + x_g^{r_3})/3 + (p_2 - p_1)(x_g^{r_1} - x_g^{r_2}) + (p_3 - p_2)(x_g^{r_2} - x_g^{r_3}) + (p_1 - p_3)(x_g^{r_3} - x_g^{r_1}), \quad (6)$$

and with probability $(1 - \tau_\mu)$ performs mutations according to Equation (2), where τ_μ is a user defined parameter. The values of p_m, $m = \{1, 2, 3\}$ and p' are obtained through the following equations:

$$p_1 = \left|f(x_g^{r_1})\right| / p', \; p_2 = \left|f(x_g^{r_2})\right| / p', \; p_3 = \left|f(x_g^{r_3})\right| / p', \text{ and}$$
$$p' = \left|f(x_g^{r_1})\right| + \left|f(x_g^{r_2})\right| + \left|f(x_g^{r_3})\right|.$$

For the rest of the chapter, we call DE_1 the differential evolution algorithm that uses Equation (1) as the mutation operator, DE_2 the algorithm that uses Equation (2), and so on.

The recombination operator is subsequently applied to further increase the diversity of the mutant individuals. To this end, the resulting individuals are combined with other predetermined individuals, called the target individuals. Specifically, for each component l ($l = 1, 2, \ldots, n$) of the mutant individual v_{g+1}^i, we randomly choose a real number r in the interval $[0, 1]$. Then, we compare this number with the *recombination constant*, ρ. If $r \leqslant \rho$, then we select, as the l–th component of the trial individual u_{g+1}^i, the l–th component of the mutant individual v_{g+1}^i. Otherwise, the l–th component of the target vector x_{g+1}^i becomes the l–th component of the trial vector. This operation yields the trial individual.

Finally, the trial vector u^i_{g+1} is accepted for the next generation only if it reduces the value of the objective function. Thus:

$$x^i_{g+1} = \begin{cases} u^i_{g+1}, & \text{if } f(u^i_{g+1}) < f(x^i_g), \\ x^i_g, & \text{otherwise.} \end{cases} \tag{7}$$

To prevent an individual from surviving indefinitely, we can employ the concept of aging [45]. To this end, each vector is randomly assigned a maximum age, i.e. an integer from the interval $[\alpha, \beta]$, where α and β are the minimum and the maximum possible age, respectively. At each iteration, the age of each vector is increased by one, and if it exceeds its maximum age then the individual "dies". This individual is then replaced by another vector randomly chosen from the current population. Note that it is desirable the best individual of the population not to be eliminated.

2.2 Implementation of Parallel Evolutionary Algorithms

Parallel processing, that is the method of having many small tasks solve one large problem, has emerged as a key enabling technology in modern computing. As a result of the demand for higher performance, lower cost, and sustained productivity, the past several years have witnessed an ever-increasing acceptance and adoption of parallel processing, both for high-performance scientific computing and for more general–purpose applications. Exploiting recent software advances [1, 20], collections of heterogeneous computers can be used as a coherent and flexible concurrent computational resource. These technologies have allowed the vast number of individual personal computers available in most scientific laboratories to be used as parallel machines at no, or at a very low, cost. Network interfaces, linking individual computers, are necessary to produce such pools of computational power.

EAs, as well as DEs, are easily parallelized [32, 43]. There are two typical models for EA parallelization. The first uses fine grained parallelism, so each individual is represented by a processor. This creates certain problems when the number of processors available is limited or when the individual's fitness to reproduce needs to be evaluated over the whole population. The second model, maps an entire subpopulation to a processor. Thus each subpopulation evolves independently toward a solution. This allows each subpopulation to develop its own solution uniquely. Then, the best individual of each subpopulation is propagated to other subpopulations, according to the selected network topology. This operation is called *migration*. This model is called the Parallel Evolutionary Algorithm (PEA).

Usually, the topology of the PEA is a ring, i.e. the best individuals from each subpopulation are allowed to migrate to the next subpopulation of the ring. This concept reduces the migration between the subpopulations and consequently the messages between the processors. The migration of the best individuals is controlled by the migration constant, $\varphi \in (0, 1)$. At each iteration, a random number from the interval $(0, 1)$ is uniformly chosen and compared with the migration constant. If the migration constant is bigger, then the best individuals of each subpopulation migrate and take the place of a randomly selected individual (different from the best one) in the next subpopulation; otherwise no migration is permitted. We have experimentally found that a migration constant, $\varphi \simeq 0.5$, is a good choice, since it allows each subpopulation to evolve for some iterations before the migration phase actually occur.

3 Function Optimization Using DE

Let us know consider the minimization problem of finding global minima of a continuous nonlinear, (possibly) nondifferentiable, multimodal objective function f. More specifically, our goal is to locate *global minimizers* x_t^* of the real–valued objective function $f\colon \mathcal{E} \to \mathbb{R}$:

$$f(x_t^*) \leqslant f(x), \quad \forall x \in \mathcal{E},$$

where $t = 1, 2, \ldots$, and the compact set $\mathcal{E} \subseteq \mathbb{R}^n$ is a n–dimensional scaled translation of the unit hypercube.

3.1 Single Objective Optimization

Firstly, we will evaluate the performance of the DE algorithms employing seven single objective test functions. We briefly describe them below.

Test Function 1: Sphere

$$f_1(x) = \sum_{j=1}^{5} x_j^2, \qquad x_j \in [-5.12, 5.12].$$

The sphere test function is a considered to be a simple minimization problem. The minimum is $f_1(0, 0, \ldots, 0) = 0$.

Test Function 2: Rosenbrock's Saddle

$$f_2(x) = 100 \cdot (x_1^2 - x_2)^2 + (1 - x_1)^2, \quad x_j \in [-2.048, 2.048].$$

This is a two–dimensional test function, which is known to be relatively difficult to minimize. The minimum is $f_2(1, 1) = 0$.

Test Function 3: Step Function

$$f_3(x) = 30 + \sum_{j=1}^{5} \lfloor x_j \rfloor, \quad x_j \in [-5.12, 5.12],$$

where the floor function $\lfloor x \rfloor$ gives the largest integer less than or equal to x. The minimum of this function is $f_3(-5 - \xi, \ldots, -5 - \xi) = 0$, where $\xi \in [0, 0.12]$. This function exhibits many flat regions that can cause search stagnation.

Test Function 4: Quartic Function

$$f_4(x) = \sum_{j=1}^{30} \left(j \cdot x_j^4 + \eta\right), \quad x_j \in [-1.28, 1.28].$$

This is test function is designed to evaluate the behavior of minimization algorithms in the presence of noise. To this end, η is a random variable following the uniform distribution in the range $[0, 1]$. The inclusion of η makes f_4 more difficult to optimize. The functional minimum of the function is $f_4(0, 0, \ldots, 0) \leqslant 30 \cdot E[\eta] = 15$, where $E[\eta]$ is the expectation of η.

Test Function 5: Shekel's Foxholes

$$f_5(x) = \frac{1}{0.002 + \psi_1(x)}, \quad x_j \in [-65.536, 65.536],$$

where, $\psi_1(x) = \sum_{i=0}^{24} \frac{1}{1+i+\sum_{j=1}^{2}(x_j - a_{ij})^6}$. The parameters for this function are:

$$a_{i1} = \{-32, -16, 0, 16, 32\}, \text{ where } i = \{0, 1, 2, 3, 4\} \text{ and } a_{i1} = a_{i \bmod 5, 1}$$
$$a_{i2} = \{-32, -16, 0, 16, 32\}, \text{ where } i = \{0, 5, 10, 15, 20\} \text{ and } a_{i2} = a_{i+k, 2},$$

and $k = \{1, 2, 3, 4\}$. The global minimum of $f_5(-32, -32) = 0.998004$.

Test Function 6: Corana Parabola

$$f_6(x) = \sum_{j=1}^{4} \begin{cases} \psi_2(x_j), & \text{if } |x_j - z_j| < 0.05, \\ \psi_3(x_j), & \text{otherwise.} \end{cases}$$

where $\psi_2(x_j) = 0.15\left(z_j - 0.05\text{sign}(z_j)\right)^2 d_j$, $\psi_3(x_j) = d_j x_j^2$, $z_j = \lfloor 5|x_j| + 0.49999 \rfloor \text{sign}(x_j) 0.2$ and $d_j = \{1, 1000, 10, 100\}$. The Corana test function defines a paraboloid with axes parallel to the coordinate axes. The function is characterized by a multitude of local minima, increasing in depth as one moves closer to the origin. The global minimum of the function is $f_6(x) = 0$, for $x_j \in (-0.05, 0.05)$.

Test Function 7: Griewangk's Function

$$f_7(x) = \sum_{j=1}^{10} \frac{x_j^2}{4000} - \prod_{j=1}^{10} \cos\left(\frac{x_j}{\sqrt{j}}\right) + 1, \text{ where } x_j \in [-400, 400].$$

This test function is riddled with local minima. The global minimum of the function is $f_7(0, 0, \ldots, 0) = 0$.

Experimental Results on Parallel Differential Evolution

Here we present results of the Parallel Differential Evolution (PARDE) algorithm [47], on the above test functions. In Table 1 the parameter setup used in the numerical experiments conducted is summarized. Specifically, *D* denotes the dimensionality of the problem, *NP* stands for the size of the subpopulation assigned to each of the processors employed, *g* is the maximum number of generations allowed, finally, μ and ρ are the values of the mutation and recombination constants, respectively.

Little effort has been devoted to the selection of the values of *NP*, μ and ρ since the scope of these experiments is to study extensively, the implications of information sharing in a parallel environment, which is controlled by the migration constant, φ. It is worth noting that further performance improvements can be achieved by further fine–tuning *NP*, μ, and ρ. The parameter τ_μ used by the trigonometric mutation strategy, Equation (6), was set to 0.1. Figure 1 illustrates the speedup achieved by assigning each subpopulation to a different processor, relative to assigning all subpopulations to

Table 1. Parameter values

Test function	D	NP	g	μ	ρ
Sphere function	5	30	1000	0.9	0.3
Rosenbrock's saddle	2	30	1000	0.9	0.5
Step function	5	20	1000	0.8	0.3
Quartic function	30	100	2000	0.8	0.5
Shekel's foxholes	2	30	1000	0.9	0.3
Corana's parabola	4	15	2000	0.4	0.2
Griewangk's function	10	50	10000	1.0	0.3

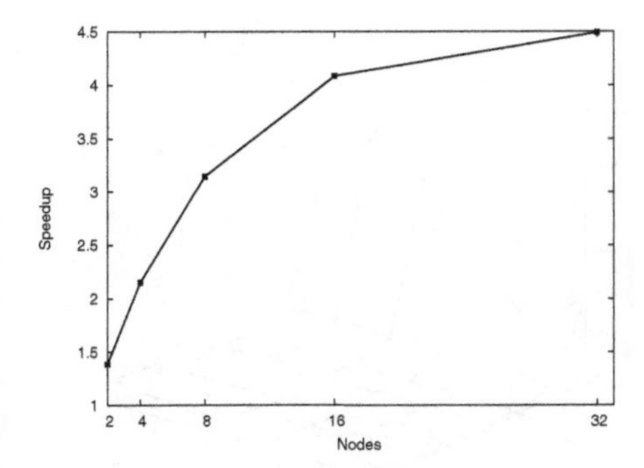

Fig. 1. Speedup using up to 32 nodes

a single processor. To obtain the plotted values, the algorithm performed 1000 generations with a migration constant equal to 0.5. Table 2 reports the mean number of generations required to locate the global minimum of each test function, averaged over all the considered mutation constants. It is clear from Table 2 that the best performing mutation strategy, for all test problems, was the first one. Furthermore, Griewangk's function appears to be the hardest to minimize. Figures 2–4 illustrate the performance of the 16–node model for all the considered mutation strategies, on all the test functions, for a particular migration constant. In all the $3D$ plots, the mutation strategies are given by the x–axis, the test functions by the y–axis, and finally, the mean number of generations required is reported in the z–axis. Concerning the overall performance of the alternative mutation strategies, the worst performance is exhibited by strategies 5 and 6. Strategies 1 and 3 appear to be the most efficient and robust.

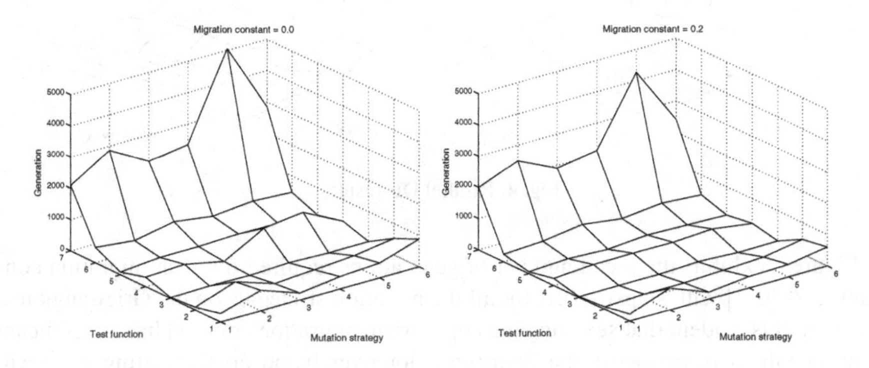

Fig. 2. Parallel DE results

Table 2. Mean number of generations for the 16–node model

Test function	Mutation Strategy					
	1	2	3	4	5	6
Sphere	**231.09**	611.58	247.02	341.08	409.79	642.02
Rosenbrock	**82.21**	386.01	130.14	220.77	271.39	371.59
Step	**21.91**	76.85	25.48	47.59	25.40	31.59
Quartic	**244.92**	249.90	260.21	406.13	454.18	244.11
Shekel	**63.39**	136.96	96.21	186.91	158.30	101.55
Corana	**282.35**	364.09	513.36	329.48	491.01	398.85
Griewangk	**1872.17**	1885.19	1975.61	2644.47	4448.66	2434.45

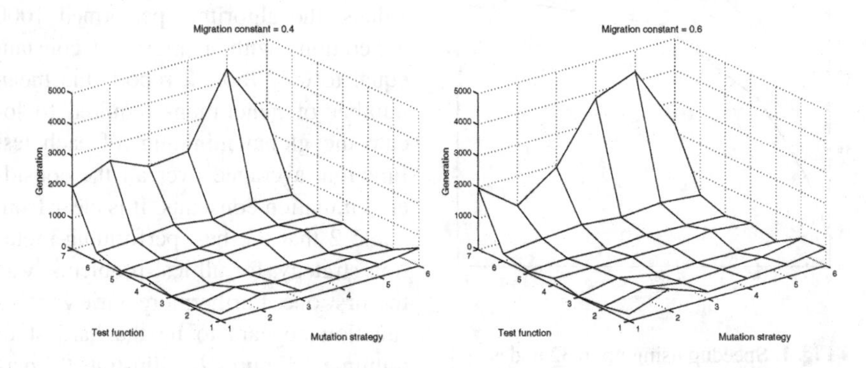

Fig. 3. Parallel DE results

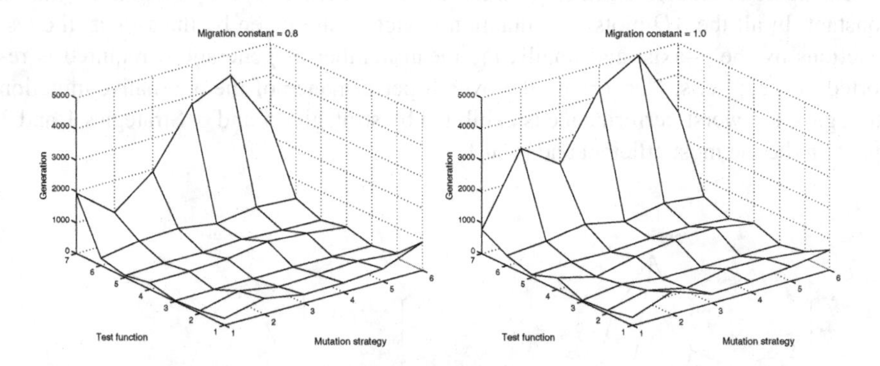

Fig. 4. Parallel DE results

Figure 5 exhibits the mean number of generations required for each migration constant, $\varphi \in [0, 1]$ with stepsize 0.1, for all the mutation strategies on the Griewangk test function. It is evident that selecting the appropriate migration constant has a significant impact on the performance of the algorithm. Moreover, it appears that setting φ close to one or to zero can lead to a substantial increase in the number of generations required.

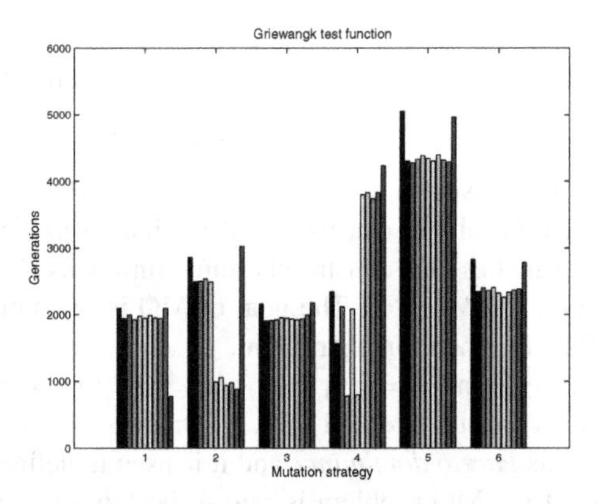

Fig. 5. Results for the Griewangk test function

A superior performance is typically obtained for intermediate values of φ. It has already been noted from the results of Table 2 and Figures 2–4 that the first mutation strategy, Equation (1), is the most efficient. From Figure 5 we can infer that this strategy also exhibits the most robust behavior with respect to the migration constant. The third mutation strategy, Equation (3), also exhibits a relatively robust behavior with respect to φ. It is worth noting, however, that the other considered mutation strategies can achieve a comparable, or even better, performance after fine–tuning the value of φ.

3.2 Multiobjective Optimization

Multiobjective Optimization (MO) problems consist of several competing and incommensurable objective functions. Such problems are frequently encountered in numerous scientific and engineering applications. The need for the concurrent minimization of more than one objective functions, renders the use of EAs particularly attractive. In contrast to traditional gradient–based techniques, EAs operate on a set of potential solutions of the problem. Thus, EAs are capable of detecting several solutions of an MO problem in a single run [11, 12, 42, 54, 59, 60]. These solutions are called *Pareto optimal*, and each corresponds to a different trade–off among the objective functions. Typically, a large number of Pareto optimal solutions exist.

Here we present a multi–population variant of DE, named *Vector Evaluated Differential Evolution* (VEDE), which is inspired by the *Vector Evaluated Genetic Algorithm* (VEGA) approach [42]. In VEDE, each population is evaluated using one of the objective functions of the problem under consideration. Information sharing among the populations takes place through the migration of the best individuals. The performance of a parallel version of VEDE, which incorporates a domination selection scheme, is investigated on widely used test problems and compared to the VEGA approach. The parallel computation of solutions of an MO problem is preferred, because besides the reduction in execution time, can also yield a better representation of the possible outcomes, thereby enhancing the performance of the algorithm [54].

Background Material

Let $\mathbb{S} \subset \mathbb{R}^n$ be an n–dimensional search space, and let k objective functions:

$$f_i(x) : \mathbb{S} \to \mathbb{R}, \quad i = 1, 2, \ldots, k,$$

be defined over S. Further assume, $g_j(x) \leqslant 0, \quad j = 1, \ldots, m$, to be m inequality constraints. Then the MO problem can be stated as finding a vector, $x^* = (x_1^*, x_2^*, \ldots, x_n^*)^\top \in S$, that satisfies the constraints and minimizes the function $\mathbf{f}(x) = [f_1(x), f_2(x), \ldots, f_k(x)] : \mathbb{R}^n \to \mathbb{R}^k$. The goal of MO is to compute a set of Pareto optimal solutions to the aforementioned problem.

Let $u = (u_1, 2, \ldots, u_k)$, and $v = (v_1, v_2, \ldots, v_k)$, be two vectors. Then, u *dominates* v if and only if, $u_i \leqslant v_i, \quad i = 1, 2, \ldots, k$, and $u_i < v_i, \quad$ for at least one i. This property is known as *Pareto dominance* and it is used to define the Pareto optimal points. A solution, x, of the MO problem is said to be *Pareto optimal* if and only if, there does not exist another solution y, such that $\mathbf{f}(y)$ dominates $\mathbf{f}(x)$. The set of all Pareto optimal solutions of an MO problem is called *Pareto optimal set* and is denoted as $\mathcal{P}^*$. The set, $\mathcal{PF}^* = \{(f_1(x), f_2(x), \ldots, f_k(x))^\top \mid x \in \mathcal{P}^*\}$, is called *Pareto front*. A Pareto front $\mathcal{PF}^*$ is *convex* if and only if, there exists $w \in \mathcal{PF}^*$, such that, $\lambda\|u\| + (1-\lambda)\|v\| \geqslant \|w\|, \quad \forall\, u, v \in \mathcal{PF}^*, \forall\, \lambda \in (0,1)$. Respectively, it is *concave* if and only if, there exists $w \in \mathcal{PF}^*$, such that, $\lambda\|u\| + (1-\lambda)\|v\| \leqslant \|w\|, \quad \forall\, u, v \in \mathcal{PF}^*, \forall\, \lambda \in (0,1)$. A Pareto front can be convex, concave or partially convex and/or concave and/or discontinuous.

The VEDE Algorithm

For VEDE a number of M subpopulations is considered in a prespecified ring topology. Each population is evaluated using as fitness function, one of the objective functions of the problem at hand. If k is the number of the objective functions, and $k < M$, then the i–th population is evaluated according to the j–th objective function, where,

$$j \equiv \begin{cases} i \bmod k, \text{ if } \ i \neq rk, \quad r = 1, 2, \ldots \\ k, \qquad \text{otherwise} \end{cases} \quad \text{and } i = 1, 2, \ldots, M.$$

In every generation, the best individual, x_g^{best}, of the i–th population, migrates to the $(i+1)$–th population of the ring. Then, the $(i+1)$–th population uses x_g^{best} as the best individual to produce its mutant vectors at generation $(g+1)$. Obviously, only the DE operators that use the best individual in the mutations, i.e. the variants described in Equations (1), (3), and (4), can take full advantage of this information exchange procedure. Moreover, a domination selection procedure, similar to that of Abbass [2], is applied, i.e. instead of using the plain DE selection operator of Equation (7), we use the following one:

$$x_{g+1}^i = \begin{cases} u_{g+1}^i, \text{ if } \mathbf{f}(u_{g+1}^i) \text{ dominates } \mathbf{f}(x_g^i), \\ x_g^i, \quad \text{otherwise,} \end{cases}$$

where $\mathbf{f}$ is the vector function defined above. This selection scheme favors non–dominated individuals in the population and it has proved to perform better in practice. VEDE can be easily parallelized. The populations can be distributed in several machines, with migrations taking place from node to node.

Experiments on Multi-Objective Optimization

Four well–known MO benchmark problems were used in the investigation of VEDE's performance. Each test problem consists of two objective functions of the form

$$\begin{aligned} f_1(x_1) &= x_1, \\ f_2(x_1, x_2, \ldots, x_n) &= g(x_2, x_3, \ldots, x_n) \times h(f_1, g). \end{aligned}$$

Specifically, we considered the following problems [60]:

Test Problem 1. This test problem is defined as:

$$\begin{aligned} f_1(x_1) &= x_1, \\ g(x_2, x_3, \ldots, x_n) &= 1 + \frac{9}{n-1} \sum_{i=2}^{n} x_i, \\ h(f_1, g) &= 1 - \sqrt{\frac{f_1}{g}}, \end{aligned}$$

with $n = 30$ and $x_i \in [0, 1]$. The Pareto front for this problem is convex.

Test Problem 2. This test problem is the non–convex counterpart to Test Problem 1. It is defined as:

$$\begin{aligned} f_1(x_1) &= x_1, \\ g(x_2, x_3, \ldots, x_n) &= 1 + \frac{9}{n-1} \sum_{i=2}^{n} x_i, \\ h(f_1, g) &= 1 - \left(\frac{f_1}{g}\right)^2, \end{aligned}$$

with $n = 30$ and $x_i \in [0, 1]$.

Test Problem 3. This test problem is defined as:

$$\begin{aligned} f_1(x_1) &= x_1, \\ g(x_2, x_3, \ldots, x_n) &= 1 + \frac{9}{n-1} \sum_{i=2}^{n} x_i, \\ h(f_1, g) &= 1 - \sqrt{\frac{f_1}{g}} - \frac{f_1}{g} \sin(10\pi f_1), \end{aligned}$$

with $n = 30$ and $x_i \in [0, 1]$. The Pareto front consists of several convex parts.

Test Problem 4. This test problem is defined as:

$$\begin{aligned} f_1(x) &= x_1, \\ g(x_2, x_3, \ldots, x_n) &= 1 + 10(n-1) + \sum_{i=2}^{n} \left(x_i^2 - 10\cos(4\pi x_i)\right), \\ h(f_1, g) &= 1 - \sqrt{\frac{f_1}{g}}, \end{aligned}$$

and it has 21^9 local Pareto fronts.

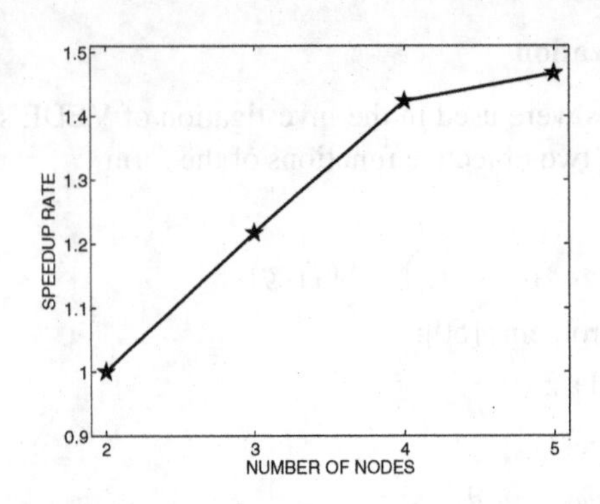

Fig. 6. Speedup gained using up to 5 nodes

All experiments were performed in parallel, using the PVM communication library. For the maintenance of the Pareto optimal set, the archiving technique described in [25], which uses an external archive, was employed. The obtained results were compared to that of VEGA algorithm. For this purpose, two established measures, namely the $\mathcal{C}$ measure [16, 60], and the $\mathcal{V}$ measure [16, 30] were employed. Metric $\mathcal{C}(A, B)$ measures the fraction of members of the Pareto front B that are dominated by members of the Pareto front A, while $\mathcal{V}(A, B)$ is the fraction of the volume of the minimal hypercube containing both fronts, that is strictly dominated by members of A but is not dominated by members of B [16]. Following the analysis presented in [60], a total number of 100 individuals divided in several populations, as well as a maximum of 250 iterations per population per run, were used. We performed 30 experiments for each test problem, using the DE with the mutation operator of Equation (1), because it suits better the migration scheme described in the previous section. This variant is denoted as VEDE1. The results are reported in the boxplots of Figures. 7–10.

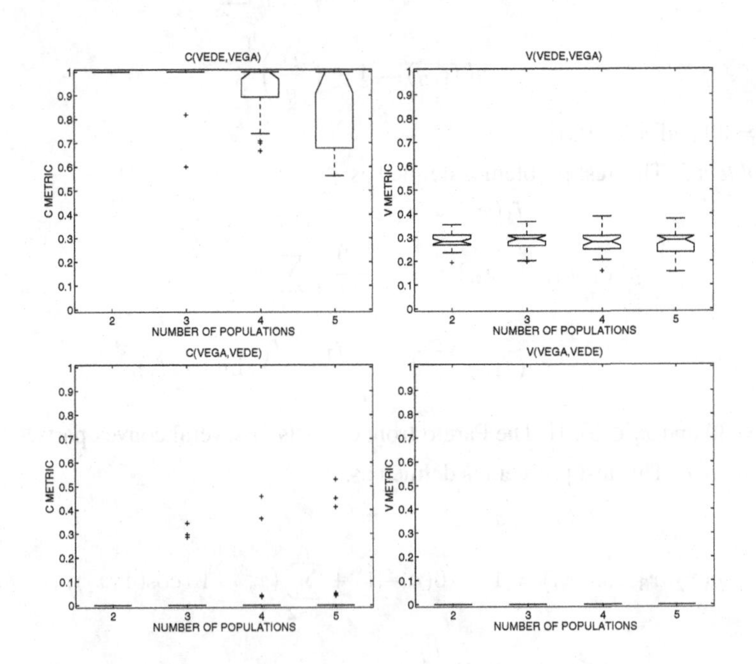

Fig. 7. Results of VEDE1 for the Test Problem 1

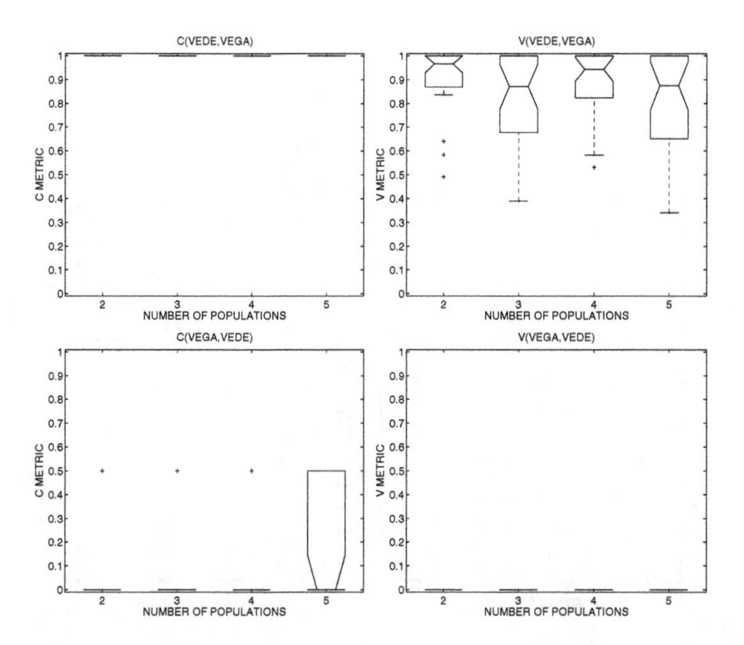

Fig. 8. Results of VEDE1 for the Test Problem 2

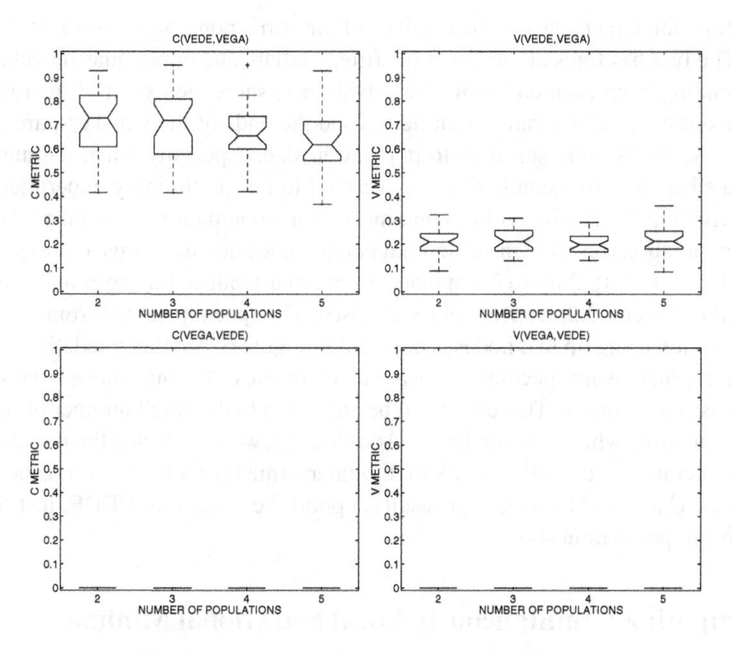

Fig. 9. Results of VEDE1 for the Test Problem 3

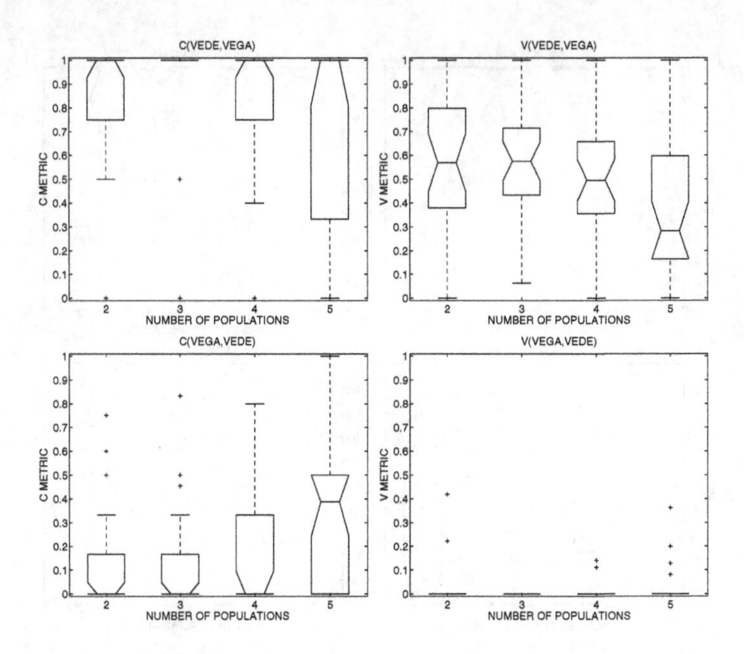

Fig. 10. Results of VEDE1 for the Test Problem 4

Each boxplot depicts the obtained values of the corresponding measure, in 30 experiments. The box has lines at the lower quartile, median, and upper quartile values. The lines extending from each end of the box (whiskers) show the extent of the rest of the data. The outliers, i.e. the values that lie beyond the ends of the whiskers, are denoted with crosses. DE is quite sensitive to population size, especially when the number of individuals becomes too small. This was verified in our preliminary experiments with VEDE. Dividing the 100 individuals into more than 5 populations (less than 20 individuals per population) resulted in substantial performance decline. Thus, our experiments were performed using 2 up to 5 populations. Standard values for the μ and ρ parameters, equal to 0.7 and 0.9, respectively, were used. The speedup gained from the parallel implementation using up to 5 nodes is depicted in Figure 6. As illustrated, there is an almost linearly increasing speedup rate using up to 4 nodes. Beyond 4 nodes, the speedup rate increases marginally. This effect can be attributed to the small number of individuals per population, which falls under 20. Additionally, we have tested the remaining DE mutation operators and in all cases, VEDE outperformed the VEGA with respect to the two metrics, $\mathcal{C}$ and $\mathcal{V}$. However, our results support the claim that VEDE, just like DE, is sensitive to population size.

4 Computing Simultaneously Local and Global Minima

In this Section the recently proposed clustering operator for Evolutionary Algorithms is described [48]. This operator utilizes already computed pieces of information regarding the search space in an attempt to discover regions containing groups of individuals located close to different minimizers. Consequently, the search is confined inside these regions and a large number of global and local minima of the objective function can be efficiently computed [48].

4.1 Exploration vs. Exploitation

The main problem when applying EAs is to find a set of control parameters which optimally balances the exploration and the exploitation capabilities of the algorithm. There is always a trade off between the efficient exploration of the search space and its effective exploitation. For example, if the recombination and mutation rates are too high, much of the space will be explored, but there is a high probability of losing good solutions. In extreme cases the algorithm has difficulty to converge to the global minimum due to insufficient exploitation of the search space. Fortunately, the convergence properties of the DE typically do not heavily depend on its control parameters. However, since not all search operators have the same impact on the exploration of the search space, the choice of the optimal mutation operator can be troublesome.

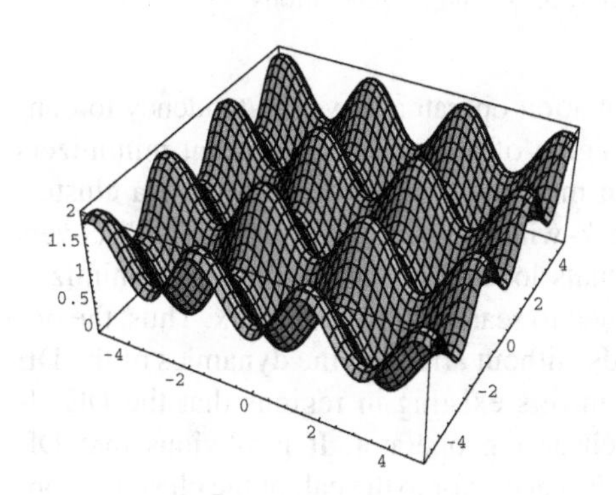

Fig. 11. Plot of $f(x) = \sin(x_1)^2 + \sin(x_2)^2$

To illustrate this we utilize the simple multimodal 2–dimensional function: $f(x_1, x_2) = \sin(x_1)^2 + \sin(x_2)^2$, where $(x_1, x_2) \in \mathbb{R}^2$. This function has an infinite number of global minimizers in $\mathbb{R}^2$, with function value equal to zero, at the points $(\kappa\pi, \lambda\pi)$, where $\kappa, \lambda \in \mathbb{Z}$. In the hypercube $[-5, 5]^2$ the function f has 9 global minimizers. In Figure 11 a surface plot of the function f is exhibited. The six DE variants described above are applied to compute the global minimizers of the objective function f. Experimental results indicate that DE_1 exhibits very fast convergence to one of the global minimizers of f. On the contrary, DE_2 explores a large portion of the search space before converging to a solution. This behavior is illustrated in Figure 12, where (for visualization purposes) a population consisting of 1000 individuals is plotted after 1, 5, 10, 20 generations of DE_2. A closer look at Equations (1) and (2) reveals that DE_1 uses the best individual as a starting point for the computation of the mutant vector, thus constantly pushing the population closer to the location of the best computed point. On the other hand, since DE_2 utilizes three randomly chosen individuals for the computation of the mutant one, its exploration capability is greatly enhanced. However, it exhibits lower convergence speed.

The performance of algorithms DE_3 and DE_4 resembles that of DE_1, due to the use of the best individual. However, DE_3 and DE_4 exhibited better exploration than DE_1, since they also incorporate randomly selected individuals. Algorithms DE_5 and DE_6 use only randomly selected individuals resulting in maximum exploration and the individuals of their populations are simultaneously attracted by more than one minimizers.

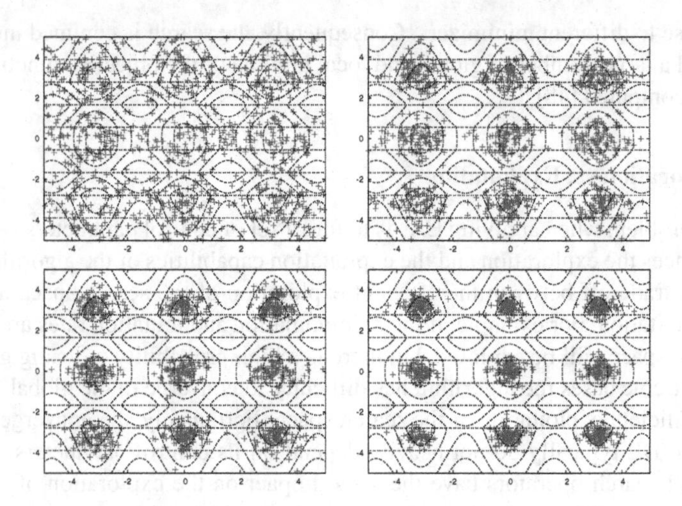

Fig. 12. DE_2 population after 1, 5, 10, and 20 generations

Experimental results show that some mutation operators have the tendency to concentrate subsets of the population in the region of attraction of different minimizers of the objective function. This observation motivated the incorporation of a clustering algorithm to identify such subsets. The k–windows clustering algorithm discovers boxes capturing subpopulations of individuals located in the region of a minimizer. Consequently, the subpopulations are confined to search within each box. Thus, the optimization of the objective function proceeds without affecting the dynamics of the DE algorithm. This process gives all the minimizers existing in regions that the DE algorithm explored before the use of the clustering operator. It is obvious that DE algorithm must adequately explore the search space prior to the call of the clustering operator. In the next section, for completeness purposes, we give a brief description of the k–windows clustering algorithm.

4.2 The Unsupervised k–Windows Clustering Algorithm

The recently proposed k–windows clustering algorithm [50] uses a windowing technique to discover the clusters present in an n–dimensional dataset. More specifically, assuming that the dataset lies in n dimensions, the algorithm initializes a number of n–dimensional windows (boxes) over the dataset. Subsequently, it iteratively perturbs these windows using the movement and enlargement procedures, in order to capture within each window patterns that belong to a single cluster.

The movement and enlargement procedures are guided by the points that lie within each window. As soon as the movement and enlargement procedures do not significantly increase the number of points within each window they terminate. The final set of windows defines the clustering result of the algorithm.

A fundamental issue in cluster analysis, independent of the particular clustering technique applied, is the determination of the number of clusters present in a dataset. The unsupervised k–windows algorithm is capable to determine the number of clusters through a generalization of the original algorithm [50]. Finally, it must be noted that no objective function evaluations are necessary during the operation of the k–windows clustering algorithm [51].

4.3 The Proposed Clustering Operator

In this section, the clustering operator is described. This operator utilizes the unsupervised k–windows algorithm and is called only once, after a user–defined number of generations. In practice, a small number of generations is sufficient for the DE algorithm to explore the search space. Afterwards, the clusters of individuals are determined and subpopulations are confined within each region. Each subpopulation has NP/β individuals, where β is the number of clusters found. If a region contains more individuals, the clustering operator selects the best NP/β. On the other hand, if less individuals exist the clustering operator initializes new ones. The result of the algorithm is the location of many minimizers in a single run, including the global one.

To better utilize the described approach it is advisable to start the DE algorithm using a mutation operator that permits adequate exploration of the search space (for example DE_2 or DE_6). Once the clusters around the minima have been determined, one can switch to a mutation operator that has faster convergence speed (for example DE_1).

For very hard optimization problems, when the objective function is defined in many dimensions and possesses multitudes of local and global minima, the clustering operator could be called more than once. The same might be true for real–life optimization tasks, where the function value of the global minimum is unknown.

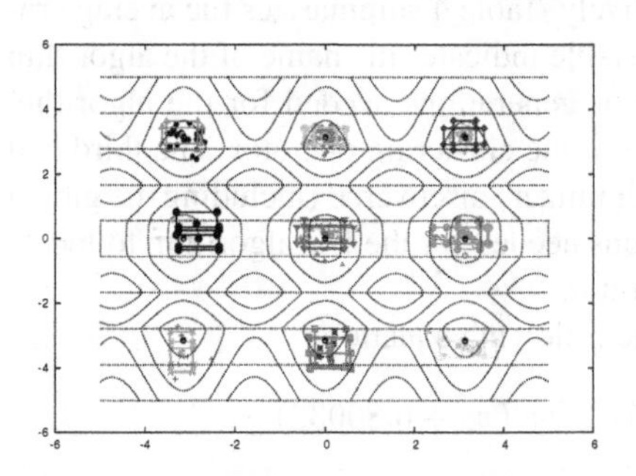

Fig. 13. Clusters and global minimizers

Each consecutive call of the clustering operator will result in more promising subregions of the original search space and will save unneeded objective function evaluations, since the subpopulations will stay focused on regions containing desirable minimizers. For all the experiments reported here, one call of the clustering operator was sufficient for the algorithm to locate the global, as well as, many local minimizers. To determine the applicability and the efficiency of the clustering operator we incorporated it to the DE algorithm and applied the new method to the multimodal test function f, which posses 9 global minimizers in the hypercube $[-5,5]^2$. The result of the application of the clustering operator on the function f is illustrated in Figure 13.

It must be noted that a number of independent experiments of the original DE algorithm gives no guarantee that all global minimizers will be detected, since the algorithm has no memory; no information concerning previously detected minimizers is kept. We performed 100 independent simulations, using each one of the six different mutation operators described above and Table 3 exhibits the average number of restarts needed

Table 3. Restarts needed to locate all the minimizers of f

	Original DE Algorithm			DE with k–win		
	Min	Mean	Max	Min	Mean	Max
DE_1	54	110.2	203	1	17.5	66
DE_2	57	119.1	294	1	1.0	1
DE_3	60	133.0	239	1	1.1	7
DE_4	52	114.9	212	1	1.0	1
DE_5	49	106.7	245	1	1.0	1
DE_6	62	111.4	221	1	1.0	1

for the DE algorithms to locate all the global minimizers of f. The modified algorithm that uses the clustering operator in most cases managed to find all the minimizers of f in a single execution.

4.4 Experimental Results on Multi-minima Discovery

We implemented and tested the clustering operator on a number of hard optimization tasks and it exhibited stable and robust performance. We report results from the Levy No. 5 test function. For each mutation operator we performed 100 independent experiments. A population consisting of 200 individuals was used and the mutation and recombination constants had values $\mu = 0.6$ and $\rho = 0.8$, respectively. The algorithm was terminated when the global minimum was located. The clustering operator was called only once for the optimization of each of the four test function considered below, after 20, 20, 10, and 200 generations, respectively. Table 4 summarizes the average results for the 100 runs. The first column of the table indicates the name of the algorithm and the second column the average number of generations needed for the algorithm to locate the global minimum without the use of the clustering operator. The third and fourth columns give the average number of minimizers discovered (including the global one) and the corresponding average generations needed for the DE algorithm to locate the global minimum using the clustering operator.

The *Levy No. 5* test function is given by the following equation:

$$f_1(\mathbf{x}) = \sigma_1\sigma_2 + (x_1 + 1.42513)^2 + (x_2 + 0.80032)^2,$$

where $x_i \in [-10, 10]$, $i = 1, 2$, and σ_1 and σ_2 are given by:

$$\sigma_1 = \sum_{i=1}^{5} i\cos[(i+1)x_1 + i], \text{ and } \sigma_2 = \sum_{j=1}^{5} j\cos[(j+1)x_2 + j].$$

There exist about 760 local minima and one global minimum with function value $f_1^* = -176.1375$ located at $\mathbf{x}^* = (-1.3068, -1.4248)$. The large number of local optimizers makes extremely difficult for any method to locate the global minimizer.

The experimental results exhibited in Table 4 indicate that generally the use of the clustering operator enhances the performance of the DE algorithms. In detail, there is an average acceleration of the algorithm's convergence speed ranging from 30% to 80%.

Table 4. Average results for the Levy function

	w/o k-win operator	with k-win operator	
	Generations	Minima located	Generations
DE_1	33.21	5.97	34.36
DE_2	70.66	20.52	54.26
DE_3	64.09	11.96	39.77
DE_4	65.11	20.22	50.25
DE_5	133.01	22.70	70.85
DE_6	64.89	19.20	50.24

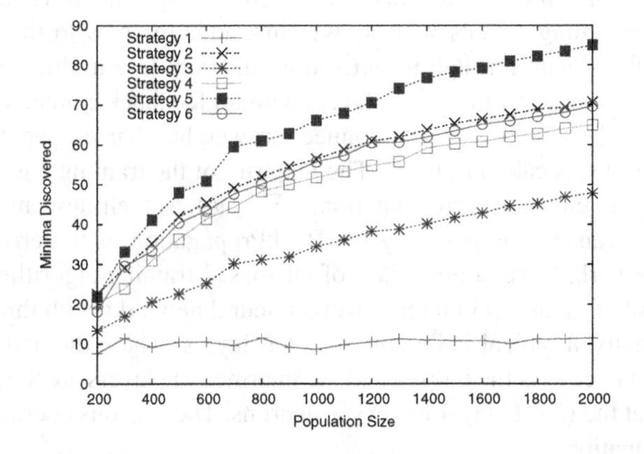

Fig. 14. The effect of the population size on the number of located minimizers

Additionally, as many as 20 minimizers (including the global one) were simultaneously computed. The only exception is DE_1 where a slight increase in the generations is observed (3%), but the modified algorithm locates the global as well as 5 local minimizers.

To better demonstrate the ability of this approach to locate many minima at once, independent runs were conducted with the number of individuals in the population gradually increasing from 200 to 2000. In general, a larger population explores better the search space and more regions containing minimizers are located by the k–windows operator. In Figure 14 we exhibit the detailed results. The algorithm DE_1 locates on average 10 minimizers regardless of the size of its populations. On the contrary, the rest of the algorithms locate more minimizers as their population is increased. DE_5 exhibited the best performance finding simultaneously up to 85 minimizers.

The above experiments show that this approach greatly accelerates the convergence speed of the DE algorithms and that in addition to the global minimum is capable to locate simultaneously many local minima without extra function evaluations. To this end, the use of the clustering operator is always suggested. In brief, the clustering operator has the following advantages:

1. locates global minimizers and local minimizers with relatively low function value,
2. in general, fewer generations are required for the DE algorithm to converge,
3. there is no need for additional function evaluations,

4. utilizes the range search algorithm for fast and reliable execution,
5. its parallel implementation is straightforward,
6. is better suited to difficult high–dimensional multimodal objective functions.

5 Neural Network Training Using DE

Artificial Feedforward Neural Networks (FNNs) have been widely used in many application areas in recent years and have shown their strength in solving hard problems in Artificial Intelligence. Although many different models of neural networks have been proposed, multilayered FNNs are the most common. FNNs consist of many interconnected identical simple processing units, called neurons. Each neuron calculates the dot product of the incoming signals with its weights, adds the bias to the resultant, and passes the calculated sum through its activation function. In a multilayer feedforward network the neurons are organized into layers with no feedback connections [23].

The incremental adaptation of the connection weights that propagate information between the neurons, is called training. The majority of the training algorithms use the negative of the gradient of the error function, $-\nabla E(w)$, as their descent direction. The gradient $\nabla E(w)$ can be computed by the BackPropagation of the error through the layers of the network. Here, a new class of DE-based training algorithms that do *not* need the gradient of E and train integer weight neural networks with threshold units is discussed. Formally, a typical FNN consists of L layers, where the first layer denotes the input, the last one, L, is the output, and the intermediate layers are the hidden layers. It is assumed that the $(l-1)$ layer has N_{l-1} neurons. The neurons operate according to the following equations

$$net_j^l = \sum_{i=1}^{N_{l-1}} w_{ij}^{l-1,l} y_i^{l-1} + \theta_j^l, \qquad y_j^l = f^l\left(net_j^l\right),$$

where $w_{ij}^{l-1,l}$ is the integer connection weight from the i-th neuron at the $(l-1)$ layer to the j-th neuron at the l-th layer, y_i^l is the output of the i-th neuron belonging to the l-th layer, θ_j^l denotes the integer bias of the j-th neuron at the l-th layer, and f is the activation function. The weights in the FNN can be expressed in vector notation. Let the weight vector have the form: $w = (w_1, w_2, \ldots, w_n)$. The weight vector, in general, defines a point in the n–dimensional real Euclidean space $\mathbb{R}^n$, where n denotes the total number of weights and biases in the network. From the optimization point of view, supervised training of an FNN is equivalent to minimizing the corresponding error function, which is a multivariate function that depends on the weights in the network. The square error over the set of input–desired output patterns with respect to every weight, is usually taken as the function to be minimized. Specifically, the error function for an input pattern t is defined as, $e_j(t) = y_j^L(t) - d_j(t)$, $j = 1, 2, \ldots, N_L$, where $d_j(t)$ is the desired response of an output neuron at the input pattern t. For a fixed, finite set of input–desired output patterns, the square error over the training set which contains T representative pairs is:

$$E(w) = \sum_{t=1}^{T} E_t(w) = \sum_{t=1}^{T} \sum_{j=1}^{N_L} e_j^2(t),$$

where $E_t(w)$ is the sum of the squares of errors associated with the pattern t. Minimization of E is attempted by using a training algorithm to update the weights.

5.1 Training Integer Weight Neural Networks with Threshold Activations

FNNs can be simulated in software, but to be utilized in real life applications, where high speed of execution is required, hardware implementation is needed. The natural implementation of an FNN – because of its modularity – is a parallel one. Hardware–friendly algorithms are essential to ensure the functionality and cost effectiveness of the hardware implementation. Moreover, the need for hardware–friendly algorithms, which have the ability to cope with time–varying problems and real–time timing constraints, has been recently increased [35].

FNNs having integer weights and biases are easier and less expensive to implement in electronics as well as in optics and the storage of the integer weights is much easier to be achieved. Additionally, the use of threshold activation functions for all the hidden and output neurons, greatly reduces the complexity of the hardware implementation, because there is no need to design and implement complicated non–linear activation functions. Another advantage of the FNNs with integer weights and threshold activation functions is that the trained neural network is to some extend immune to noise in the training data. Such networks only capture the main feature of the training data. Low amplitude noise that possibly contaminates the training set cannot perturb the discrete weights, because those networks require relatively large variations to "jump" from one integer weight value to another.

Mathematical operations that are easy to implement in software might often be very burdensome in the hardware and therefore more costly. This property reduces the amount of memory required for weight storage in digital electronic implementations. Additionally, it simplifies the digital multiplication operation. Finally, if inputs are restricted to the set $\{-1, 1\}$ (bipolar inputs), the neurons in the first hidden layer require only sign changes during multiplication operations, and only integer additions.

To apply DE to neural network training with integer weights, we initialize the individuals with N–dimensional integer weight vectors, following a uniform probability distribution, and evolve them over time. The only problem is that the mutation operator results in real weight vectors. As our aim is to maintain an integer weight population at each generation, each component of the mutant weight vector is rounded to the nearest integer.

It must be noted that the evolutionary class of algorithms does not need the activation function to be differentiable and is suitable for training with threshold units [34]. In the first phase of our approach, the DE algorithms are used to train a neural network "off–line", using sigmoid activation functions, such as:

$$f_1(x) = \tanh(\lambda x) \equiv \frac{2}{1 + e^{-\lambda x}} - 1, \qquad f_2(x) = \frac{1}{1 + e^{-\lambda x}},$$

where λ is the gain parameter. This seems to be a good practice since the network is trained much faster with sigmoid functions. In the second phase we alter the gain of the sigmoid function in such a way that allows a mapping to a threshold unit network.

Specifically, when the inputs are correctly classified and the network error is relatively small, the value of λ is increased in the sequence $(1, 10, 20, 30, 40, 50, \infty)$. Additional training might be necessary after each increase of λ. That justifies the additional iterations needed to train an FNN, using only threshold activation functions. This procedure is analogous to taking the limit of the sigmoid function as the gain parameter λ goes to infinity. Finally, the trained network uses only threshold activation functions and thus the complexity of the hardware implementation is greatly reduced. If new input data are introduced, training *can* be continued sequentially or in parallel "on–chip".

5.2 Experiments on Neural Network Training

Next, we exhibit results of the DE-based algorithms on the Encoder/Decoder FNN training problem. For all the simulations bipolar input and output vectors have been used. Table 5 summarizes the performance of the DE algorithms using different mutation rules when sigmoid activation functions are used. Hyperbolic tangent activation functions in both the hidden and output layer neurons have been used. In Table 6 we exhibit the performance of the DE algorithms, when the training has been performed as described above in order to lead to a network that uses only threshold activation functions.

We must note here that a key feature of the DE algorithms is that *only* error function values are needed. No gradient information is required, so there is no need of backward passes. For the test problems considered, we made no effort to tune the mutation, recombination and migration constants, μ, ρ and φ respectively, to obtain optimal or at least nearly optimal convergence speed. Default fixed values ($\mu = 0.5$, $\rho = 0.7$ and $\varphi = 0.3$) have been used instead. Smaller values of φ can further reduce the messages between the processors, but may result in rare and inefficient migrations. It is obvious that one can try to fine–tune the μ, ρ, φ and *NP* parameters to achieve better results, i.e. less error function evaluations and/or exhibit higher success rates. The weight subpopulations have been initialized with random integers from the interval $[-3, 3]$ and the total population size $3NP$ has been divided equally to 3 subpopulations, each having *NP* individuals. Regarding the total population size, experimental results have shown that a good choice is $2n \leqslant 3NP \leqslant 4n$, where n denotes the dimensionality of the error function, i.e. the total number of weights and biases. It is obvious that the exploitation of the weight space is more effective for large values of *NP*, but sometimes more error function evaluations are required. On the other hand, small values of *NP* render the algorithm inefficient and more generations are required to converge.

4–2–4 Encoder/Decoder

Here, we consider the 4–2–4 encoder/decoder (sixteen weights and six biases, dimension of the problem $n = 22$). The network is presented with 4 distinct input patterns, each having only one bit turned on. The task is to duplicate the input pattern in the output units. Since all information must flow through the hidden units, the network must develop a unique encoding for each of the 4 patterns in the 2 hidden units and a set of connection weights performing the encoding and decoding operations. This particular encoding is considered to be "tight", since the number of the hidden nodes equals the

Table 5. Results for the encoder/decoder problem using sigmoid activation functions

Algorithm	min	mean	max	s.d.	Success
DE_1	330	1614.8	4686	868.5	100%
DE_2	3960	8160.6	13376	2160.5	100%
DE_3	308	1428.2	4004	660.9	100%
DE_4	660	4540.5	8514	1505.4	100%
DE_5	7260	13110.9	20636	3092.4	100%

Table 6. Results for the encoder/decoder problem using threshold activation functions

Algorithm	min	mean	max	s.d.	Success
DE_1	990	2520.8	23260	2326.4	100%
DE_2	4796	8724.9	16588	2264.6	100%
DE_3	1034	2104.5	4664	680.0	100%
DE_4	1870	4778.1	9724	1278.0	100%
DE_5	6072	14070.3	20746	2795.4	100%

base 2 logarithm of the input nodes ($\log_2 4 = 2$). This problem has been selected because it is quite close to real world pattern classification tasks, where small changes in the input pattern cause small changes in the output pattern. The size of each subpopulation was $NP = 20$. The low and high bound of the age of each individual, were $\alpha = 50$ and $\beta = 200$ respectively. A typical 3–bit weight vector is $w = (0, 2, -2, 3, -3, -3, 2, 3, -3, -3, 2, -3, -2, 2, 3, 2, 1, 0, -3, -3, -2, -2)$ and the corresponding value of the error function is $E = 0.0459$. Simulation results are exhibited in Tables 5 and 6.

6 Data Mining Using DE

Although originally designed for global optimization the DE algorithm is versatile enough to be applied in a variety of scientific tasks. For example here we present an application of DE for data clustering. Evolutionary clustering is a recent trend in cluster analysis, that has the potential to yield high partitioning accuracy results. Traditional evolutionary techniques applied in clustering are typically hindered by the high cost involved in the computation of the objective function. In this section we demonstrate how DE can be employed to evolve cluster solutions. Furthermore we present how recent advances in clustering can be employed to estimate the number of clusters from this evolutionary technique. Finally, by employing real world datasets, we exhibit the high quality clustering results that this scheme can provide.

6.1 Data Clustering

Clustering is a fundamental step in the process of transforming data to knowledge. It aims at discovering groups (clusters) in a set of objects such that similarity among the objects in the same group is higher than that of objects belonging to different clusters.

The application domain of clustering techniques is very wide including data mining, text mining, statistical data analysis, compression and vector quantization, global optimization and web personalization [7, 15, 38, 52].

Clustering algorithms are traditionally categorized into three main categories, Hierarchical, Partitioning [44] and Distance-based. Hierarchical clustering algorithms construct hierarchies of clusters in a top-down (agglomerative) or bottom-up (divisive) fashion. Hierarchical clustering algorithms have proved to yield high quality results especially for applications involving clustering text collections. Nonetheless, their high computational requirements, usually limits their applicability in real life applications, where the number of samples and their dimensionality is typically high (the cost is quadratic to the number of samples).

Partitioning clustering algorithms, start from an initial clustering (that can be randomly formed) and create partitionings by iteratively adjusting the clusters based on the distance of the data points from a representative member of the cluster. The most commonly used partitioning clustering algorithm is k-means. This algorithm initializes k centers and iteratively assigns each data point to the cluster whose centroid minimizes the Euclidean distance from the data point. Algorithms of this type can give good clustering results at low cost, since their running time is proportional to kN, where N is the number of patterns present in the dataset. However, their results rely heavily on their initialization and they can converge to arbitrary local optima.

Distance based clustering algorithms create a partitioning by considering neighbors of data points. DBSCAN [41] is a distance-based clustering algorithm that has proved quite effective for spatial databases. Clusters are considered as high density neighborhoods of data points. Although the density parameter is critical for the successful application of DBSCAN, recently proposed heuristics appear to yield high quality results. The computational complexity of DBSCAN comes up to $O(N \log(N))$ under the assumption that the data are organized in a spatial index (R^*-tree).

In evolutionary clustering, a solution to the clustering problem is typically encoded as a chromosome. By employing evolutionary operators and a population of solutions the algorithm probes the search space to find a globally optimum partition of the data. In early approaches [9, 26], chromosomes encoded the partition of n objects into K clusters and Genetic Algorithms were employed to identify the best partition. However, the sensitivity of GAs to the selection of the various parameters like population size, and crossover and mutation probabilities, as well as, the difficulties associated with the representation scheme, presented a major problem. Better results were obtained through hybrid approaches [5].

However, it is possible to represent the clustering procedure as an optimization problem of locating the optimal centroids of clusters. Thus, all evolutionary techniques can be employed since a clustering solution can be represented as a real-valued vector of the centroids. Previous approaches employed Evolutionary Strategies [6], Evolutionary Programming [18], and recently Particle Swarm Optimization [53]. All these approaches demonstrated that it is possible to obtain high quality partitions, but at a high computational cost. Here we attempt to tackle the high computational cost of traditional evolutionary techniques by introducing a new fitness criterion. This criterion is based on a windowing technique already employed in other clustering algorithms [51, 55].

A critical and open issue in cluster analysis, is the determination of the number of clusters present in a dataset. The evolutionary clustering techniques proposed so far, with the exception of [22], require from the user to specify the number of clusters present in the data prior to the execution of the algorithm. The described approach can provide an approximation to the number of clusters present in a dataset.

6.2 Designing an Efficient Clustering Fitness Criterion

Let the data set comprise a set $X = \{x_1, x_2, \ldots, x_n\}$, where x_j is a data vector in the d dimensional Euclidean space $\mathbb{R}^d$. A k clustering of X is a partition C of X into k disjoint groups C_i, for $i = 1, 2, \ldots, k$. The clustering problem amounts to the determination of a partition of X which is optimal with respect to a function f that quantifies the goodness of the partition.

Different statistical functions have been proposed for f [31, 58]. But in most approaches at least a full scan over the dataset is necessary to compute the function value for a specific instance. Evolving a population using such a fitness criterion can be expensive in terms of computational cost, compared to k-means like approaches that typically do not require more than 10 to 20 scans of the dataset. However an efficient clustering fitness criterion can be constructed by utilizing computational geometry techniques. In detail, let us define structure in the form of axis parallel hyper-rectangles (d-ranges).

Definition 1: Let a d-range of size $a \in \mathbb{R}$ and center $z \in \mathbb{R}^d$ be the orthogonal range $[z_1 - a, z_1 + a] \times \cdots \times [z_d - a, z_d + a]$. Assume further, that the set $S_{a,z}$, with respect to the set X, is defined as $S_{a,z} = \{y \in X : z_i - a \leqslant y_i \leqslant z_i + a, \quad \forall i = 1, 2, \ldots, d\}$. Then the *Window Density Function* WDF for the set X, with respect to a given size $a \in \mathbb{R}$, is defined as: $\text{WDF}_a(z) = |S_{a,z}|$, where $|\cdot|$ denotes the cardinality of the set, i.e. a measure of the number of elements of the set.

In other words, WDF represents the number of points from the dataset X, that reside in a window of size a centered at z. WDF is a meaningful clustering objective function, since as the center of a d-range, z, moves to the center of the cluster the number of points around it should increase. As it is obvious the size a, is critical to the procedure as it

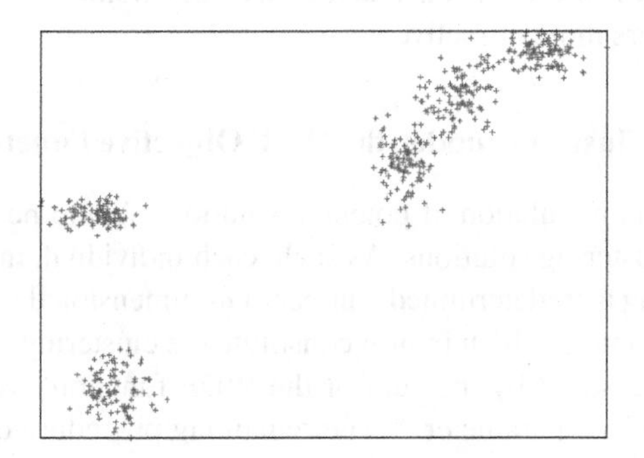

Fig. 15. Dataset $DSet_1$

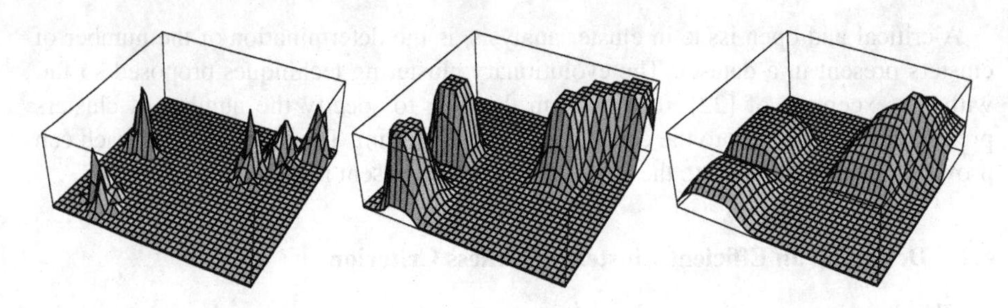

Fig. 16. 3D-Plot of WDF for $Dset_1$ and $a = 1, 5, 10$ (left to right)

determines the location of the minimizers of the objective function. To illustrate this we employ the dataset $Dset_1$, exhibited in Figure 15. This dataset contains 500 points organized in 5 clusters with 100 points each. Each cluster is constructed by sampling 100 points from a two dimensional Gaussian distribution. The mean of each distribution was randomly scattered in the range $[0, 200]^2$, and the covariance matrices were randomly generated by obtaining for each element of the matrix a random number between 1 and 2.

In Figure 16, the 3D plots of WDF are provided to visualize the impact of the parameter a. As the value of a increases, the extreme points of WDF tend to merge. When $a = 1$ there are five maxima, equal to the number of clusters. On the other hand, when $a = 10$, the three maxima corresponding to the three closest clusters previously identified merge to a single one.

The most important feature of the density function is that it is not necessary to scan the entire dataset to obtain a fitness value for a specific object. In particular, the computation of WDF is the well studied Computational Geometry *Orthogonal Range Search Problem*. Numerous Computational Geometry techniques have been proposed to address this problem. All these techniques employ a preprocessing stage at which they construct a data structure storing the patterns. This data structure allows them to answer range queries fast. For applications of very high dimensionality, data structures like the Multidimensional Binary Tree [37], and Bentley and Maurer [8] seem more suitable. On the other hand, for low dimensional data with a large number of points the approach of Alevizos [3] appears more attractive.

6.3 Evolutionary Clustering under the WDF Objective Function

In the DE settings, the population of potential solutions should be properly defined to represent nominal clustering solutions. As such, each individual, in the clustering context, is expressed using a predetermined number of d–dimensional vectors that represent the centers of the d–ranges, which in turn constitute the clustering result. The fitness of each individual is measured by the sum of the WDF function over all the d–ranges, under a fixed value of the parameter a. The remaining procedure of the DE algorithm remains unchanged.

As it is obvious an evolutionary optimization procedure using the above described characteristics aims at discovering the set of d-ranges that include as many points from

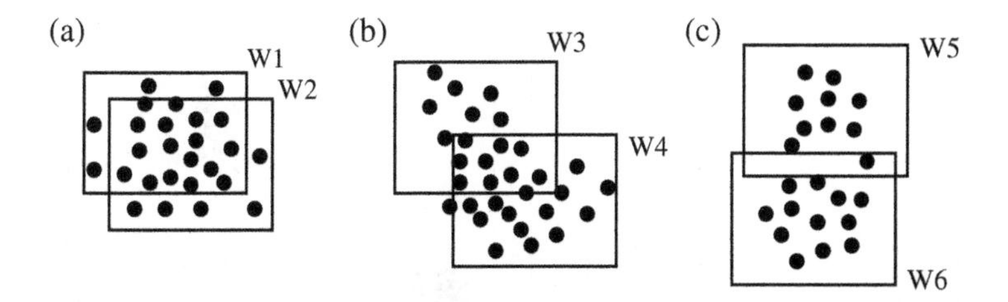

Fig. 17. (a) $W1$ and $W2$ satisfy the similarity condition and $W1$ is deleted. (b) $W3$ and $W4$ satisfy the merge operation and are considered to belong to the same cluster. (c) $W5$ and $W6$ capture two different clusters.

the dataset as possible. Thus a single execution is able to determine a clustering result. Note that in the final clustering solution empty d-ranges may appear, or even d-ranges that overlap may exist. By employing the *merge* operation of the unsupervised k-windows clustering algorithm [51], the number of clusters can be approximated. During this step, for each pair of overlapping windows, the number of patterns that lie in their intersection is computed. With respect to the proportion of this number to the total number of points contained in each window, the algorithm can decide whether to either:

(a) Ignore one window if the proportion is very high.
(b) Consider the windows to contain parts of the same cluster if the proportion is relatively high.
(c) Consider the windows to capture different clusters, if the proportion is low.

An example of this operation is exhibited in Figure 17 and a high level description of the described algorithmic scheme follows:

DEEC algorithm
Step 1. Construct a data structure for the storing of the data.
Step 2. Set the parameter a of WDF function.
Step 3. Repeat
Step 4. Execute the DE algorithm.
Step 5. Until a sufficient part of the dataset is covered
or a maximum number or iterations is performed.
Step 6. Merge the resulting d–ranges
Step 7. Report the final clusters.

6.4 Evolutionary Clustering Results

To demonstrate the applicability of this approach we firstly employ $Dset_1$, exhibited in Figure 15, which is two dimensional and allows the visual inspection of the results. Note that in all the experiments reported in this section the population size was set to 20 individuals, and a maximum of 200 epochs was allowed. The DE parameters μ and p were set to 0.6, and 0.8, respectively, in all experiments. Moreover, if the d–ranges of the best individual discovered contain more than 90% of the total points the execution

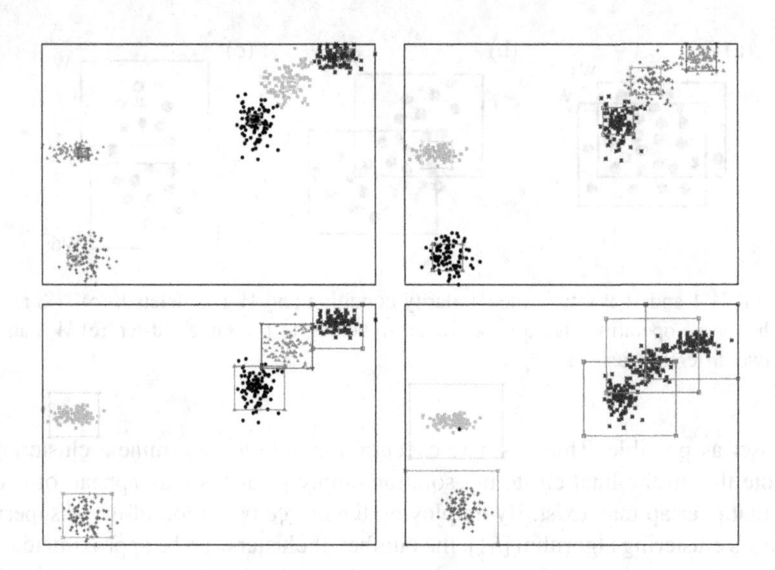

Fig. 18. The clustering result of DEUC for $a = 1, 3, 5, 10$

of DE terminates. The application of the DEUC algorithm over the $Dset_1$ dataset with the parameter a obtaining values $1, 2, 5, 10$ is exhibited in Figure 18. These results were obtained, by stopping the iterative executions of DE when more than 90% of the dataset was covered. Each individual encoded the center of five d-ranges. Comparing the clustering result, with the 3D plots of the WDF function in Figure 16, it is obvious that DEUC is able to detect the extrema of WDF and form a clustering result that is in accordance with the form of WDF. The colors in the plots correspond to the different cluster labels of the points that were assigned to the closest d-range under the Euclidean metric. It is obvious that DEUC is able to provide visually optimal clustering results when a ranges between 1 and 5. On the other hand, when a is too large the adjacent clusters are merged to a single cluster by the merging procedure.

Comparing the results of DEUC involves the usage of a clustering algorithm that can approximate the number of clusters. To compare the results of DEUC with other approaches we employ the DBSCAN clustering algorithm [41]. This choice is motivated by the fact that DBSCAN computes the number of points ($MinPts$) that reside in a hypersphere of size *Eps*. Thus, the *Eps* parameter of DBSCAN is strongly related to the a parameter of the WDF function. The execution of DBSCAN on $Dset_1$, setting $MinPts = 5$, (anything with less than 5 points in an *Eps* neighborhood around it is considered noise), and for *Eps* obtaining the values $Eps = 1, 3, 5, 10$ is exhibited in Figure 19. Similarly in this case the colors designate different cluster labels, and the red crosses represent points recognized as noise. From the plots we can see that DBSCAN is more sensitive to the value of *Eps* than DEUC is on the value of a. Moreover, for DBSCAN to be able to recognize the three different adjacent clusters a very careful selection of *Eps* and $MinPts$ is needed.

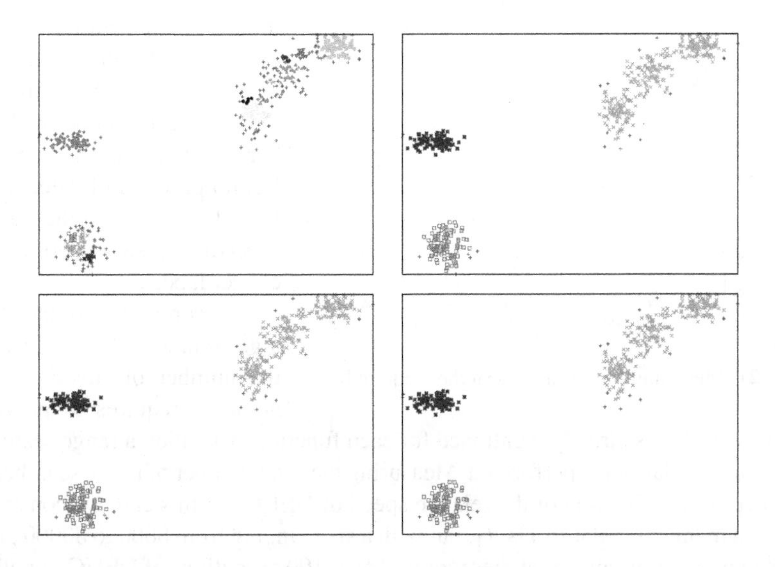

Fig. 19. The clustering result of DBSCAN for $Eps = 1, 3, 5, 10$

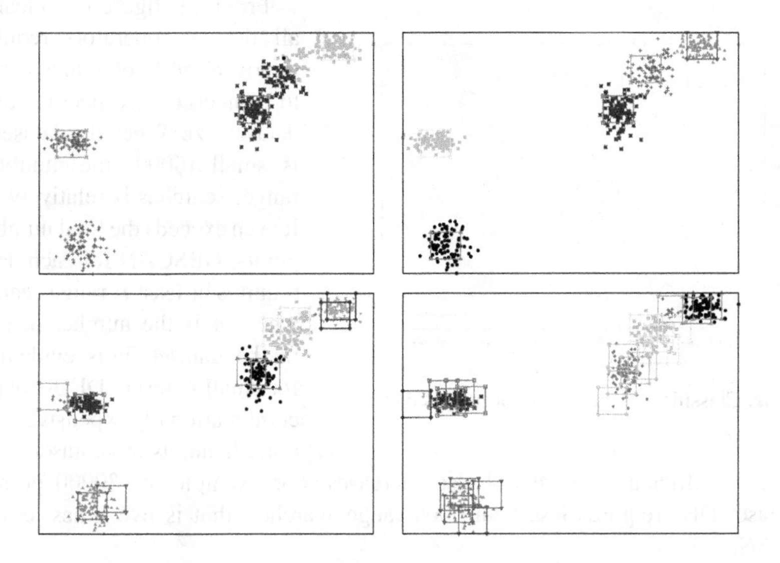

Fig. 20. The impact on the clustering result of different number of d-ranges (3,5,10 and 15) when $a = 3$

Next, in Figure 20, we investigate the ability of DEUC to approximate the number of clusters. To this end we apply DEUC using 3, 5, 10 and 15 windows. As illustrated, when the number of d-ranges is less than the true number of clusters, each d-range is located over a minimum of WDF, but due to the inability to cover all the

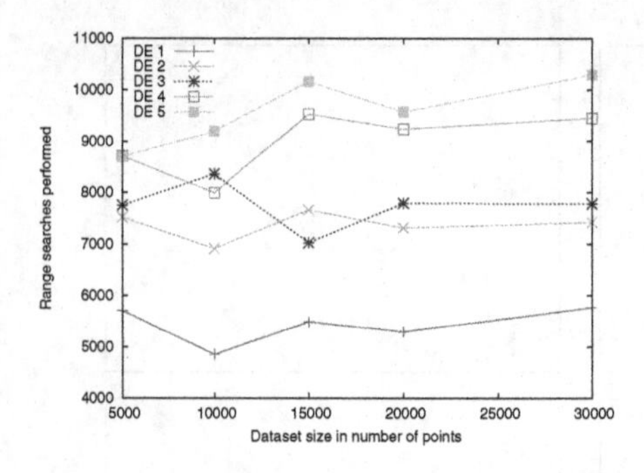

Fig. 21. Mean number of range searches required

minimizers the cluster labels are incorrect. On the other hand, as the number of d-ranges becomes larger than the real number of clusters, the algorithm has no problem of detecting the five clusters, since the merging procedure assigns correctly the cluster labels.

The complexity of the DEUC algorithm can be analyzed by the number of function evaluations it requires to provide a clustering result. As already mentioned for each function evaluation a range search operation over the dataset is performed. Measuring the total number of range searches that are needed is an indication of the relative speed of DEUC. To this end, we constructed $Dset_2$ in a manner similar to $Dset_1$, but with a size ranging from 5000 to 30000 points. The mean number of range searches required over 100 executions of DEUC, for all mutation operators is depicted in Figure 21.

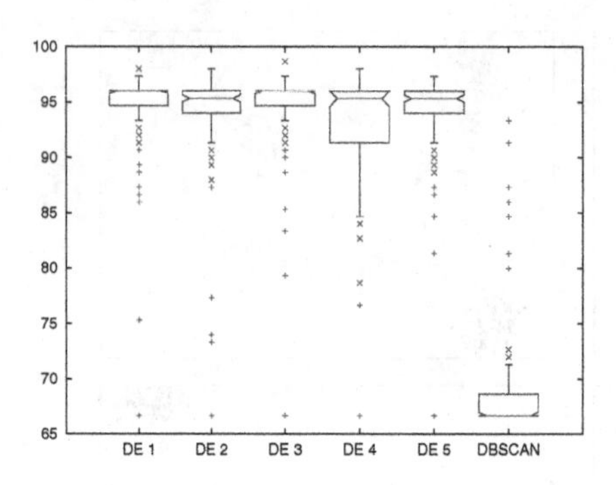

Fig. 22. Classification accuracy for the Iris dataset

From this figure it is clear that all the DE operators require a steady number of range searches to converge, irrespective of the dataset size. When the dataset size is small (5000) the number of ranges searches is relatively high. It even exceeds the total number of points. DBSCAN for each dataset requires at least n range searches, where n is the number of points in the dataset. It is evident that for small datasets DEUC appears computationally expensive. On the other hand, as the dataset size increases, the efficiency of DEUC also increases. For example for 30000 points in the dataset DE_1 requires less than 6000 range searches, that is five times less than DBSCAN.

To demonstrate the quality of the partitioning results we employ the four dimensional Iris dataset $Dset_{iris}$ from the UCI Machine Learning Repository [10]. This dataset is among the best known databases to be found in the pattern recognition literature. It contains 150 records of four features. The features are measurements of the sepal and petal length and width of three different types of the iris plant (Setosa, Versicolour and Virginica). The 150 records are equally distributed in three classes, each corresponding to a different type of the plant. To evaluate the clustering result we resolve to the correspondence they have to the true cluster labels of the patterns.

Ideally, each cluster should contain patterns that belong to only one type of the Iris plant. After normalizing the data in the $[10, 100]^4$ range, DEUC was executed 100 times, using a population of 20 individuals while each individual encoded 5 d–ranges. In most cases 3 clusters were recognized by the algorithm, but there were also cases that resulted in 4 and 5 clusters. Moreover, as a comparison measure we executed DBSCAN using all the combinations of values in $[1, 10]$ with a step of 1, for the Eps and $MinPts$ parameters, yielding 100 different clustering results. In the box-plots exhibited in Figure 22, we summarize the results with respect to the partitioning accuracy. As it is obvious from Figure 22, all the different DE operators are able to capture the dynamics of the dataset and result in high partitioning accuracy. Among all the operators DE_3 exhibits the most robust behavior and is able to provide the best results even with respect to outliers. On the other hand, DBSCAN is unable to provide highly accurate results since in this dataset two of the classes are somewhat close and DBSCAN tends to merge them to a single cluster, thus destroying its classification accuracy.

In conclusion, it seems that it is possible using clustering criteria as the WDF density function, to design efficient and effective evolutionary clustering techniques. This is achieved by utilizing Computational Geometry data structures. Moreover, such an evolutionary procedure has the ability to approximate the number of clusters and yield high quality partitions as it is evident from the experimental results.

7 Real Life Application: DNA Microarrays

To understand a biological processes that a living cell undergoes, one has to measure the *gene expression levels* in different developmental phases, different body tissues, and different clinical conditions. Compared to the traditional approach to genomic research, which has been to examine and collect data for a single gene locally, DNA microarray technologies have rendered possible the simultaneous monitoring of the expression pattern of thousands of genes. Unfortunately, the original gene expression data are contaminated with noise, missing values and systematic variations due to the experimental procedure. Several methodologies can be employed to alleviate these problems, such as Singular Value Decomposition based methods, weighted k–nearest neighbors, row averages, replication of the experiments to model the noise, and/or normalization, which is the process of identifying and removing systematic sources of variation. Discovering the patterns hidden in the gene expression microarray data and subsequently using them to classify the various conditions is a tremendous opportunity and a challenge for functional genomics and proteomics. A promising approach to address this task is to utilize computational intelligence techniques, such as EAs and Feedforward Neural Networks (FNNs). Unfortunately, employing FNNs (or any other classifier) directly to classify the samples is almost infeasible due to the *curse of dimensionality* (limited number of samples coupled with very high feature dimensionality). One solution is to preprocess the expression matrix using a dimension reduction technique.

Here, we follow a different approach. DE and FNNs are employed to discover subsets of informative genes that accurately characterize all the samples [49]. Generally, the aim is to reduce the initial gene pool from several thousand genes (5,000–10,000 or more) to 50–100. Several gene selection methods based on statistical analysis have been developed to select these predictive genes and perform dimension reduction. Those methods include t-statistics, information gain theory, and principal component analysis (PCA). It is evident that the choice of feature selection is difficult and bears a significant effect on the overall classification accuracy. Typically, accuracy on the training data can be quite high, but not replicated on the testing data.

7.1 Algorithms and Methodology

To classify samples using microarray data, it is necessary to decide which genes, from the ones assayed, should be included in the classifier. Including too few genes and the test data will be incorrectly classified. On the other hand, having too many genes is not desirable either, as many of the genes will be irrelevant, mostly adding noise. This is particularly severe with a noisy data set and few subjects, as is the case with microarray data.

In the literature, both supervised and unsupervised classifiers have been used to build classification models from microarray data. This study addresses the supervised classification task where data samples belong to a known class. EAs are applied to microarray classification to determine the optimal, or near optimal, subset of predictive genes on complex and large spaces of possible gene sets. Although a vast number of gene subsets are evaluated by the EA, selecting the most informative genes is a non trivial task. Common problems include the existence of:

(a) relevant genes that are not included in the final subset, because of the insufficient exploration of the gene pool,
(b) significantly different subsets of genes being the most informative as the evolution progresses, and
(c) many subsets that perform equally well, as they all predict the test data satisfactorily.

From a practical point of view, the lack of a unique solution does not seem to present a problem.

The EA approach we describe maintains a population of trial gene subsets; imposes random changes on the genes that compose those subsets; and incorporates selection (driven by a neural network classifier) to determine which are the most informative ones. Only those genes are maintained in successive generations; the rest are removed from the trial pool. At each iteration, every subset is given as input to an FNN classifier and the effectiveness of the FNN determines the fitness of the subset of genes. The size of the population and the number of features in each subset are parameters that we explore experimentally.

For the outlined system, each population member represents a subset of genes, so a special representation must be designed. When seeking subsets containing n genes, each individual consists of n integers. The first integer is the index of the first gene to be included in the subset, the second integer denotes the number of genes to skip until the second gene to be included is reached, the third integer component denotes the number of genes to skip until the third included gene, and so on. This representation was necessary in order to avoid multiple inclusion of the same gene. Moreover, a version of DE that uses integer vectors has been thoroughly studied in previous Section.

FNNs were used as a classifier to evaluate the fitness of each gene subset. One third of the data set is used as a training set for the FNN and one third is used to measure the classification accuracy of the FNN classifier. The remaining patterns of the data set are kept to estimate the classification capability of the final gene subset. All the FNNs were trained using the well known and widely used Resilient backpropagation (Rprop) [40] training algorithm. Rprop is a fast local adaptive learning scheme, performing supervised training. To update each weight of the FNN, Rprop exploits information concerning the sign of the partial derivative of the error function. In our experiments, the five parameters of the Rprop method were initialized using values commonly employed in the literature. In particular, the increase factor was set to $\eta^+ = 1.2$; the decrease factor was set to $\eta^- = 0.5$; the initial update value is set to $\Delta_0 = 0.1$; the maximum step, which prevents the weights from becoming too large, was $\Delta_{\max} = 50$; and the minimum step, which is used to avoid too small weight updates, was constantly fixed to $\Delta_{\min} = 10^{-6}$.

7.2 Presentation of Experiments in Evolutionary Dimension Reduction

Next, we report the experimental results. We have tested and compared the performance of the described system on many publicly available microarray data sets. Here we report results from the following two data sets:

(a) The COLON data set [4] consists of 40 tumor and 22 normal colon tissues. For each sample there exist 2000 gene expression level measurements. The data set is available at `http://microarray.princeton.edu/oncology`.
(b) The PROSTATE data set [13] contains 52 prostate tumor samples and 50 nontumor prostate samples. For each sample there exist 6033 gene expression level measurements. It is available at `http://www.broad.mit.edu/cgi-bin/cancer/datasets.cgi`.

Since the appropriate size of the most predictive gene set is unknown, DE was employed for various gene set sizes ranging from 10 to 100 with a step of 10. The FNN used at the fitness function consisted of 2 hidden layers with eight and seven neurons, respectively. The input layer contained as many neurons as the size of the gene set. One output neuron was used at the output layer whose value for each sample determined the network classification decision. Since both problems had two different classes for the patterns, a value lower than 0.5 regarded the pattern to belong to class 1 otherwise regarded it to belong to class 2.

For each different gene set size the data was partitioned randomly into a learning set consisting of two-thirds of the whole set and a test set consisting of the remaining one third, as already mentioned. The one third of the training set was used by the Rprop algorithm to train the FNNs, and the performance of the respective gene set was measured in the other one third. The test set was only used to evaluate the classification accuracy that can be obtained using the final gene set discovered by the DE algorithm. To reduce the variability, the splitting was repeated 10 times and 10 independent runs were performed each time, resulting in a total of 100 experiments, for gene set size.

The classification accuracy of the system is illustrated using boxplots in Figure 23. Each boxplot depicts the obtained values for the classification accuracy, in the 100

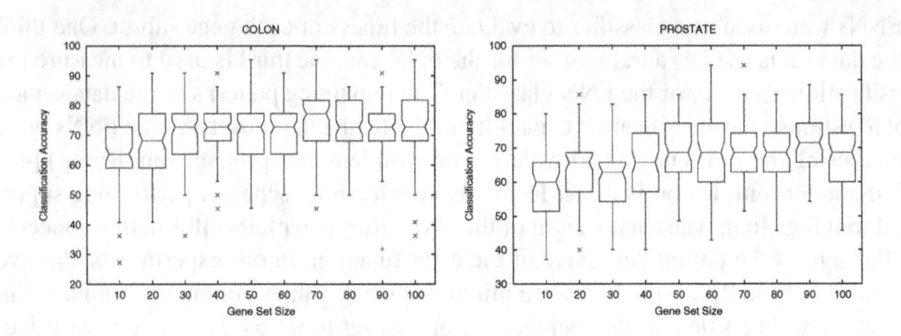

Fig. 23. Classification accuracy obtained by FNNs trained using the DE selected gene set for the COLON (left) and PROSTATE (right) datasets

experiments. As demonstrated, using a gene set size of 50–80 for the COLON dataset the algorithm managed to achieve the best results; comparable to those obtained by other approaches. The same is achieved for the PROSTATE dataset for a gene set size ranging from 40 to 60.

8 Future Directions: What Lies Ahead?

More than ten years have passed since the initial proposal of DE by Storn and Price, and DE has been accepted as a strong the robust global optimization algorithm capable to handle nondifferentiable, nonlinear and multimodal objective functions. Although the driving force of DE are the mutation operators, little progress has been made to the extension of the algorithm by introducing new operators. Here, we utilize Genetic Programming (GP) to evolve novel DE mutation operators.

GP is a method for automatically creating working computer programs employing principles of Darwinian evolution, and having as input a high-level statement of the problem [28]. GP aspires to induce a population of computer programs that gradually improve as they evolve and experience the data on which they are evaluated. In this Section, we present a comparison between already known human-designed mutation operators and new genetically programmed ones. Our experimental results indicate that the performance of the genetically programmed operators is comparable and in some cases is considerably better than the already existing human designed ones. A genetically evolved operator also exhibited the most robust performance. Additionally, the genetic evolution resulted in parameter free mutation operators.

8.1 Genetic Programming

GP is an extension of Genetic Algorithms in which individuals are no longer fixed-length strings but rather computer programs expressed as *syntax trees*. GP individuals consist of function and terminal nodes. Terminal nodes store a value which they return as an output, while functions process their inputs to compute an output. The terminal set, T, is comprised of the inputs, the constants supplied, and the zero-argument functions.

Thus, terminal nodes have an arity of zero. On the other hand, the function set, F, is composed of the statements and functions available to GP.

The primary GP search operators are *crossover* and *mutation*. In crossover, a randomly selected subtree from each of the two selected parents is exchanged between them to form two new individuals (offsprings). The idea is that useful building blocks for the solution of a problem are accumulated in the population and crossover permits the aggregation of good building blocks into even better solutions to the problem [29]. Crossover is the predominant search operator in GP [56]. Mutation operates on a single individual by altering a random subtree. Next, we briefly describe the GP initialization and the GP operators used.

The GP Initialization

The individuals in the GP population are initialized by recursively generating syntax trees composed of random function and terminal nodes. Two established GP initialization methods are the grow and the full method. Both methods require from the user to specify the maximum initial tree depth. According to the grow method, nodes are selected randomly from the function and the terminal sets. The grow method, therefore, produces trees of irregular shape, since once a terminal node is inserted the path ending with this node cannot be extended, even if the maximum initial depth has not been reached. On the other hand, in the full initialization method only function nodes are selected until the maximum initial depth is reached. Beyond that depth only terminal nodes are chosen to end the branches. This method results in a balanced tree, every branch of which reaches the maximum initial depth.

The GP Selection Algorithm

To derive the individuals that will comprise the population of the next generation, GP initially selects individuals from the current generation. The selection operators that have been proposed for Genetic Algorithms are also applicable to GP. In this study, we employed the most commonly encountered one, namely *roulette wheel selection*. Define the fitness of the ith individual as E_i, where E is the error function we wish to minimize. Then the probability of selecting individual i as a parent of an individual of the next generation is equal to $E_i^n / \sum_{j=1}^{N} E_j^n$; where $E_i^n = 1/(1 + E_i)$.

The GP Crossover Operator

The crossover operator combines the genetic material of two parents chosen by the selection operator to yield two offsprings. In particular, a real number r is randomly chosen in the interval $[0.1]$. Crossover takes place only if $r \leqslant C$, where C is the predefined crossover constant. In this case, a random node in each parent is chosen and the subtrees rooted at these nodes are exchanged between the parents to yield the offsprings. If an offspring exceeds the maximum depth it is discarded and the corresponding parent individual takes its place in the population of the next generation. Thus, crossover produces offsprings by swapping a part of one parent with a part of the other. If crossover does not take place ($r > C$) the offsprings are exact copies of the parents.

The GP Mutation Operator

After the crossover operator has finished, each offspring produced undergoes mutation. The probability of mutation is a user defined parameter. The mutation operator in GP randomly selects a node of the tree. If the node is a function then it is replaced by another function. If the node is terminal, another randomly selected terminal is used instead (point mutation) [56]. The mutated individual is then placed back into the population.

8.2 Genetically Programmed Differential Evolution Mutation Operators

As previously mentioned, in this section we investigate the discover of new efficient DE mutation operators using GP [33]. This is possible since, mutation operators are simply the composition of elementary functions such as addition, subtraction, and multiplication, operating on the vectors that represent individuals of the DE population. To this end, the terminal set used for GP, included two numerical constants, the vector of the best so far DE individual, x_g^{best}, three vectors of different randomly selected DE individuals, and the fixed mutation constant μ employed by the DE mutation operators. In detail, the terminal set used in this study was, $T = \{0.5, 1, \mu, x_g^{\text{best}}, x_g^{r1}, x_g^{r2}, x_g^{r3}\}$. The function set was $F = \{+, -, \odot, \oslash\}$, where $\odot$ and $\oslash$ are defined as follows:

$$x \odot y = x^{\top} \cdot \text{diag}\{y_1, y_2, \ldots, y_n\} = (x_1 y_1, x_2 y_2, \ldots, x_n y_n)^{\top}$$
$$x \oslash y = x^{\top} \cdot \text{diag}\{1/y_1, 1/y_2, \ldots, 1/y_n\} = (x_1/y_1, x_2/y_2, \ldots, x_n/y_n)^{\top},$$

where the vectors $x, y \in \mathbb{R}^n$, with $x^{\top} = (x_1, x_2, \ldots, x_n)$ and $y^{\top} = (y_1, y_2, \ldots, y_n)$. Note that the operator $\oslash$ utilizes a protected division; if the absolute value of the denominator is less 0.0001, then $\oslash$ returns 1.

The presentation of the problem and the fitness function typically define the space of candidate solutions for each particular problem. At present, more than one performance measure are applicable. One approach is to use the distance of the discovered minimizer from the global one to measure the operator's performance [36]. However, in many real life applications the location of the global minimizer is unknown. Conversely, the value of the global minimum could be known (for example when minimizing the sum of squares, a chemical or physical process, etc.).

In this study, we defined a fitness function, suitable for general optimization tasks, which utilizes three benchmark optimization problems discussed in previous Section: the Shekel's Foxholes, the Corana Parabola and the Levy No. 5 test problems. More specifically, the performance of each operator was measured through the sum of the generations required to locate the global minimum on each benchmark function, plus the minimum function values that were discovered. It is known that the performance of the DE algorithm (like the performance of every other EA) can vary with the initial random individuals. To reduce the effect of the stochastic nature of DE, 10 independent evaluations were performed, and the final fitness was averaged. If the global minimum was not found after 100 generations DE terminated. Using this fitness function, we strain GP evolution towards obtaining DE operators capable of locating the global optimum, within a minimum number of generations.

8.3 Experimental Discovery of Genetically Programmed Operators

The computational experiments were performed utilizing a novel GP–DE interface. We employed the full GP initialization method with a maximum initial tree depth of 3. Another critical GP parameter is the maximum allowed depth for the trees. The maximum depth parameter is the largest allowed depth between the root node and the outermost terminals. The maximum depth during the GP execution was 100. GP population size was 40, while the maximum number of generations was set to 1000. The mutation and crossover probabilities for GP were set to 0.6 and 0.1, respectively. The values for the parameters μ and ρ employed by the DE algorithm (irrespective of the mutation operator), were set to 0.6 and 0.8, respectively.

We conducted 100 independent GP experiments. The five best performing DE mutation operators discovered are the following:

$$v_{g+1}^{i} = (x_g^{r1} + x_g^{r2}) \oslash (1 + (x_g^{r3} \oslash x_g^{\text{best}})), \tag{8}$$

$$v_{g+1}^{i} = x_g^{\text{best}} + 0.5(x_g^{r1} - x_g^{r2}), \tag{9}$$

$$v_{g+1}^{i} = \left(x_g^{r1} + x_g^{r3}\right) \oslash \left((x_g^{r1} \oslash x_g^{\text{best}}) + (x_g^{r2} \oslash x_g^{\text{best}})\right), \tag{10}$$

$$v_{g+1}^{i} = \left(x_g^{r3} + x_g^{\text{best}}\right) \oslash \left((x_g^{r3} \oslash x_g^{r1}) + (x_g^{r2} \oslash x_g^{\text{best}})\right), \tag{11}$$

$$v_{g+1}^{i} = \left((x_g^{r1} \odot x_g^{r3}) \oslash (x_g^{r1} + x_g^{r3})\right) \odot \left((x_g^{\text{best}} \oslash x_g^{r3}) + (x_g^{\text{best}} \oslash x_g^{r2})\right). \tag{12}$$

Throughout the remaining chapter, we call $\text{GPDE}_1, \text{GPDE}_2, \ldots, \text{GPDE}_5$ the DE algorithm that uses Equation (8), Equation (9), ..., Equation (12) as the mutation operator, respectively. It is evident that this methodology allows us to routinely "invent" new specialized DE operators, which are optimal or near-optimal for a specific problem. Notice that although the mutation constant μ was included in the terminal set, all the above mentioned GP derived DE mutation operators are parameter free. This is a considerable advantage since it alleviates the need for parameter tuning by the user.

The original DE algorithm exploits the information from the differences between pairs of individuals to guide its search in the solution domain. Although, in all the mutation operators discovered here, individuals interact in pairs, pairwise differences are not encountered in any GPDE operator but GPDE_2. Indeed, GPDE_2 is equivalent to DE_1 for the special case that $\mu = 0.5$. The experimental results reported below suggest that this particular setting is more effective than a typical value of μ for the benchmark problems considered. Note that DE_1 has been documented as one of the most effective and robust mutation operators. It is also interesting to note that the best individual of the current generation appears at least once in all GPDE operators.

To measure the efficiency and effectiveness of the newly discovered GPDE operators, we tested them on the three previously mentioned optimization benchmark functions, as well as on two additional functions; namely the Griewangk's and the Rosenbrock's Saddle test functions.

The performance of the human-designed and the genetically programmed DE mutation operators is presented in Table 7. In particular, for each mutation operator and for each benchmark function, Table 7 reports the mean number of generations required to locate a global minimizer (Gen.), as well as, the percentage of times the algorithm was

Table 7. Human Designed vs. Genetically Programmed Differential Evolution Operators

	Training Phase						Testing Phase			
	Problem 1		Problem 2		Problem 3		Problem 1		Problem 2	
	Gen.	(%)	Gen.	(%)	Gen.	(%)	Gen.	(%)	Gen.	(%)
DE_1	95.6	5	87.5	42	64.9	50	29.8	**100**	—	0
DE_2	83.8	**95**	—	0	76.8	**100**	56.8	**100**	—	0
DE_3	97.5	13	97.9	35	76.9	63	43.7	**100**	—	0
DE_4	—	0	39.8	79	89.2	16	52.3	74	96.8	12
DE_5	94.6	20	—	0	92.9	22	81.2	85	—	0
DE_6	81.2	64	—	0	72.1	**100**	63.1	96	—	0
$GPDE_1$	—	0	29.9	94	43.7	96	55.9	68	93.6	44
$GPDE_2$	84.1	23	98.8	3	**34.0**	88	**25.1**	**100**	—	0
$GPDE_3$	**78.5**	25	58.2	57	52.8	65	28.0	**100**	76.9	53
$GPDE_4$	—	0	**12.1**	**100**	82.6	32	38.8	**100**	**20.3**	**100**
$GPDE_5$	—	0	25.7	**100**	57.1	59	38.4	93	82.2	28

(— denotes that the algorithm failed to find the global minimum in all runs).

successful in locating a global minimizer (%). The reported results are averages over 100 independent experiments for each mutation operator. Note, that in the cases DE was unable to identify a global minimizer, the maximum allowed number of generations was added to the sum used to compute the mean number of generations required to locate a global minimizer. The entry "—" in the table suggests that the success rate of a mutation operator for the corresponding benchmark was zero. Finally, bold faced entries are used to indicate the mutation operator with the lowest mean number of generations to detect a global minimizer and the one with the highest success rate.

With respect to the mean number of generations required to detect a global minimizer, the best performing mutation operator is in all cases derived by Genetic Programming. The best performing mutation strategy in this respect is $GPDE_2$, which is a special case of DE_1. For two out the five optimization problems (Train Problem 3 and Test Problem 1), DE_1 requires the lowest mean number of generations to compute a global minimizer among the original DE operators. On the same two problems $GPDE_2$ is the overall best performing strategy in this respect, but it performs badly on Test Problem 2. $GPDE_4$ is by far the best performing strategy on Train Problem 2 and Test Problem 2, for which most operators performed badly. With respect to the percentage of times a minimizer was located, the two types of operators perform similarly well on Test Problem 1. Last but not least, it is important to note that the most robust operator with respect to both criteria is $GPDE_3$. It is the best performing operator with respect to mean number of generations on Train Problem 1, the second best on Test Problem 1, and the third best performing on Train Problem 3 and Test Problem 2. Furthermore, it is the only operator that achieved a positive percentage of locating a minimizer on all the test functions. Our experience is that $GPDE_3$ is stable and effective, and can be used to optimize an unknown function with good results.

In accordance to the "no free lunch theorem" [57], it is impossible to find a single DE operator that outperforms all the other in every test problem. Instead, here we try to discover new DE operators better suited for general optimization problems, or classes of problems. The experimental results indicate that the best performing DE mutation operator is in all cases GP derived. GP has been able to automatically evolve a variety of new DE mutation operators that operate as well or considerably better, for the considered problems, than the already existing human-designed ones. It is interesting to note that all the new DE mutation operators are parameter free, in the sense that no mutation constant is needed.

9 Synopsis

In this chapter we presented an overview of the major applications areas of differential evolution. The DE algorithms have shown their strength in tackling many difficult problems from diverse scientific areas, including single and multiobjective function optimization, neural network training, clustering, and real life DNA microarray classification.

All the experiments presented in this chapter have been performed using distributed computing environments, since DE can be easily parallelized in a virtual parallel environment so as to improve both its speed and performance. The results indicate that the extent of information exchange among subpopulations assigned to different processor nodes, aids the algorithm to converge faster and find better solutions. To demonstrate that we have introduced the parallel, multi–population DE algorithm for single and multiobjective optimization.

Next, we presented a case where DE can be utilized to perform data clustering. Additionally, clustering algorithms can also aid DE to locate simultaneously multiple local and global minimizers of an objective function. This can be accomplished by the new clustering operator. This operator incorporates the unsupervised k–windows clustering algorithm, utilizing already computed pieces of information regarding the search space in an attempt to discover regions containing groups of individuals located close to different minimizers. Then, the search is confined inside these regions and a large number of global and local minimizers of the objective function can be efficiently computed.

The real life DE applications presented here include the training of integer weight neural networks with threshold activations and the selection of genes of DNA microarrays in order to obtain high predictive accuracy subsets. In both cases, the DE addressed problems of very high dimensionality successfully.

We closed this chapter with a discussion on promising future extensions of the algorithm by the incorporation of genetically programmed mutation operators. These operators improve the quality of the solutions and accelerate the execution of the algorithm. It must be noted that the genetic evolution resulted in parameter free DE operators. This is a considerable advantage since it alleviates the need for parameter tuning by the user. The results indicate that the performance of the genetically programmed operators is comparable and in some cases is considerably better than the already existing human designed ones. We feel that this can be a very interesting future research direction.

References

1. MPI the message passing interface standard, `http://www-unix.mcs.anl.gov/mpi/`
2. Abbass, H.: Self–adaptive pareto differential evolution. In: Proceedings of the IEEE 2002 Congress on Evolutionary Computation, Honolulu, Hawaii, pp. 831–836. IEEE Press, Los Alamitos (2002)
3. Alevizos, P.: An algorithm for orthogonal range search in $d \geqslant 3$ dimensions. In: Proceedings of the 14th European Workshop on Computational Geometry. Barcelona (1998)
4. Alon, U., Barkai, N., Notterman, D.A., Gish, K., Ybarra, S., Mack, D., Levine, A.J.: Broad patterns of gene expression revealed by clustering analysis of tumor and normal colon tissues probed by oligonucleotide array. Proc. Natl. Acad. Sci. USA 96(12), 6745–6750 (1999)
5. Babu, G.P., Murty, M.N.: A near optimal initial seed value selection in k-means algorithm using a genetic algorithm. Pattern Recogn. Lett. 14(10), 763–769 (1993)
6. Babu, G.P., Murty, M.N.: Clustering with evolution strategies. Pattern Recogn. 27, 321–329 (1994)
7. Becker, R.W., Lago, G.V.: A global optimization algorithm. In: Proceedings of the 8th Allerton Conference on Circuits and Systems Theory, pp. 3–12 (1970)
8. Bentley, J.L., Maurer, H.A.: Efficient worst-case data structures for range searching. Acta Informatica 13, 1551–1568 (1980)
9. Bhuyan, J.N., Raghavan, V.V., Venkatesh, K.E.: Genetic algorithm for clustering with an ordered representation. In: Fourth International Conference on Genetic Algorithms, pp. 408–415 (1991)
10. Blake, C.L., Merz, C.J.: UCI repository of machine learning databases (1998)
11. Coello Coello, C.A., Van Veldhuizen, D.A., Lamont, G.B.: Evolutionary Algorithms for Solving Multi–Objective Problems. Kluwer, New York (2002)
12. Deb, K.: Multi–objective genetic algorithms: Problem difficulties and construction of test problems. Evolutionary Computation 7(3), 205–230 (1999)
13. Singh, D., et al.: Gene expression correlates of clinical prostate cancer behavior. Cancer Cell 1, 203–209 (2002)
14. Fan, H.Y., Lampinen, J.: A trigonometric mutation operation to differential evolution. Journal of Global Optimization 27, 105–129 (2003)
15. Fayyad, U.M., Piatetsky-Shapiro, G., Smyth, P.: Advances in Knowledge Discovery and Data Mining. MIT Press, Cambridge (1996)
16. Fieldsend, J.E., Everson, R.M., Singh, S.: Using unconstrained elite archives for multiobjective optimization. IEEE Trans. Evol. Comp. 7(3), 305–323 (2003)
17. Fogel, D.: Evolutionary Computation: Towards a New Philosophy of Machine Intelligence. IEEE Press, Piscataway (1996)
18. Fogel, D.B., Simpson, P.K.: Evolving fuzzy clusters. In: International Conference on Neural Networks, pp. 1829–1834 (1993)
19. Fogel, L.J., Owens, A.J., Walsh, M.J.: Artificial intelligence through simulated evolution. Wiley, Chichester (1966)
20. Geist, A., Beguelin, A., Dongarra, J., Jiang, W., Manchek, R., Sunderam, V.: PVM: Parallel Virtual Machine. A Users Guide and Tutorial for Networked Parallel Computing. MIT Press, Cambridge (1994)
21. Goldberg, D.: Genetic Algorithms in Search, Optimization, and Machine Learning. Addison Wesley, Reading (1989)
22. Handl, J., Knowles, J.: Evolutionary multiobjective clustering. In: Yao, X., Burke, E.K., Lozano, J.A., Smith, J., Merelo-Guervós, J.J., Bullinaria, J.A., Rowe, J.E., Tiňo, P., Kabán, A., Schwefel, H.-P. (eds.) PPSN 2004. LNCS, vol. 3242, pp. 1081–1091. Springer, Heidelberg (2004)

23. Haykin, S.: Neural Networks. Macmillan College Publishing Company, New York (1999)
24. Holland, J.H.: Adaptation in natural and artificial system. University of Michigan Press (1975)
25. Jin, Y., Olhofer, M., Sendhoff, B.: Evolutionary dynamic weighted aggregation for multiobjective optimization: Why does it work and how? In: Proceedings GECCO 2001 Conference, San Francisco, CA, pp. 1042–1049 (2001)
26. Jones, D., Beltramo, M.A.: Solving partitioning problems with genetic algorithms. In: Fourth International Conference on Genetic Algorithms, pp. 442–449 (1991)
27. Kennedy, J., Eberhart, R.C.: Particle swarm optimization. In: Proceedings IEEE International Conference on Neural Networks, Piscataway, NJ, vol. IV, pp. 1942–1948. IEEE Service Center (1995)
28. Koza, J.R.: Hierarchical genetic algorithms operating on populations of computer programs. In: Proceedings of the Eleventh International Joint Conference on Artificial Intelligence, pp. 768–774 (1989)
29. Koza, J.R.: Genetic Programming: On the Programming of Computers by Means of Natural Selection. MIT Press, Cambridge (1992)
30. Laumanns, M., Zitzler, E., Thiele, L.: A unified model for multiobjective evolutionary algorithms with elitism. In: Proc. IEEE Congr. Evol. Comp., Piscataway, NJ, pp. 46–53. IEEE Press, Los Alamitos (2000)
31. Marriott, F.H.C.: Optimisation methods of cluster analysis. Biometrics 69(2), 417–422 (1982)
32. Michalewicz, Z., Fogel, D.B.: How to solve it: Modern Heuristics. Springer, Heidelberg (2000)
33. Pavlidis, N.G., Tasoulis, D.K., Plagianakos, V.P., Vrahatis, M.N.: Human designed vs. genetically programmed differential evolution operators. In: IEEE Congress on Evolutionary Computation, pp. 1880–1886 (2006)
34. Plagianakos, V.P., Vrahatis, M.N.: Training neural networks with threshold activation functions and constrained integer weights. In: IEEE International Joint Conference on Neural Networks (IJCNN 2000), Como, Italy (2000)
35. Plagianakos, V.P., Vrahatis, M.N.: Parallel evolutionary training algorithms for 'hardware-friendly' neural networks. Natural Computing 1, 307–322 (2002)
36. Poli, R., Langdon, W.B., Holland, O.: Extending particle swarm optimisation via genetic programming. In: Keijzer, M., Tettamanzi, A.G.B., Collet, P., van Hemert, J.I., Tomassini, M. (eds.) EuroGP 2005. LNCS, vol. 3447. Springer, Heidelberg (2005)
37. Preparata, F., Shamos, M.: Computational Geometry. Springer, New York (1985)
38. Ramasubramanian, V., Paliwal, K.: Fast k-dimensional tree algorithms for nearest neighbor search with application to vector quantization encoding. IEEE Transactions on Signal Processing 40(3), 518–531 (1992)
39. Rechenberg, I.: Evolution strategy. In: Zurada, J.M., Marks II, R.J., Robinson, C. (eds.) Computational Intelligence: Imitating Life. IEEE Press, Piscataway (1994)
40. Riedmiller, M., Braun, H.: A direct adaptive method for faster backpropagation learning: The rprop algorithm. In: Proceedings of the IEEE International Conference on Neural Networks, San Francisco, CA, pp. 586–591 (1993)
41. Sander, J., Ester, M., Kriegel, H.-P., Xu, X.: Density-based clustering in spatial databases: The algorithm gdbscan and its applications. Data Mining and Knowledge Discovery 2(2), 169–194 (1998)
42. Schaffer, J.D.: Multiple Objective Optimization With Vector Evaluated Genetic Algorithms. PhD thesis, Vanderbilt University, Nashville, TN, USA (1984)
43. Schwefel, H.-P.: Evolution and Optimum Seeking. Wiley, New York (1995)
44. Steinbach, M., Karypis, G., Kumar, V.: A comparison of document clustering techniques. In: KDD Workshop on Text Mining (2000)

45. Storn, R.: System design by constraint adaptation and differential evolution. IEEE Transactions on Evolutionary Computation 3, 22–34 (1999)
46. Storn, R., Price, K.: Differential evolution – a simple and efficient adaptive scheme for global optimization over continuous spaces. Journal of Global Optimization 11, 341–359 (1997)
47. Tasoulis, D.K., Pavlidis, N.G., Plagianakos, V.P., Vrahatis, M.N.: Parallel differential evolution. In: IEEE Congress on Evolutionary Computation (CEC 2004) (2004)
48. Tasoulis, D.K., Plagianakos, V.P., Vrahatis, M.N.: Clustering in evolutionary algorithms to efficiently compute simultaneously local and global minima. In: IEEE Congress on Evolutionary Computation, pp. 1847–1854 (2005)
49. Tasoulis, D.K., Plagianakos, V.P., Vrahatis, M.N.: Differential evolution algorithms for finding predictive gene subsets in microarray data. In: Artificial Intelligence Applications and Innovations. IFIP International Federation for Information Processing, vol. 204, pp. 484–491 (2006)
50. Tasoulis, D.K., Vrahatis, M.N.: Novel approaches to unsupervised clustering through the k-windows algorithm. In: Sirmakessis, S. (ed.) Knowledge Mining. Studies in Fuzziness and Soft Computing, vol. 185, pp. 51–78. Springer, Heidelberg (2005)
51. Tasoulis, D.K., Vrahatis, M.N.: Unsupervised clustering on dynamic databases. Pattern Recognition Letters 26(13), 2116–2127 (2005)
52. Torn, A., Zilinskas, A.: Global Optimization. Springer, Berlin (1989)
53. van der Merwe, D.W., Engelbrecht, A.P.: Data clustering using particle swarm optimization. In: Congress on Evolutionary Computation, Canberra, Australia, pp. 215–220 (2003)
54. Van Veldhuizen, D.A., Zydallis, J.B., Lamont, G.B.: Considerations in engineering parallel multiobjective evolutionary algorithms. IEEE Trans. Evol. Comp. 7(2), 144–173 (2003)
55. Vrahatis, M.N., Boutsinas, B., Alevizos, P., Pavlides, G.: The new k-windows algorithm for improving the k-means clustering algorithm. Journal of Complexity 18, 375–391 (2002)
56. Wolfgang, B., Nordin, P., Keller, R.E., Francone, F.D.: Genetic programming: An Introduction: on the automatic evolution of computer programs and its applications. Morgan Kaufmann Publishers Inc., San Francisco (1998)
57. Wolpert, D.H., Macready, W.G.: No free lunch theorems for optimization. IEEE Transactions on Evolutionary Computation 1(1), 67–82 (1997)
58. Yang, M.-S., Wu, K.-L.: A similarity-based robust clustering method. IEEE Transactions on Pattern Analysis and Machine Intelligence 26(4), 434–448 (2004)
59. Zitzler, E.: Evolutionary Algorithms for Multiobjective Optimization: Methods and Applications. PhD thesis, Swiss Federal Institute of Technology Zürich, Switzerland (1999)
60. Zitzler, E., Deb, K., Thiele, L.: Comparison of multiobjective evolution algorithms: Empirical results. Evolutionary Computation 8(2), 173–195 (2000)

The Differential Evolution Algorithm as Applied to Array Antennas and Imaging

A. Massa[1], M. Pastorino[2], and A. Randazzo[2]

[1] Department of Information and Communication Technology, University of Trento,
Via Sommarive 14, I-38050 Trento, Italy
Phone: +390461882057
Fax: +390461881696
andrea.massa@ing.unitn.it
[2] Department of Biophysical and Electronic Engineering, University of Genoa,
Via Opera Pia 11A, I-16145, Genova, Italy
Phone: +39010352242
Fax: +390103532245
{pastorino,randazzo}@dibe.unige.it

Summary. The application of the differential evolution method in two important areas of applied electromagnetics is discussed in this chapter. The first one refers to the synthesis and design of array antennas, for which differential evolution, as well as other evolutionary algorithms, is now considered a fundamental design tool. The second one concerns the diagnostic applications faced as a result of using radiofrequency and microwave imaging techniques. Being based on the inverse scattering problem, these techniques suffer from nonlinearity and ill posedness. The differential evolution method has been successfully proposed for optimizing this multimodal and complex inverse problem.

The chapter includes a brief review of some results recently published in the scientific literature concerning the application of differential evolution to the above-mentioned problems. Moreover, the main contributions of the authors in these areas are reviewed and discussed. Finally, some new results are reported.

1 Introduction

Evolutionary algorithms are now very common for the solution of complex problems in the field of applied electromagnetics [1-8]. The flexibility, the accuracy, and the possibility of obtaining the global optimum of an optimization problem, which very often correspond to the "best" solution, are the most appreciated features of evolutionary algorithms. A number of electromagnetic problems for which real time is not a requirement have greatly benefited from the development of this kind of approaches, and, more generally, from the introduction of global optimization methods, including "serial" methods (e.g. the simulated annealing) [9] and population-based method (e.g., the genetic algorithm (GA).) [10].

Among the various applications that, in recent year, have been treated by using stochastic global optimization methods in the field of applied electromagnetics, array antenna design and radiofrequency and microwave imaging represent significant examples.

U.K. Chakraborty (Ed.): Advances in Differential Evolution, SCI 143, pp. 239–255, 2008.
springerlink.com

In the following, the application of the differential evolution (DE) algorithm to these research areas is reviewed. In the first case, the DE method has been applied to solve several different synthesis problem. Among them (which are discussed in Section 2), the synthesis of monopulse antennas is described in details. The other case concerns the complex problem of the inspection and imaging of unknown targets by inverting field-scattered data. The problem is multimodal, nonlinear, and ill posed. In several works reported in the scientific literature, which are briefly reviewed in Section 3, the differential evolution method has yielded very accurate reconstruction results. The case of tomographic imaging for inspecting inhomogeneous dielectric targets is described in details. Some new results, concerning both the applicative areas of antenna synthesis and electromagnetic imaging, are reported for illustrative purposes.

2 The DE Method in Antenna Applications

The differential evolution method has been intensively applied in the field of antenna synthesis and design, as well as most of the new evolutionary algorithms that have opened new grounds in this field. In its basic formulation, the array synthesis problem concerns the definition of some parameters of a given array configurations (e.g., number and/or positions of the antenna elements and their excitation coefficients, which are usually complex numbers, although, in some cases, the amplitudes or the phases are assumed to be given.) The unknown parameters are optimized in order to fulfill prescribed constrains concerning, for example, the shape of the array pattern, the beamwidth, the level of the side lobes, etc.

The way in which the DE is applied to these synthesis problems is schematized in Fig. 1.

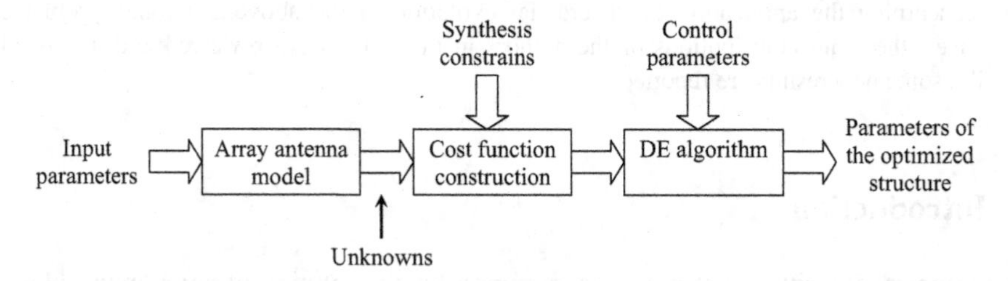

Fig. 1. Schematic representation of the application of the DE algorithm to the array synthesis problem

Usually, one chooses the array configuration (e.g., linear, circular, planar array) and the other fixed elements of the design. Then a suitable cost function is constructed on the bases of the prescribed constrains (which are in general related, as previously mentioned, to some radiation properties of the array.) The unknown parameters are the "arguments" of the cost function, and the DE algorithm is used to optimize the cost function versus the unknown parameters. The result is of course an optimized design of the array.

Several applications have been recently proposed in the scientific literature. In [11], the DE method has been used to suppress the sideband radiation patterns in time modulated linear array antennas. The DE algorithm has been found to be a very effective tool in optimizing the static excitation amplitudes and the "switch-on" time intervals of each element. In this application, the DE algorithm has been applied to optimize 32 variables. Moreover, the authors of [11] have found the DE method to be "more powerful" than the standard GA for the present application.

The DE algorithm has been also efficiently used for the synthesis of uniform amplitude arrays [12]. In particular, two classes of arrays have been optimized, i.e., unequally spaced arrays with equal phases (position-only synthesis) and unequal phases (position/phase synthesis). For the proposed application, the authors have devised some guidelines for the choice of the parameters of the DE method based on the reported numerical simulations. In particular, following the implemented version of the DE algorithm reported in reference [19] of [12], they suggest to choice the probability of generating the trial members (c_r) in the range $0.5 \delta c_r \delta 1$, the probability of mutation (c_m) in the range $0 < c_m \delta 0.2$ and the constant function F controlling the differential variations between 0.4 and 1.

In designing an array antenna, however, the mutual coupling among the antenna elements must be often taken into account properly. In [13], a method to compensate the mutual coupling effects in time-modulated arrays has been reported. In that work, the coupling effects are compensated by making the broadside beam of the antenna to match a standard low-sidelobe pattern of Taylor type [14]. In particular, the weights of the "compensated" elements, as well as the time sequences of the time-modulated array, have been optimized by the DE method, which has resulted to be very effective for this optimization problem. In particular, a L-band antenna with 16 elements (equally-spaced printed dipoles linearly aligned) has been designed.

The DE algorithm has been also used in [15], in order to perform the power-pattern synthesis, which has several applications both in telecommunications and in the development of electronic countermeasure systems. In this field, the use of stochastic approaches has represented a significant advance over the approaches based on classical synthesis methods (e.g., the Woodward-Lawson method [14].)

Another antenna synthesis problem that has been successfully solved by using the DE algorithm concerns the design of monopulse array antennas [14,16-18]. Monopulse antennas are able to generate sequentially the so-called sum pattern (which has a maximum in the broadside direction) and difference pattern (which has a null in the same direction.) Although several methods for implementing monopulse antennas have been proposed, in the last years a growing interest has been drawn to methods that use proper feed networks in order to avoid the need for the design of two completely independent feeds for the sum and different patterns. A challenging approach is the one proposed by Lopez et al. [18] that is based on a subarray configuration in which one of the excitation sets (for the sum or difference pattern) is assumed to be known and optimum, whereas the other one is realized by using a subarray configuration to reduce the feeding complexity. The objective of the synthesis is to construct a reduced subarray configuration able to synthesize as better as possible this pattern. In particular, the problem can be recast as an optimization problem in which a functional is constructed and optimized in order to define, for each array element, the corresponding subarray, the weights of all the subarrays and, consequently, the excitation sets of the pattern to be constructed. Clearly, the goal of the synthesis is to realize a good compromise between the feed network complexity and the quality of the patterns.

To show the approach, we refer here to a linear array of $M=2N$ equally-spaced elements whose array factor $F(\theta)$ is given by [14]

$$F(\theta)=\sum_{n=-N}^{-1} a_n e^{j\left(n+\frac{1}{2}\right)kd\cos\theta}+\sum_{n=1}^{N} a_n e^{j\left(n-\frac{1}{2}\right)kd\cos\theta} \tag{1}$$

where a_n are the complex excitation coefficients, k is the wavenumber, and θ defines the angle at which F is calculated with respect to the broadside direction. Finally, d is the inter-element distance.

The sum pattern is constructed by using assigned and symmetric coefficients, $a_{-n}^{s}=a_{n}^{s}$. Equation (1) then reduces to

$$F_s(\theta)=\sum_{n=1}^{N} a_n^s \cos\left[\frac{1}{2}(2n-1)kd\cos\theta\right] \tag{2}$$

In order to construct the difference pattern, the M elements of the array are grouped in P subarrays. Each subarray has a weighting coefficient, g_p, $p = 1,..,P$, and the group membership of the antennas must be optimized in order to create a difference pattern fulfilling the prescribed requirements. To achieve this goal, a positive integer number, c_n $(0 \le c_n \le P)$, is associated to each element of the array and denotes the subarray at which the antenna element must be connected. In particular, if $c_n = p$, the n-th element is to be connected to the p-th subarray. If $c_n = 0$, the element is not considered in the synthesis process. The set of elements associated to the p-th subarray is indicated by $\Gamma(p)$. Once the memberships of the subarrays have been defined, the excitation coefficients of the difference pattern can be obtained by multiplying each coefficient of the sum pattern for the coefficient of the corresponding subarray. Formally,

$$a_n^d = a_n^s \sum_{p=1}^{P} \delta_{c_n p} g_p \text{ , } n=1,\ldots,N \tag{3}$$

where $\delta_{c_n p}$ denotes the Kronecker function, i.e., $\delta_{c_n p}=1$ if $c_n = p$, $\delta_{c_n p}=0$ elsewhere. Since the excitations of the difference pattern must be antisymmetric, i.e. $a_{-n} = -a_n$, only one half of the array is considered in the synthesis problem. The array pattern is then given by

$$F_d(\theta)=\sum_{n=1}^{N} a_n^d \sin\left[\frac{1}{2}(2n-1)kd\cos\theta\right] \tag{4}$$

The previous approach has been followed in [18] and the obtained functional has been optimized by a standard binary GA. However, in [19] the same optimization problem has been faced by using a DE algorithm with hybrid chromosomes (constituted by real and integer genes), which allow to avoid coding and decoding processes for the real variables (excitation coefficients). In [19], the cost function has been constructed in order to obtain a side lobe level (*SLL*) with a prescribed value

$$f(\mathbf{x}_k(i)) = [SLL_k(i) - SLL_d]^2 H([SLL_k(i) - SLL_d]) \tag{5}$$

where SLL_d is the desired side lobe level, $SLL_k(i)$ is the *SLL* value corresponding to $\mathbf{x}_k(i)$, which is the *i*-th element of the population at the *k*-th iteration, and $H(\cdot)$ denotes the Heaviside step function. However, other types of constraints can be used in the synthesis process (as previously discussed).

The application of the DE algorithm has been found to be particularly suitable in this case. Several different configurations have been analyzed in [19]. In particular, the choice of the weighting factor *F* and of the probability that control the crossover operator has been made after a large numerical assessment. It resulted that for the present application good results can be obtained by assuming $c_r = 0.7$ and $F = 0.5$. In particular, these values avoid a premature convergence to local minima or a slow convergence rate.

For the design of monopulse antennas with the subarray configuration, the application of a standard real-coded GA has been discussed, too [19]. By using the same initial population and the same cost function (equation (5)) the DE method has been found to be superior in terms of cost functions evaluations needed to obtain the same level of accuracy in determining the weights of the subarrays and the various group memberships of the antenna elements.

The DE algorithm has shown excellent capabilities also in another synthesis problem again related to monopulse antennas, i.e., the maximization of the directivity of the difference pattern [20]. To this end, according with reference [5] of [20], the following fitness function has been used (directivity)

$$f = 2\frac{\mathbf{a}_d^t \mathbf{A}_d \mathbf{a}_d}{\mathbf{a}_d^t \mathbf{B}_d \mathbf{a}_d} \tag{6}$$

where $\mathbf{A}_d$ and $\mathbf{B}_d$ are $N \times N$ matrices whose elements are given by

$$\begin{aligned} a_{ij} &= \sin\left(\frac{1}{2}(2i-1)kd\cos\theta\right)\sin\left(\frac{1}{2}(2j-1)kd\cos\theta\right), \\ b_{ij} &= \operatorname{sinc}((j-i)kd) - \operatorname{sinc}((j+i-1)kd), \\ i,j &= 1,\ldots,N, \end{aligned} \tag{7}$$

and $\mathbf{a}_d$ denotes the array containing the values of the excitation coefficients of the difference pattern. For the above synthesis problems, accurate results have been

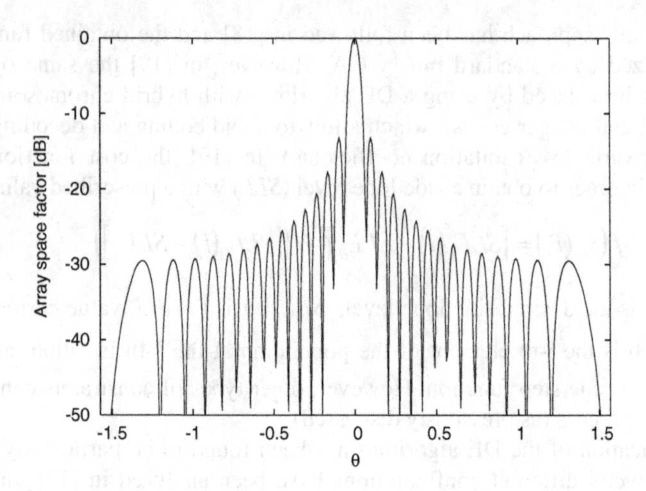

Fig. 2. Array space factor for the sum pattern

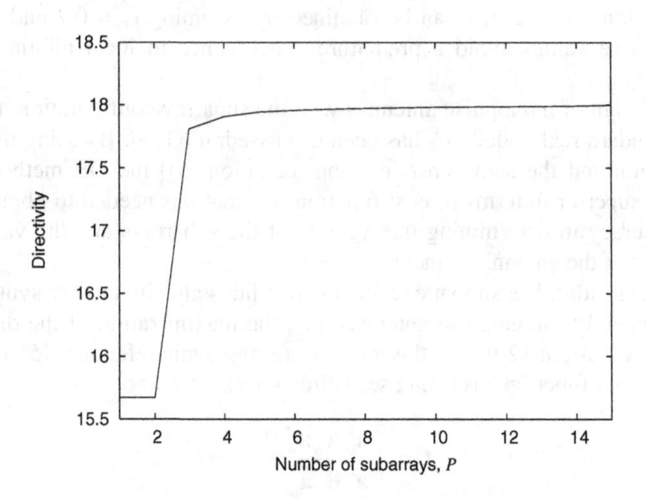

Fig. 3. Directivity of the difference pattern versus the number of subarrays

obtained by setting $c_r = 0.8$ and the parameter F in the range [0.5, 2] (a random choice has been performed in [20]).

An example is reported in the following. An array of $2N = 30$ elements with spacing $d = 0.5\lambda$ has been considered. The sum pattern has been obtained by using uniform excitations a_n^s, $n = 1,\ldots,N$. The corresponding array space factor is shown in Fig. 2.

The difference pattern has been computed by using the DE algorithm. The population size has been set equal to ten times the number of unknowns, i.e., $N_p = 10NP$.

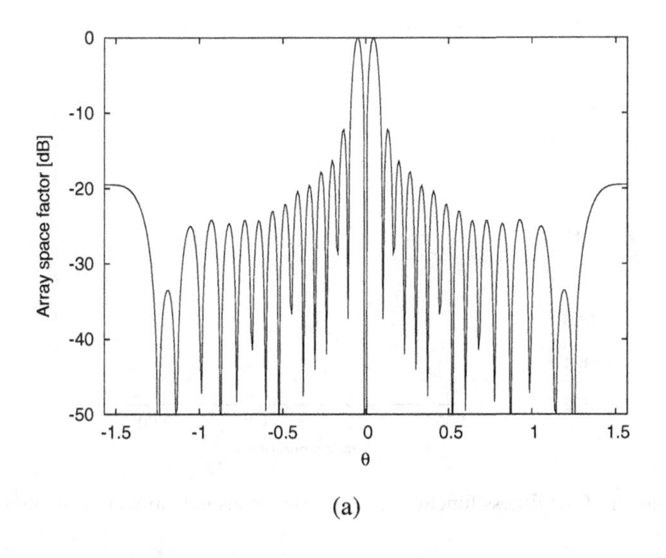

(a)

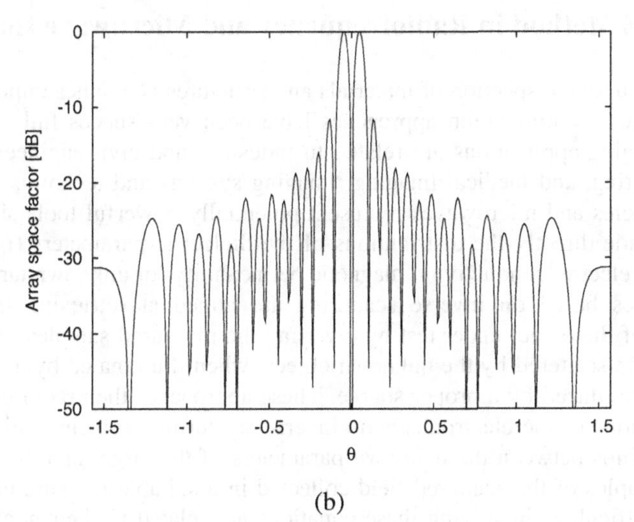

(b)

Fig. 4. Array space factor for the difference pattern. (a) $P = 5$. (b) $P = 10$.

Moreover, the maximum number of generations has been set equal to $k_{max} = 1000$. The number of subarrays has been changed in the range [1, 15].

The behavior of the directivity obtained by the DE-based approach versus the number of subarrays is reported in Fig. 3. Fig. 4 shows two examples of the obtained difference-mode array space factor (for the cases in which $P = 5$ and $P = 10$). Moreover, the behavior of the fitness function versus the number of iterations for the case in which $P = 10$ is reported in Fig. 5.

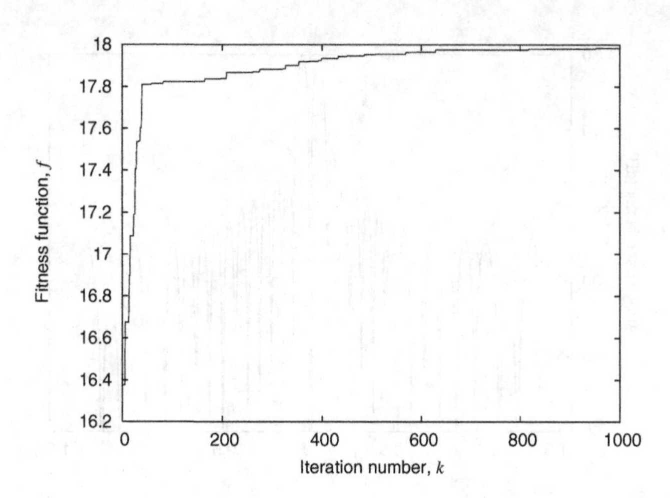

Fig. 5. Behavior of the fitness function (directivity) versus the number of iterations. $P = 10$.

3 The DE Method in Radiofrequency and Microwave Imaging

The nondestructive inspection of materials and structures is another important area in which stochastic optimization approaches have been very successfully applied [21-25]. Challenging applications are related to industrial and civil engineering, subsurface prospecting, and medical imaging. Imaging systems and techniques working at radiofrequencies and microwaves represent potentially powerful tools since they are able to provide directly the distributions of the dielectric parameters (i.e., dielectric permittivity, electric conductivity, magnetic permeability) of unknown targets.

Approaches based on inverse scattering are aimed at retrieving the dielectric parameters of the object under test by inverting the measured samples of the electromagnetic field scattered by the unknown objects when illuminated by a known "incident" wave produced by a proper source. These approaches then require the solution of the equations of the electromagnetic inverse scattering problem, which constitute the relationships between the unknown parameters of the target and the "input" data (i.e., the samples of the scattered field collected in a suitable measurement domain.) The main difficulties in solving these equations are related to their nonlinear nature and their severe ill-posedness, which require the use of "regularizations" techniques. In addition, the computational load is usually a limiting factor for the application of these methods.

The use of evolutionary algorithms for solving the inverse scattering problems is a very suitable choice, since these methods are able to find the global minimum of the problem, which corresponds to the "correct" solution, whereas local minima correspond to false solutions, which, in imaging modalities, often result in "artifacts" in the final images.

However, since evolutionary algorithms are computationally expensive, they can be difficulty applied in a straightforward way to discretized problems in which the object (and, more precisely, the investigation region where the object is assumed to be present) is represented by grids of pixels (in two-dimensional imaging) or voxels (in three-dimensional imaging, which, as a matter of fact, is still in its "infancy".) Consequently, evolutionary algorithms can be efficiently applied only with reduced parameterization or combined with other inversion strategies in order to devise hybrid approaches (e. g., those combining deterministic and stochastic methods.)

In the recent literature, some very interesting imaging approaches based on inverse scattering have been successfully solved by using the DE algorithm, which has been particularly appreciated for some peculiarities that will be discussed in the following.

To describe the inverse scattering formulation, we refer to the case in which the body under test is assumed to be of cylindrical shape (i.e., the cross section is uniform along a given direction, the cylindrical axis). Moreover, the illuminating field is polarized along this axis (in the following, the z axis), i.e., it has only the component parallel to the cylinder axis (transverse magnetic illumination conditions [26].) Under these hypotheses, which are usually followed in tomography, the problem to be solved is now a two-dimensional and scalar one. Let the cross section of the object be characterized by the dielectric parameters $\varepsilon = \varepsilon(x, y)$ (dielectric permittivity) and $\sigma = \sigma(x, y)$ (electric conductivity), where (x, y) denotes the transversal coordinates. Non magnetic materials are assumed ($\mu = \mu_0$), although the formulation could be easily extended to these materials.

The equation that relates the incident interrogating field, the dielectric properties of the object under test and the field "scattered" by the target is the so-called Lippmann-Schwinger integral equation [26]

$$E^{tot}(x,y) = E^{inc}(x,y) + \iint_A \tau(s,t) E^{tot}(s,t) G^{2D}(x,y/s,t) ds dt \tag{8}$$

where $E^{inc}(x, y)$ and $E^{tot}(x, y)$ denote the z-components of the incident and total electric fields, $\tau(x, y) = j\omega[\varepsilon(x, y) - \varepsilon_0]$ is the *object function*, which includes the information on the scatterer (ε_0 is the dielectric permittivity of vacuum), A is the investigation area (i.e., the area that can be investigated by the imaging system and that includes, by hypothesis, the cross section of the cylindrical target); finally, $G^{2D}(x, y/s,t)$ is the known Green's function for free space [26] and is given by

$$G^{2D}(x,y/s,t) = -\frac{j}{4} H_0^{(2)}\left(k\sqrt{(x-s)^2 + (y-t)^2}\right) \tag{9}$$

where $k = \omega\sqrt{\varepsilon_0 \mu_0}$ is the wavenumber in the propagation medium (it is assumed here to be vacuum, but different media could be simply assumed by modifying the propagation constants), ω is the operating angular frequency, and $H_0^{(2)}$ is the Hankel function of second kind and zero-th order. For lossy dielectrics, the relative dielectric permittivity is complex and given by $\tilde{\varepsilon}(x, y) = \varepsilon(x, y) - j\omega^{-1}\sigma(x, y)$.

By using proper sensors, the total electric field, which is the sum of the incident field and the field scattered by the unknown target, is collected in a certain region outside the investigation area A. Usually, the same field cannot be measured inside the target region A and, consequently, it is an unknown quantity. The incident field is known everywhere. It results that the inverse scattering problem is nonlinear, since both $\tau(x,y)$ and $E^{tot}(x,y)$ for points $(x,y)\in A$ must be determined.

In the past, several approximate methods have been proposed to overcome the nonlinearity of this problem. In particular, linearized approaches, in which the unknown internal total electric field is essentially approximated by the know incident field, have been proposed. These methods are valid for weak scatterers only. However, to inspect very strong scatterers, the original nonlinear equation should be solved. In that case, one can resort to numerical methods, which "discretize" the continuous model and result in algebraic (nonlinear) equations to be solved. Evolutionary algorithms can play a key role in solving these kind of problems as it will be discussed in the following.

It should be mentioned that, since the information content of the data (which are measured in the observation domain) may not be sufficient to retrieve the unknown distributions of the dielectric parameters and also due to the severe ill-posedness of the inverse scattering problem, another relation is usually employed, in addition to equation (8). In general, the scattering equation for the internal field is used. This equation is formally equal to equation (8), but in this case $(x,y)\in A$. It imposes that the retrieved dielectric properties of the object and the internal total electric field must be consistent with the known incident field inside *A*.

The nonlinear inverse scattering problem has been solved in the past by using deterministic techniques (e.g., conjugate gradient methods, Newton methods, etc.) [27-40]. The main limitation of these techniques is that they are local methods and can be trapped in local minima corresponding to false solutions. In general, they require that the starting point of the iterative search must be very close to the right solution. In practical applications, this requires "a priori" information on the configuration to be inspected that is not always available.

It is almost evident that this challenging nonlinear problem can notably benefit from the application of stochastic optimization methods. As a matter of fact, since they introduction in the field of computational electromagnetics, stochastic approaches have been exploited for inverse scattering based diagnostic approaches. The DE algorithm has been found to be particularly suitable for this kind of applications.

In order to apply optimization techniques (a schematic representation is reported in Fig. 6), the problem solution is first recast as an optimization problem by defining a suitable cost function (often called *fitness* function.)

The fitness function can contain different terms, e.g.,

$$f\{\tau,E^{tot}\}=c_1 f_{ext}\{\tau,E^{tot}\}+c_2 f_{\text{int}}\{\tau,E^{tot}\}+c_3 f_{penalty}\{\tau,E^{tot}\} \qquad (10)$$

where $f_{ext}\{\tau,E^{tot}\}$ is a term related to the minimization of the residual of the discretized version of the scattering equation for the measured data in the external region(observation domain); $f_{\text{int}}\{\tau,E^{tot}\}$ is an analogous term concerning the equation for the field inside the investigation area *A*; finally, $f_{penalty}\{\tau,E^{tot}\}$ represents a term in which all the available information on the scatterer under test can be included as a penalty function. This term can also play the role of a regularization function. In

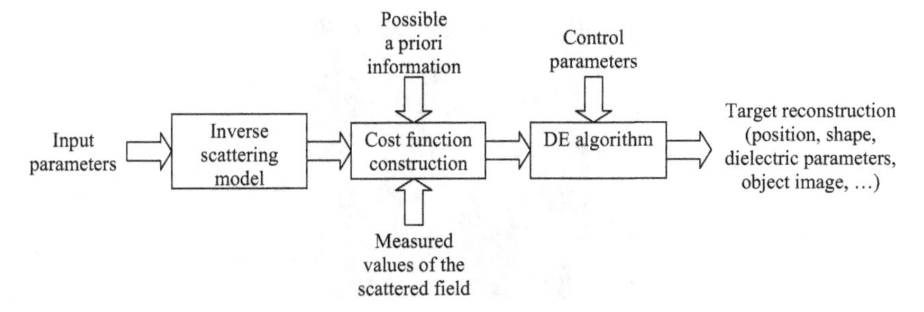

Fig. 6. Schematic representation of the application of the DE to the electromagnetic imaging problem

equation (10) each term is multiplied by a constant coefficient that can be used to properly "balance" the different terms.

An example of the results that can be obtained by the previous approach is reported for illustration. The electric field data have been collected by using $V = 8$ views, and for each view, the scattered field has been collected in $M = 51$ measurement points. In particular, in the v-th view, the source, modeled as an infinite z-directed line-current source, is located at position $\mathbf{s}_v = (2.4\lambda_0, 0.25(v-1)\pi)$, $v = 1,...,V$, whereas the measurement points are located on an arc of circumference at positions $\mathbf{r}_m^v = (2.4\lambda_0, 0.25v\pi + 0.027(m-1)\pi), m = 1,...,M$, $v = 1,...,V$.

The investigation area is a square domain of side $L = 1.5\lambda_0$, centered in the origin of the coordinate system. In the reconstruction phase, this domain is discretized by using a mesh of $N = 14 \times 14 = 196$ square subdomains. The input electric field data have been obtained by using a numerical simulator based on the method of moments [26] in which a finer mesh (with respect to that considered in the inversion phase) is used. Furthermore, the electric field values have been corrupted by a Gaussian noise (with zero mean value) with a signal-to-noise ratio of 25 dB.

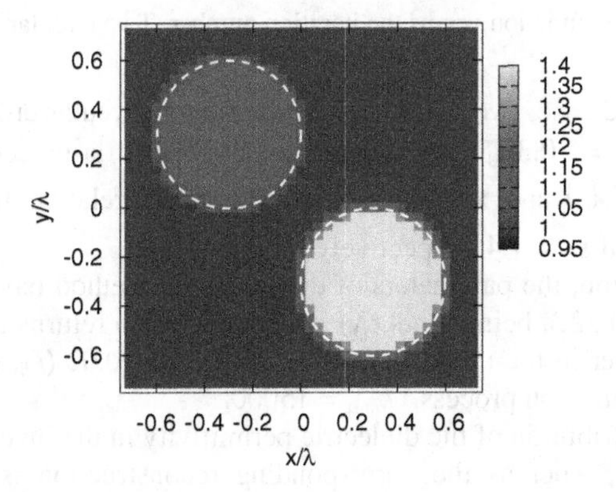

Fig. 7. Actual distribution of the dielectric permittivity in the investigation domain. Two separated circular cylinders. $N = 196$.

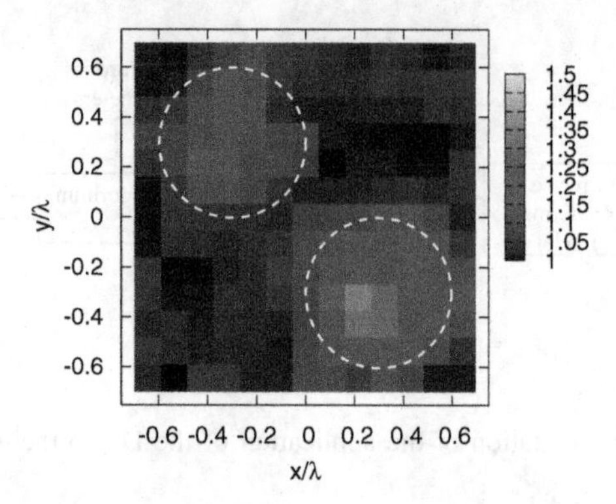

Fig. 8. Reconstructed distribution of the dielectric permittivity in the investigation domain. Two separated circular cylinders. $N = 196$. V = 8. M = 51. SNR = 25 dB.

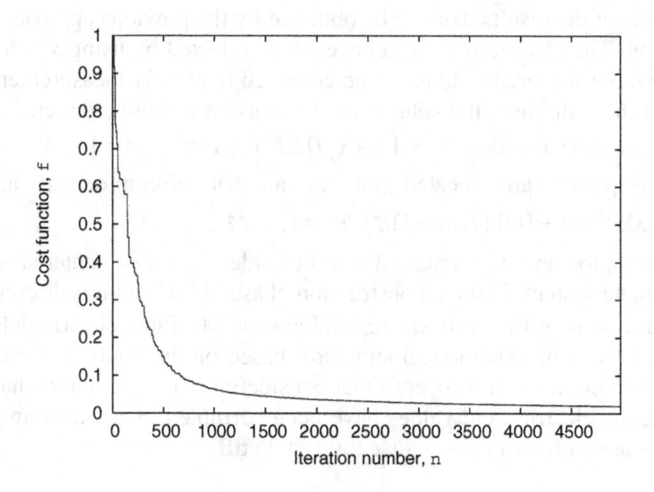

Fig. 9. Cost function versus the iteration number. Two circular cylinders.

In the considered case, two dielectric cylinders with circular cross sections of radiuses $r_1 = r_2 = 0.3\lambda_0$ are considered. They are centered at points $\mathbf{x}_c^{(1)} = (-0.3\lambda_0, 0.3\lambda_0)$ and $\mathbf{x}_c^{(2)} = (0.3\lambda_0, -0.3\lambda_0)$. The relative dielectric permittivities are $\varepsilon_{r1} = 1.2$ and $\varepsilon_{r2} = 1.4$, respectively.

For this simulation, the parameters of the inversion method have been set equal to $c_r = 0.8$, F = rand(0., 2.), being rand(x,y) a function which returns a uniformly distributed random number in the range [x,y], $N_p = 2000$, $f_{th} = 0.02$ (f_{th}, threshold value for stopping the minimization process,) $k_{max} = 15000$.

The original distribution of the dielectric permittivity in the investigation domain is reported in Fig. 7, whereas the corresponding reconstruction is shown in Fig. 8. Moreover, Fig. 9 shows the behavior of the cost function versus the iteration number.

It should be noted that we refers here to a case in which the investigation area is represented by a pixel grid, being the objective of this example the evaluation of the

DE algorithm capabilities in solving the inverse problem. However, being computational intensive, the DE method can only be applied with a reduced parameterization. Consequently, focusing or hybrid approaches should be used. In the first case, the inspection starts with a coarse grid and the grid is successively refined as well as the position of the scatterer is retrieved (iterative approach). In the former case, once the scatterer has been localized by the DE, it is further inspected by a deterministic procedure (e.g., a conjugate gradient method) with a finer discretization.

A focused approach has been considered for example in [41], where the DE algorithm has been applied for the detection of buried structures. Essentially, the formulation is the same of the one reported in this section, but the Green's function for free space $G^{2D}(x,y/s,t)$ is replaced by the Green's function for the half space, which is given by [42]

$$G_{2D}^{12}(x,y/x',y')=\frac{1}{2\pi}\int_{-\infty}^{+\infty}\frac{j}{\gamma_1+\gamma_2}e^{-j\lambda(x-x')}e^{-j(\gamma_1 y-\gamma_2 y')}d\lambda \tag{11}$$

when it is used in the "external" equation (equation (8), in the case $(x,y)\notin A$), and by

$$G_{2D}^{22}(x,y/x',y')=\frac{1}{2\pi}\int_{-\infty}^{+\infty}\frac{j}{2\gamma_2}\frac{\gamma_2-\gamma_1}{\gamma_1+\gamma_2}e^{-j\lambda(x-x')}e^{-j\gamma_2(y+y')}d\lambda+ \\ \frac{1}{2\pi}\int_{-\infty}^{+\infty}\frac{j}{2\gamma_2}e^{-j\lambda(x-x')}e^{-j\gamma_2|y+y'|}d\lambda \tag{12}$$

for the equation for the internal total electric field (equation (8) in the case $(x,y)\in A$). Equations (11) and (12) are valid for a half space geometry with homogeneous half spaces [42]. In these relations, $\gamma_1=\sqrt{(\lambda^2-k_u^2)}$ and $\gamma_2=\sqrt{(\lambda^2-k_l^2)}$, being k_u and k_l the propagation constants in the upper and lower regions, respectively.

In [41] the DE algorithm has been applied by combining two of the various possible implementing strategies for this evolutionary approach. In particular, the *DE/1/best/bin* version [43] is used until the cost function has reached a predefined value; successively, the *DE/1/rand/bin* strategy [43] is applied. It has been found that the *DE/1/best/bin* strategy is quite able to rapidly locate the "attraction basin" of a minimum, but, since it uses the best individual of the population to perform the mutation, it can sometimes be trapped in a local minimum. This drawback is overcame by switching, after a predefined threshold, to the *DE/1/rand/bin* strategy, which is able to explore more efficiently the search space, without modifying the previous best solution if it is inside the correct attraction basin. Concerning the choice of the control parameters, F has been chosen in the range [0.5,1.0], whereas good reconstructions have been obtained with $c_r = 0.8$.

As previously mentioned, the application of the DE to imaging problems requires a reduced parameterization. This has been obtained in [44-45] by approximating the cylindrical scatterers with canonical objects with circular and elliptical cross sections.

The proposed application concerns the inspection of tunnels and pipes in a cross-borehole configuration [46] in which the effects of the interface between the upper and lower media have been neglected assuming deeply buried objects. In the inverse problem, the problem unknowns are represented by the cylinder center and the radius (circular cross section) or the major semiaxis, the eccentricity, and the tilt angle (elliptic cross section). The DE algorithm has been applied to a cost function simply based on the "external" equation only (equation (8) for $(x, y) \notin A$). Excellent reconstructions have been obtained with the following control parameters: $N_p = 25$, $k_{max} = 40$, $c_r = 0.9$ and $F = 0.7$.

Finally, the DE algorithm has been further applied to inspecting single and multiple PEC cylinders of arbitrary shapes [47-49]. The DE-based approach has been proven to be able to obtain good reconstructions of the profiles of the PEC cylinders (both with synthetic and real data) by using the following values of the control parameters: $c_r = 0.9$ and $F = 0.7$. For this specific application, the DE method has been compared with a standard real-coded GA under the same operating conditions. In this case, too, the DE algorithm has been found to outperform the GA due to the need for a smaller population. It should be mentioned that the author used in [49] the so-called "dynamic DE strategy," in which an additional competition is introduced between the "resultant vector" and the current optimal individual. The current optimal element is replaced if the new element corresponds to a better solution and the updated element is immediately included in the new population.

4 Conclusions

In this chapter, the application of the differential evolution algorithm to antenna synthesis and microwave imaging has been reviewed. These two areas constitute important examples in the framework of the computational electromagnetics, which, due to the complex mathematical problems involved, greatly benefit from the introduction of evolutionary algorithms. The differential evolution method, in particular, has been found very suitable in various antenna synthesis problems, including the design of monopulse antennas, for which a recently proposed approach based on a subarray configuration has been outlined and discussed with the help of new results. Furthermore, the application of the differential evolution method for retrieving unknown targets has been discussed, too. Recently proposed solutions, both for perfectly conducting objects and penetrable materials, have been briefly reviewed. The formulation of the inverse-scattering based inverse problem has been reported. Following an approach based on the differential evolution method (previously applied for detecting buried objects in a half space) an example showing the reconstruction capabilities of the algorithm (in a tomographic configuration) has been included. Clearly, as previously discussed, differential evolution being computationally expensive can be applied in the imaging field only with reduced parameterizations. Consequently, other strategies can be followed for high-resolution imaging. However, the differential evolution has shown excellent capabilities in finding the "attraction basin" for the related optimization problem. Analogously, very efficient array antennas have been designed for different applications by applying the differential evolution method. Finally, in both the applications, comparative results have been reported in the scientific literature suggesting the superiority of the differential evolution method with respect to the standard genetic algorithm.

References

1. Johnson, J.M., Ramat-Samii, Y.: Genetic algorithms in engineering electromagnetics. IEEE Antennas Propagat. Mag. 39(4), 7–21 (1997)
2. Haupt, R.L.: An introduction to genetic algorithms for electromagnetics. IEEE Antennas Propagat. Mag. 37(2), 7–15 (1995)
3. Weile, D.S., Michielssen, E.: Genetic algorithm optimization applied to electromagnetics: a review. IEEE Trans. Antennas Propagat. 45, 343–353 (1997)
4. Rahmat-Samii, Y., Michielssen, E.: Electromagnetic Optimization by Genetic Algorithms. Wiley, New York (1999)
5. Price, K.: An introduction to differential differential evolution. In: Corne, D., Dorigo, M., Glover, F. (eds.) New Ideas in Optimization. McGraw-Hill, New York (1999)
6. Storn, R., Price, K.: Differential evolution - a simple and efficient heuristic for global optimization over continuous spaces. J. Global Optimization 11, 341–359 (1997)
7. Moscato, P.: On evolution, search, optimization, genetic algorithms and martial arts towards memetic algorithms. Tech. Rep. Caltech Concurrent Computation Program, Report. 826, California Institute of Technology, Pasadena, California, USA (1989)
8. Robinson, J., Rahmat-Samii, Y.: Particle swarm optimization in electromagnetics. IEEE Trans. Antennas Propagat. 52(2), 397–407 (2004)
9. Davis, L.: Genetic Algorithms and Simulated Annealing. Morgan Kaufmann Publishers Inc., San Francisco (1987)
10. Goldberg, D.E.: Genetic Algorithms in Search, Optimization and Machine Learning. Addison-Wesley, Boston (1989)
11. Yang, S., Gan, Y.B., Qing, A.: Sideband suppression in time-modulated linear arrays by the differential evolution algorithm. IEEE Antennas Wireless Propagat. Lett. 1, 173–175 (2002)
12. Kurup, D.G., Himdi, M., Rydberg, A.: Synthesis of uniform amplitude unequally spaced antenna array using the differential evolution algorithm. IEEE Trans. Antenna Propagat. 51(9), 2210–2217 (2003)
13. Yang, S., Nie, Z.: Mutual coupling compensation in time modulated linear antenna array. IEEE Trans. Antennas Propagat. 53(12), 4182–4185 (2005)
14. Balanis, C.A.: Antenna theory: analysis and design. Wiley, New York (1982)
15. Yang, S., Gan, Y.B., Tan, P.K.: A new technique for power-pattern synthesis in time-modulated linear arrays. IEEE Antennas Wireless Propagat. Lett. 2, 285–287 (2005)
16. McNamara, D.A.: Synthesis of sum and difference patterns for two-section monopulse arrays. Inst. Elect. Eng. Proc. pt H 135(6), 371–374 (1996)
17. Ares, F., Rodrìguez, J.A., Moreno, E.: Optimal compromise among sum and difference patterns. J. Electromagnetic Waves and Appl. 10, 1543–1555 (1996)
18. Lòpez, P., Rodrìguez, J.A., Ares, F., Moreno, E.: Subarray weighting for the difference patterns of monopulse antennas: Joint optimization of subarray configurations and weights. IEEE Trans. Antennas Propagat. 49(11), 1606–1608 (2001)
19. Caorsi, S., Massa, A., Pastorino, M., Randazzo, A.: Optimization of the difference patterns for monopulse antennas by an hybrid real/integer-coded differential evolution method. IEEE Trans. Antennas Propagat. 53(1), 372–376 (2005)
20. Massa, A., Pastorino, M., Randazzo, A.: Optimization of the directivity of a monopulse antenna with a subarray weighting by an hybrid differential evolution method. IEEE Antennas Wireless Propagat. Lett. 5, 155–158 (2006)
21. Chiu, C.C., Liu, P.T.: Image reconstruction of a perfectly conducting cylinder by the genetic algorithm. IEE Proc. Microwave Antennas Propag. 143 (1996)

22. Kent, S., Gunel, T.: Dielectric permittivity estimation of cylindrical objects using genetic algorithm. J. Microwave Power and Electromagn. Energy 32, 109–113 (1997)
23. Caorsi, S., Massa, A., Pastorino, M., Raffetto, M., Randazzo, A.: Detection of buried inhomogeneous elliptic cylinders by a memetic algorithm. IEEE Trans. Antennas Propagat. 51, 2878–2884 (2003)
24. Qian, Z.P., Hong, W.: Image reconstruction of conducting cylinder based on FD-MEI and genetic algorithm. In: Proc. IEEE APS Int. Symp., vol. 2, pp. 718–721 (1998)
25. Caorsi, S., Massa, A., Pastorino, M.: A computational technique based on a real-coded genetic algorithm for microwave imaging purposes. IEEE Trans. Geosci Remote Sensing, special issue on Computational Wave Issues in Remote Sensing, Imaging and Target Identification, Propagation, and Inverse Scattering 38, 1697–1708 (2000)
26. Balanis, C.A.: Advanced engineering electromagnetics. Wiley, New York (1989)
27. Dourthe, C., Pichot, C., Dauvignac, J.Y., Cariou, J.: Inversion algorithm and measurement system for microwave tomography of buried object. Radio Sci. 35, 1097–1108 (2000)
28. Cui, T.J., Chew, W.C.: Diffraction tomographic algorithm for the detection of three-dimensional object buried in a lossy half-space. IEEE Trans. Geosci. Remote Sensing 50, 42–49 (2002)
29. Ramananjaona, C., Lambert, M., Lesselier, D., Zolésio, J.P.: Shape reconstruction of buried obstacles by controlled evolution of a level set: from a min-max formulation to numerical experimentation. Inverse Problems 17, 1087–1111 (2001)
30. Cui, T.J., Aydiner, A.A., Chew, W.C., Wright, D.L., Smith, D.W.: Three-dimensional imaging of buried object in very lossy earth by inversion of VETEM data. IEEE Trans. Geosci. Remote Sensing 41, 2197–2209 (2003)
31. Smith, G.S., Petersson, L.E.R.: On the use of evanescent electromagnetic waves in the detection and identification of object buried in lossy soil. IEEE Trans. Antennas Propagat. 48, 1295–1300 (2000)
32. Micolau, G., Saillard, M., Borderies, P.: DORT method as applied to ultrawideband signals for detection of buried objects. IEEE Trans. Geosci. Remote Sensing 41, 1813–1820 (2003)
33. Ferrayé, R., Dauvignac, J.Y., Pichot, C.: An inverse scattering method based on contour deformation by means of a level set method using frequency hopping technique. IEEE Trans. Antennas Propagat. 51, 1100–1112 (2003)
34. Zhang, Z.Q., Liu, Q.H.: Two nonlinear inverse methods for electromagnetic induction measurements. IEEE Trans. Geosci. Remote Sensing 39, 1331–1339 (2001)
35. Lambert, M., Lesselier, D.: Binary-constrained inversion of a buried cylindrical obstacle from complete and phaseless magnetic fields. Inverse Problems 16, 563–576 (2000)
36. Chommeloux, L., Pichot, C., Bolomey, J.C.: Electromagnetic modeling for microwave imaging of cylindrical buried inhomogeneities. IEEE Trans. Microwave Theory Tech. 34, 1064–1076 (1986)
37. Hughes, D., Zoughi, R.: A method for evaluating the dielectric properties of composites using a combined embedded modulated scattering and near-field microwave nondestructive testing technique. In: Proc. 18th IEEE Instrum. Meas. Technol. Conf., pp. 1882–1886 (2001)
38. Kleinman, R.E., van den Berg, P.M.: Two-dimensional location and shape reconstruction. Radio Sci. 29, 1157–1169 (1994)
39. Tijhuis, A.G., Belkebir, K., Litman, A.C.S., de Hon, B.: Theoretical and computational aspects of 2-D inverse profiling. IEEE Trans. Geosci. Remote Sensing 39, 1316–1330 (2001)

40. Franza, O., Joachimowicz, N., Bolomey, J.C.: SICS: A sensor interaction compensation scheme for microwave imaging. IEEE Trans. Antennas Propagat. 50, 211–216 (2002)
41. Massa, A., Pastorino, M., Randazzo, A.: Reconstruction of Two-Dimensional Buried Objects by a Differential Evolution Method. Inverse Problems, special session on Electromagnetic Characterization of Buried Obstacles 20(6), S135–S150 (2004)
42. Sommerfeld, A.: Partial Differential Equations in Physics. Academic Press, New York (1949)
43. Storn, R.: On the Usage of Differential Evolution for Function Optimization. In: Proc. 1996 Biennial Conference of the North American Fuzzy Information Processing Society (NAFIPS 1996), pp. 519–523 (1996)
44. Michalski, K.A.: Electromagnetic imaging of elliptical-cylindrical conductors and tunnel using a differential evolution algorithm. Microwave Opt. Technol. Lett. 28(3), 164–169 (2001)
45. Michalski, K.A.: Electromagnetic imaging of circular-cylindrical conductors and tunnels using a differential evolution algorithm. Microwave Opt. Technol. Lett. 27(5), 330–334 (2000)
46. Bonnard, S., Vincent, P., Saillard, M.: Cross-borehole inverse scattering using a boundary finite-element method. Inverse Problems 14, 521–534 (1998)
47. Qing, A.: Electromagnetic inverse scattering of multiple two-dimensional perfectly conducting objects by the differential evolution strategy. IEEE Trans. Antennas Propagat. 51(6), 1251–1262 (2003)
48. Qing, A.: Electromagnetic inverse scattering of multiple perfectly conducting cylinders by differential evolution strategy with individuals in groups (GDES). IEEE Trans. Antennas Propagat. 52(5), 1223–1229 (2004)
49. Qing, A.: Dynamic differential evolution strategy and applications in electromagnetic inverse scattering problems. IEEE Trans. Geosci. Remote Sensing 44(1), 116–125 (2006)

Applications of Differential Evolution in Power System Optimization

L. Lakshminarasimman and S. Subramanian

Department of Electrical Engineering, Annamalai University,
Annamalainagar – 608002, India
llnarasimman@gmail.com

Summary. Modern power systems are very large, complex and widely distributed. Scarcity in energy resources, increasing power generation cost and ever-growing demand for electric energy necessitates optimal operation of power systems. Even a small reduction in production cost may lead to a large savings. Hence efficient algorithms for solving the power system scheduling are needed. New optimization methods based on evolutionary computation that abstract the principle of natural selection and genetics are employed for scheduling problems. They are easy to implement and have the capability to converge to global optimum at a relatively lesser computational effort. Differential Evolution (DE), a numerical optimization approach is simple, easy to implement, significantly faster and robust. It has been verified as a promising candidate for solving real-valued engineering optimization problems. This chapter is concerned with the applications of differential evolution and its variants for various power system scheduling problems like economic dispatch, dynamic economic dispatch and unit commitment. Different case studies have been conducted including nonlinearities such as the valve-point effects, prohibited operating zones and transmission losses. This chapter enumerates the advantages of differential evolution to determine the most economic conditions of the electric power system.

1 Introduction

Electricity is the indispensable form of energy in the modern world. The modern economy is totally dependent on the electricity as a basic input. The load demand is increasing year by year. The increasing energy demand and decreasing energy resources have necessitated the optimum use of available resources. One of the requirements of power system operation is to supply power to the customers economically. In recent years, more stringent requirements have been imposed on electric utilities in order to supply high quality electrical energy. Interconnections between systems are also increasing to enhance reliability and economy. Therefore, optimum scheduling of power plant generation is of great importance to electric utility systems and the optimal operating strategies are to be determined to satisfy versatile operational constraints.

The application of optimization techniques to power system planning and operation has been an active research in the recent past. Power system optimization problems are very difficult to solve because power systems are very large, complex, geographically widely distributed and are influenced by many unexpected events. It is therefore

U.K. Chakraborty (Ed.): Advances in Differential Evolution, SCI 143, pp. 257–273, 2008.
springerlink.com

necessary to employ most efficient optimization methods to take full advantages in simplifying the formulation and implementation of the problem.

A wide variety of mathematical optimization techniques have been applied to solve the power system operation and control problems. However, they are beset by weak convergence, unrealistic assumptions and inadequate modeling of power systems. The traditional optimization techniques have the possibility of getting trapped at local optima, depending upon the degree of non-linearity and the initial guess. Therefore, there is a need to develop new optimization techniques that can deal with the highly non-linear characteristics of power system components, and are able to determine the global optimum solution.

In the recent past non-traditional optimization techniques called evolutionary computation techniques such as genetic algorithms (GA), evolutionary programming (EP) and differential evolution (DE), are employed for power system optimization problems. Evolutionary algorithms are powerful optimization techniques based on the principle of natural selection. These algorithms are easy to implement and have the capability to converge to global optimum at a relatively lesser computational effort. The advantages of evolutionary computation includes conceptual simplicity, broader domain applications, efficiency in solving real world problems, versatility in incorporating domain knowledge, hybridization with conventional techniques, parallelism, robustness, self adoption and requirement of least human expertise. These techniques do not require any in-depth mathematical understanding of the problems to which they are applied. Among the evolutionary computation techniques DE is catching up fast and is being applied to a wide range of complex power system optimization problems.

Differential evolution (DE) developed by Storn and Price, (1997) is a numerical optimization approach that is simple, easy to implement, significantly faster and robust. The fittest of an offspring competes one-to-one with that of corresponding parent, which is different from the other evolutionary algorithms. This one-to-one competition gives rise to faster convergence rate. DE is the real coded GA combined with an adaptive random search using a normal random generator. DE uses floating point numbers that are more appropriate than integers for representing points in a continuous space. This method has been verified as a promising candidate for solving real-valued optimization problems.

The DE has been successfully applied for various power system optimization problems such generation expansion planning (Kannan et al. 2005) and hydrothermal scheduling (Lakshminarasimman and Subramanian, 2006). The hybrid differential evolution (HDE) has been employed for the solution of large capacitor placement problem (Chiou et al. 2004). The mixed integer hybrid differential evolution (MIHDE) has been employed for hydrothermal coordination (Lakshminarasimman and Subramanian, 2007), hydrothermal optimal power flow (Lakshminarasimman and Subramanian, 2007a) and network reconfiguration problem (Su and Lee, 2003).

This chapter focuses on the applications of DE algorithms to various power system optimization problems such as economic dispatch (ED), dynamic economic dispatch (DED) and unit commitment (UC). The ED problem, one of the fundamental issues in power system operation, is solved using the differential evolution algorithm. DED that determines the optimal operation of units with predicted load demand over a scheduling period has been solved using hybrid differential evolution. The optimal unit commitment solution is an essential factor in planning and operation of power systems. It is a combinatorial optimization problem involving continuous and discrete variables and has been solved using mixed integer hybrid differential evolution (MIHDE).

2 Economic Dispatch

Economic dispatch is an optimization problem, which is to distribute the total required generation among the units in operation so as to minimize the total cost of generation and transmission for a prescribed schedule of loads. It has complex and nonlinear characteristics with heavy equality and inequality constraints, such as load and operational constraints (Wood AJ and Wollenberg, 1984).

2.1 Problem Objective

The objective of economic dispatch problem can be defined as

$$\text{Minimize} \quad F = \sum_{i=1}^{N_g} FC(i) \tag{1}$$

where N_g is number of thermal units and F is the total cost of generation. The production cost, FC in terms of decision variables of generated powers is expressed as

$$FC(i) = a_i + b_i P_i + c_i P_i^2 \tag{2}$$

where P_i is the real power generation and a_i, b_i and c_i are the fuel cost coefficients of the ith generator.

2.2 Problem Constraints

Power Balance Constraint
While minimizing the total generation cost, the total generation should be equal to the total system demand plus the transmission network loss. Therefore, the power system equality constraint is expressed as

$$\sum_{i=1}^{N_g} P_i - P_D - P_L = 0 \tag{3}$$

where P_D is the total load of the system and P_L is the total transmission losses of the system. P_L is a function of unit power outputs that can be represented using B coefficients as given by

$$P_L = \sum_{i=1}^{N_g} \sum_{j=1}^{N_g} P_i B_{ij} P_j + \sum_{i=1}^{N_g} B_{0i} P_i + B_{00} \tag{4}$$

where B_{ij}, B_{0i} and B_{00} are the loss coefficients, which are constants under certain assumed operating conditions.

Generator Capacity Constraints
For stable operation, the real power generation of each generator should be restricted between its lower and upper limits of generation. The generator capacity constraints are expressed as

$$P_i^{min} \le P_i \le P_i^{max} \tag{5}$$

where min and max represent the minimum and maximum values.

Plant operators, to avoid shortening the life of their equipment, try to keep thermal gradients inside the turbine within safe limits. This mechanical constraint is usually translated into a limit on the rate of increase of electrical power output. Therefore, the operating range of all online units is restricted by their ramp rate limits as given by

$$P_i - P_i^0 \le UR_i \tag{6}$$

$$P_i^0 - P_i \le DR_i \tag{7}$$

where P_i^0 is the previous output power, UR_i is the up-ramp limit and DR_i is the down-ramp limit of the ith generator.

The prohibited operating zone constraints avoid the operation of units in the prohibited zones. The prohibited operating zones of a unit divide the operating range between its minimum and maximum generation limits into several disjoint convex sub-regions. The valve points of thermal units also generate many prohibited zones. Therefore, the feasible operating zones of unit *i* can be described as

$$P_i \in \begin{cases} P_i^{min} \le P_i \le P_{i,1}^{l} \\ P_{i,j-1}^{u} \le P_i \le P_{i,j}^{l} \quad j = 2,3,\ldots,n_i \\ P_{i,n_i}^{u} \le P_i \le P_{i,}^{max} \end{cases} \tag{8}$$

where n_i is the number of prohibited zones of ith unit.

2.3 Economic Dispatch Using Differential Evolution

The differential evolution algorithm employed for economic dispatch problem is briefly discussed as follows:

Initialization
The initial population of N_p individuals is randomly selected based on uniform probability distribution for all variables to cover the entire search space uniformly. The real power generation of ith plant is expressed as

$$P_i = P_i^{min} + \rho\left(P_i^{max} - P_i^{min}\right) \tag{9}$$

where $\rho \in [0,1]$ is uniformly distributed random number.

A penalty function approach is used to handle the power balance constraint. The penalty factor reduces the fitness of the vector according to the magnitude of constraint violation. The objective function of the ED problem, which is to be minimized, is given by

$$\psi = \sum_{i=1}^{N_g}\left(a_i + b_i P_i + c_i P_i^2\right) + \sum_{z=1}^{N_c} \lambda_z \left|VIOL_z\right| \tag{10}$$

where λ is the penalty factor, N_c represents the number of constraints and VIOL is the constraint violation.

Mutation

DE generates new parameter vectors by adding the weighted difference vector between two population members to a third member. A perturbed individual is therefore generated on the basis of the parent individual in the mutation process by

$$\hat{Z}_i^{G+1} = Z_p^G + F \times \left(Z_j^G - Z_k^G\right) \tag{11}$$

where F is a scaling factor and j and k are randomly selected. The scaling factor $F \in [0,1]$ ensures the fastest possible convergence and G represents generation number.

If the new decision variable is out of the limits (lower and upper) by an amount, this amount is subtracted or added to the limit violated to shift the value inside the limits and appropriate adjustments are made to satisfy the prohibited operating zone constraints.

Crossover

In the crossover operation, the gene of an individual at the next generation is produced from the perturbed individual and the present individual by a binomial distribution to perform the crossover operation to generate the offspring.

$$\hat{Z}_i^{G+1} = \begin{cases} Z_{ji}^G, & \text{if a random number} > C_R \\ \hat{Z}_{ji}^{G+1}, & \text{otherwise} \end{cases} \tag{12}$$

where $i = 1, \ldots, N_p$, $j = 1, \ldots, n$ and the crossover factor $C_R \in [0,1]$ is assigned by the user.

Evaluation and Selection

In the evaluation process an offspring competes one-to-one with the parent. The parent is replaced by its offspring if the fitness of the offspring is better than that of its parent. Contrarily, the parent is retained in next generation if the fitness of offspring is worse than the parent as expressed by

$$Z_i^{G+1} = \arg\min\left\{\psi\left(Z_i^G\right),\ \psi\left(\hat{Z}_i^{G+1}\right)\right\} \tag{13}$$

$$\hat{Z}_b^{G+1} = \arg\min\left\{\psi\left(Z_i^{G+1}\right), i=1,\ldots N_P\right\} \tag{14}$$

where $i = 1,\ldots N_P$ and arg min means the argument of the minimum.

Then the vector with lesser cost replaces the initial population. With the members of the next generation thus selected, the cycle repeats until the maximum number of generations or no improvement is seen in the best individual after many generations.

Different strategies can be adopted in DE algorithm depending on the vector to be perturbed, number of difference vectors considered for perturbation, and the type of crossover. In this study, DE with random vector perturbation and binominal crossover is employed. DE control parameters are the population size N_P, weight applied to the random differential F and crossover constant C_R.

2.4 Computer Simulation

In this study, the performance of the DE based economic dispatch algorithm is implemented using C++ code on a PIV 2.4GHz personal computer and is evaluated using an illustrative test system consisting of 15 generators (Gaing ZL, 2003).

The generating unit characteristics are given in Table 1 and 2. The loss coefficients matrix can be taken from the reference. The dimension of the problem is 15 variables representing the real power generation and the control parameters are $N_p = 78$, $F = 1.0$ and $CR = 0.9$. The comparison of the optimal system costs obtained from the DE based approach with that of particle swarm optimization (PSO) and genetic algorithm is given in Table 3. The proposed approach yields better results than PSO and GA.

Table 1. Generating unit data

Unit	P_i^{min} (MW)	P_i^{max} (MW)	a_i (MW)	b_i (\$/MW)	c_i (\$/MW^2)	UR_i (MW/h)	DR_i (MW/h)	P_i^0
1	150	455	671	10.1	0.000299	80	120	400
2	150	455	574	10.2	0.000183	80	120	300
3	20	130	374	8.8	0.001126	130	130	105
4	20	130	374	8.8	0.001126	130	130	100
5	150	470	461	10.4	0.000205	80	120	90
6	135	460	630	10.1	0.000301	80	120	400
7	135	465	548	9.8	0.000364	80	120	350
8	60	300	227	11.2	0.000338	65	100	95
9	25	162	173	11.2	0.000807	60	100	105
10	25	160	175	10.7	0.001203	60	100	110
11	20	80	186	10.2	0.003586	80	80	60
12	20	80	230	9.9	0.005513	80	80	40
13	25	85	225	13.1	0.000371	80	80	30
14	15	55	309	12.1	0.001929	55	55	20
15	15	55	323	12.4	0.004447	55	55	20

Table 2. Prohibited zones of generating units

Unit	Prohibited zones (MW)
2	[185 225] [305 335] [420 450]
5	[180 200] [305 335] [390 420]
6	[230 255] [365 395] [430 455]
12	[30 40] [55 65]

The success rate, a best measure for the performance of the technique, is defined the ratio of the total number of times the optimal solution is found to the total number of test runs. DE is found to have a success rate of hundred percent. The proposed approach converges to the optimal solution in 0.25 seconds. The proposed algorithm has been demonstrated to have superior features including high-quality solution and computational efficiency.

Table 3. Comparison of optimal solution for 15-generator system

Generation Schedule	DE	PSO	GA
P_1, MW	455.0000	439.1162	415.3108
P_2, MW	420.0000	407.9727	359.7206
P_3, MW	130.0000	119.6324	104.4250
P_4, MW	130.0000	129.9925	74.9853
P_5, MW	270.0000	151.0681	380.2844
P_6, MW	460.0000	459.9978	426.7992
P_7, MW	430.0000	425.5601	341.3164
P_8, MW	60.0000	98.5699	124.7867
P_9, MW	25.0000	113.4936	133.1445
P_{10}, MW	63.0498	101.1142	89.2567
P_{11}, MW	80.0000	33.9116	60.0572
P_{12}, MW	79.9349	79.9583	49.9998
P_{13}, MW	25.0000	25.0042	38.7713
P_{14}, MW	15.0000	41.4140	41.9425
P_{15}, MW	15.0000	35.6140	22.6445
Total power output (MW)	2658.0	2662.4	2668.4
P_L (MW)	27.9778	32.4306	38.2782
Total cost ($/h)	32589	32858	33113

3 Dynamic Economic Dispatch

Dynamic economic dispatch (DED), an extension of the economic dispatch problem, is a method of scheduling the online generators with a predicted load demand over a certain period of time taking into account the various constraints imposed on the system operation. The DED problem has been recognized as not only a more accurate formulation of the economic dispatch but also a difficult dynamic optimization problem because of its large dimensionality (Han et al. 2001).

3.1 Objective Function

The objective of the dynamic economic dispatch problem is to schedule the committed units economically over a scheduling period T as given by

$$\text{Minimize} \quad F = \sum_{t=1}^{T} \sum_{i=1}^{N_g} FC(i,t) \tag{15}$$

The production cost is expressed in a quadratic form with valve point loading effect is given as

$$FC(i,t) = a_i + b_i P_{it} + c_i P_{it}^2 + \left| e_i \times \sin\left\{ f_i \times \left(P_{it}^{min} - P_{it} \right) \right\} \right| \tag{16}$$

where e_i and f_i represent the cost coefficients of ith unit valve point effects.

3.2 System Constraints

Power Balance Constraint

The power system equality constraint is expressed as

$$\sum_{i=1}^{N_g} P_{it} - P_{Dt} - P_{Lt} = 0 \tag{17}$$

where $t = 1,2,\ldots,T$, P_{Dt} is the forecasted total power demand at time t and P_{Lt} is the total transmission losses of the system at time t. The general form of loss formula using B coefficients is

$$P_{Lt} = \sum_{i=1}^{N_g} \sum_{j=1}^{N_g} P_{it} B_{ij} P_{jt} \tag{18}$$

Generator Capacity Constraints

The generator capacity constraints are expressed as

$$P_i^{min} \le P_{it} \le P_i^{max} \tag{19}$$

The ramp rate limits are given by

$$P_{it} - P_{i(t-1)} \le UR_i \tag{20}$$

$$P_{i(t-1)} - P_i \le DR_i \tag{21}$$

3.3 DED Using Hybrid Differential Evolution

The dynamic economic dispatch has to determine optimal scheduling of generators over specified intervals of time period. Therefore, the number of decision variables will be number of generating units multiplied by the number of time intervals. The population size should be 5-10 times the value of the dimension of the problem in order to avoid premature convergence.

Hybrid differential evolution (HDE) has overcome the usage of large population, which results in lesser computation time. The hybrid version employs two additional operations, acceleration operation that improves the fitness from one generation to another and migration operation to upgrade the exploration of the search space. The acceleration phase is used to accelerate convergence, although this faster convergence generally leads to obtaining a local optimum. The migration phase is used to escape this local optimum point since the new candidate individuals are regenerated on the basis of the best individual at the current generation. Correspondingly, the diversity can still be retained by such a regeneration procedure.

The DED problem with large number of decision variables is well suited for the application of hybrid differential evolution. The mutation, crossover and evaluation operations are same as explained in the preceding sections. Only the initialization, acceleration and migration phases of the HDE algorithm are enumerated as follows:

Initialization

The initial population of N_p individuals is randomly selected based on uniform probability distribution for all variables to cover the entire search space uniformly. The real power generation of ith plant at time t is expressed as

$$P_{it} = P_i^{min} + \rho\left(P_i^{max} - P_i^{min}\right) \tag{22}$$

where $\rho \in [0,1]$ is uniformly distributed random number. The objective function of the DED problem, which is to be minimized, is given by

$$\psi = \sum_{t=1}^{T} \sum_{i=1}^{N_g} FC(P_{it}) + \sum_{z=1}^{N_c} \lambda_z |VIOL_z| \tag{23}$$

where λ is the penalty factor, N_c represents the number of constraints and VIOL is the constraint violation.

Acceleration Operation

If the best fitness at the present generation is not further improved by the mutation and crossover operations, then the present best individual is pushed towards a better point. Thus, the accelerated phase is represented as

$$Z_b^{G+1} = \begin{cases} \hat{Z}_b^{G+1}, & \text{if } \psi\left(\hat{Z}_b^{G+1}\right) < \psi\left(\hat{Z}_b^{G}\right) \\ \hat{Z}_b^{G+1} - \alpha \nabla \psi, & \text{otherwise} \end{cases} \tag{24}$$

where Z_b^{G+1} is the best individual. The gradient of the objective function $\nabla\psi$ can be calculated with finite variation. The step size $\alpha \in [0,1]$ is determined by the descent property. Initially α is set a value of one to obtain the new individual Z_b^N. If the descent property is satisfied, i.e.,

$$\psi\left(Z_b^N\right) < \psi\left(Z_b^{G+1}\right) \tag{25}$$

then the Z_b^N becomes a candidate in the next generation and is added to this population replacing the worst individual. If the descent property is not satisfied, then step size is lowered a little. The descent method is repeated to search Z_b^N until $\alpha\nabla\psi$ is sufficiently small or a specified number of iterations are performed. This faster decent results in a premature convergence and the migration phase regenerates a new population.

Migration Operation

A migration phase is introduced to regenerate a newly diverse population of individuals to enhance the investigation over the search space, and thus, reduce the pressure of selection from a small population. The new populations are obtained based on the best individual Z_b^{G+1}. The hth gene of the ith individual is regenerated as

$$Z_{hi}^{G+1} = \begin{cases} Z_{hb}^{G+1} + \delta_{hi}\left(Z_{h\,min} - Z_{hb}^{G+1}\right) & \text{if } \tilde{\delta}_{hi} < \dfrac{Z_{hb}^{G+1} - Z_{h\,min}}{Z_{h\,max} - Z_{h\,min}} \\ Z_{hb}^{G+1} + \delta_{hi}\left(Z_{h\,max} - Z_{hb}^{G+1}\right) & \text{otherwise} \end{cases} \qquad i = 1, \ldots, Np, \quad h = 1, \ldots, n \tag{26}$$

where δ and $\tilde{\delta}$ denote uniformly distributed random numbers. This diversified population is then used as the initial decision parameters to escape the local optimum points. The migration operation is performed only if the population diversity ρ is smaller than the desired tolerance of population diversity ε_1.

$$\rho = \frac{\left\{\sum\limits_{\substack{i=1 \\ i\neq b}}^{N_p}\left(\sum\limits_{h=1}^{n}\eta_Z\right)\right\}}{n\left(N_p - 1\right)} < \varepsilon_1 \tag{27}$$

where

$$\eta_Z = \begin{cases} 1, & \text{if } \left|\dfrac{Z_{hi} - Z_{hb}}{Z_{hb}}\right| > \varepsilon_2 \\ 0, & \text{otherwise} \end{cases} \tag{28}$$

Parameter ε_2 expresses the gene diversity with respect to the best individual. η_Z is the scale index. Degree of population diversity is between zero and one. A value of zero implies that all genes gather around the best individual. On the other hand, the value of one implies that the current candidate individuals are a diversified population. Therefore, the tolerance of population diversity is accordingly assigned within this region.

With the members of the next generation thus selected, the cycle repeats until there is no improvement in the best individual. In this study also HDE with random vector perturbation and binominal crossover is employed.

3.4 Simulation Results

In this study, the performance of the proposed HDE based dynamic economic dispatch algorithm is evaluated using a five-unit test system. The scheduling horizon is chosen as one day with 24 intervals of one hour each. In order to illustrate the robustness of the proposed algorithm the effect of valve point loading is also included in the fuel cost characteristics. This case study does not consider the prohibited discharge zones (Panigrahi et al. 2006). The thermal generator data and the load demand are summarized in Tables 4 and 5. The B-coefficient data are as same as given in reference.

The number of decision variables is 120 (5×24), which represents the generations over the entire scheduling period. The best production costs obtained by using the proposed HDE based DED approach is found to be $44,235 as against $47,356

Table 4. Generating unit data

Unit	P_i^{min} (MW)	P_i^{max} (MW)	a_i (MW)	b_i ($/MW)	c_i ($/MW^2)	e_i ($/h)	f_i (1/MW)	UR_i (MW/h)	DR_i (MW/h)
1	10	75	20	2.0	0.0080	100	0.042	30	30
2	20	125	60	1.8	0.0030	140	0.040	30	30
3	30	175	100	2.1	0.0012	160	0.038	40	40
4	40	250	120	2.0	0.0010	180	0.037	50	50
5	50	300	40	1.8	0.0015	200	0.035	50	50

Table 5. Load demand for 24 hours

Time (h)	Load (MW)	Time (h)	Load (MW)	Time (h)	Load (MW)	Time (h)	Load (MW)
1	410	7	626	13	704	19	654
2	435	8	654	14	690	20	704
3	475	9	690	15	654	21	680
4	530	10	704	16	580	22	605
5	558	11	720	17	558	23	527
6	608	12	740	18	608	24	463

Table 6. Optimal generation schedules in MW for 5 units system

Time (h)	P_1 (MW)	P_2 (MW)	P_3 (MW)	P_4 (MW)	P_5 (MW)
1	10.0000	105.0000	30.0000	40.0000	229.4899
2	10.0000	95.0000	30.0000	90.0000	229.4697
3	10.0000	95.0000	30.0000	140.0000	229.5456
4	10.0000	95.0000	30.0000	190.0000	227.3372
5	10.0000	97.7046	30.0000	200.0000	229.4675
6	19.7666	89.8540	70.0000	210.0000	227.8791
7	10.0000	95.0000	110.0000	204.4402	229.1949
8	12.9776	101.2120	110.0000	210.7557	230.2335
9	40.0000	110.0487	111.3030	210.0000	229.5979
10	70.0000	101.2120	106.2030	207.0073	230.2335
11	75.0000	105.0000	112.8693	211.3861	228.8660
12	75.0000	101.1819	119.7351	209.7317	246.0311
13	49.0258	101.1819	121.5641	212.7384	230.0821
14	46.9723	97.6674	115.5194	211.9302	230.2335
15	25.2711	98.4512	112.3599	200.0000	227.5293
16	10.0000	95.0000	115.5194	150.0000	228.8709
17	10.0000	97.5175	112.5328	123.6323	227.5293
18	10.0000	97.3031	111.8747	170.2601	229.5031
19	29.0263	93.2956	109.9234	209.1855	229.1019
20	59.0263	102.2993	118.0707	206.3134	229.1094
21	75.0000	97.4094	109.9234	194.1341	229.1019
22	45.0000	98.4975	114.1504	144.1341	228.3510
23	15.0000	98.4012	112.7850	117.6849	226.4812
24	10.0000	95.0000	114.1504	67.6849	229.3537

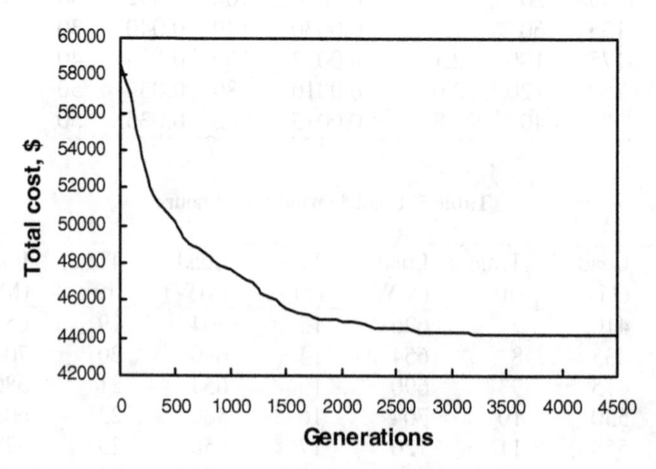

Fig. 1. Convergence characteristics of HDE based dynamic economic dispatch

obtained using simulated annealing (SA) technique. The best generation schedule is given in Table 6. The convergence characteristics of the HDE based DED is shown in Figure 1. From the results it is quite evident that the proposed HDE based dynamic economic dispatch algorithm provides better solution than those reported in literature. The best HDE parameters are $N_p = 32$, $F = 1.0$ and $C_R = 0.75$. The optimum results are obtained in 12 seconds.

4 Unit Commitment

The goal of optimal unit commitment (UC) problem is to properly schedule the on/off state of all the units in the system such that the total cost is minimum while satisfying a large number of constraints. The UC problem is formulated as a combinatorial optimization problem with 1/0 variables that represent on /off states and continuous variables that represent the power generations. The number of combinations of 1/0 variables grows exponentially as being a large-scale problem. Therefore, this problem is known as one of the problems that are most difficult to solve in power systems.

4.1 Objective Function

The objective of the unit commitment problem is to minimize the thermal cost of committed units over a scheduling period T, and is expressed as

$$\text{Minimize} \qquad F = \sum_{t=1}^{T} \sum_{i=1}^{N_g} \{u_{it} FC(i,t) + SC(i,t)\} \tag{29}$$

where u_{it} represents the operating status of the ith thermal unit at a time t and SC is the start-up cost. The production cost can be expressed in a general quadratic form using cost coefficients of the thermal plant.

4.2 System Constraints

Power Balance Constraint

The load balance constraint expressed as

$$\sum_{i=1}^{N_g} u_{it} P_{it} = P_{Dt} \tag{30}$$

The spinning reserve constraint is given by

$$\sum_{i=1}^{N_g} u_{it} P_i^{max} - P_{Dt} \geq R_t \tag{31}$$

where R_t is the reserve requirement at time t.

The reserve requirement was 10% of the hourly load and the start-up cost was calculated as

$$SC(i,t) = \begin{cases} HSC(i), & \text{if } md(i) \le X_i^{off} \le md(i) + CST(i) \\ CSC(i), & \text{if } X_i^{off} > md(i) + CST(i) \end{cases} \quad (32)$$

where CST is the cold starting hour.

Unit Constraints

The generator capacity constraints are expressed as

$$u_{it} P_i^{min} \le P_{it} \le u_{it} P_i^{max} \quad (33)$$

where min and max represent the minimum and maximum values. The minimum uptime *mu* and minimum down time, *md* constraints are represented as

$$mu_i \le X_i^{on} \quad (34)$$

$$md_i \le X_i^{off} \quad (35)$$

where X^{on} and X^{off} represent the duration for which the unit is continuously on and off respectively.

4.3 UC Using Mixed Integer Hybrid Differential Evolution

The unit commitment problem is a mixed integer non-linear optimization problem that includes continuous variables (X) representing the generation schedules and discrete variables (Y) representing the operating status of the thermal units. The HDE algorithm is modified to handle both these variables. Therefore unit commitment problem is well suited for the application of mixed integer hybrid differential evolution (MIHDE). The different steps in the MIHDE algorithm for the unit commitment problem are briefly discussed as follows:

Initialization

$$(X_i^0, Y_i^0) = (X_{min}, Y_{min}) + \rho_i (X_{max} - X_{min}) + \text{round}\left(\rho_i (Y_{max} - Y_{min})\right) \quad (36)$$

where round represents the nearest integer to the real number.

The fitness function of the unit commitment problem in terms of thermal generations and their operating status is given by

$$\psi = \sum_{t=1}^{T} \sum_{i=1}^{N_s} \{u(i,t) FC(i,t) + SC(i,t)\} + \sum_{z=1}^{N_c} \phi_z |VIOL_z| \quad (37)$$

Mutation

The mutation process is explained as

$$\left(U_i^{G+1}, V_i^{G+1}\right) = (X_i^0, Y_i^0) + F(X_j^G - X_k^G) + \text{round}(F(Y_j^G - Y_k^G)) \quad (38)$$

Crossover

The perturbed individual $\left(U_i^{G+1}, V_i^{G+1}\right)$ and the current individual (X_i^G, Y_i^G) are selected by a binomial distribution to perform the crossover operation to generate the offspring.

$$U_{hi}^{G+1} = \begin{cases} X_{hi}^{G}, & \text{if a random number} > C_R \\ U_{hi}^{G+1}, & \text{otherwise} \end{cases} \tag{39}$$

$$V_{gi}^{G+1} = \begin{cases} Y_{gi}^{G}, & \text{if a random number} > C_R \\ V_{gi}^{G+1}, & \text{otherwise} \end{cases} \tag{40}$$

Evaluation and selection

The steps involved in the evaluation process are expressed as

$$\left(X_i^{G+1}, Y_i^{G+1}\right) = \arg\min\left\{\psi\left(X_i^G, Y_i^G\right), \psi\left(U_i^{G+1}, U_i^{G+1}\right)\right\} \tag{41}$$

$$\left(X_b^{G+1}, Y_b^{G+1}\right) = \arg\min\left\{\psi\left(X_i^{G+1}, Y_i^{G+1}\right), i = 1, \ldots N_P\right\} \tag{42}$$

Acceleration Operation

The accelerated phase is represented as:

$$\left(X_b^{G+1}, Y_b^{G+1}\right) = \begin{cases} \left(X_b^{G+1}, Y_b^{G+1}\right), & \text{if } \psi\left(X_b^{G+1}, Y_b^{G+1}\right) < \psi\left(X_b^{N}, Y_b^{G+1}\right) \\ \left(X_b^{G+1}, Y_b^{G+1}\right) - \varepsilon_1 \nabla\psi, & \text{otherwise} \end{cases} \tag{43}$$

Migration Operation

The migration phase is represented as:

$$X_{hi}^{G+1} = \begin{cases} X_{hb}^{G+1} + \delta_{hi}\left(X_{h\min} - X_{hb}^{G+1}\right), & \text{if } \tilde{\delta}_{hi} < \dfrac{X_{hb}^{G+1} - X_{h\min}}{X_{h\max} - X_{h\min}} \\ X_{hb}^{G+1} + \delta_{hi}\left(X_{h\max} - X_{hb}^{G+1}\right), & \text{otherwise} \end{cases} \tag{44}$$

$$Y_{gi}^{G+1} = \begin{cases} Y_{gb}^{G+1} + round\left(\delta_{gi}\left(Y_{h\min} - Y_{hb}^{G+1}\right)\right), & \text{if } \tilde{\delta}_{gi} < \dfrac{Y_{gb}^{G+1} - Y_{h\min}}{Y_{g\max} - Y_{h\min}} \\ Y_{gb}^{G+1} + round\left(\varepsilon_4\left(Y_{h\max} - Y_{hb}^{G+1}\right)\right), & \text{otherwise} \end{cases} \tag{45}$$

The migration operation is performed only if the population diversity ρ is smaller than the desired tolerance of population diversity ε_1.

$$\rho = \frac{\left\{ \sum_{\substack{i=1 \\ i \neq b}}^{N_p} \left(\sum_{h=1}^{N_c} \eta_X + \sum_{g=1}^{N_d} \eta_Y \right) \right\}}{\left(N_c + N_d\right)\left(N_p - 1\right)} < \varepsilon_1 \tag{46}$$

where

$$\eta_X = \begin{cases} 1, & \text{if} \left| \dfrac{X_{ji} - X_{jb}}{X_{jb}} \right| > \varepsilon_2 \\ 0, & \text{otherwise} \end{cases} \tag{47}$$

$$\eta_Y = \begin{cases} 0, & \text{if } Y_{gi}^{G+1} = Y_{bi}^{G+1} \\ 1, & \text{otherwise} \end{cases} \tag{48}$$

4.4 Simulation Results

The MIHDE algorithm is tested on a 20 generating units test system. The algorithm is implemented under the same conditions as taken from reference (Kazarlis et al. 1996). The total number of decision variables is 960 (20×24+20×24) representing the generations of units and their operating status.

Table 7 provides comparison of the optimal production costs obtained using the proposed approach with that of the conventional techniques like LR and other evolutionary computation techniques like EP (Juste et al. 1999), integer coded genetic algorithm -ICGA (Damousis et al. 2004) and matrix real coded genetic algorithm – MRCGA(Sun et al. 2005). From the results it is quite evident that the proposed MIHDE based algorithm provides better solution for the large scale unit commitment of thermal plants. The MIHDE parameters are $N_p = 30$, $F = 1.0$ and $CR = 0.95$. The proposed approach converges to the optimal solution in 17 seconds.

Table 7. Comparison of production cost for 20 units system

Technique	Best cost ($)
Proposed approach	1124959
MRCGA	1125035
ICGA	1127244
GA	1126243
EP	1127257
LR	1130660

5 Conclusion

This chapter presented the application of differential evolution and its variants for solution of power system optimization problems. The algorithms have been devised to efficiently according to the problem dimensionality and the constraints. It is quiet evident from the comparison against other evolutionary algorithms that the differential evolution approach provides a competitive performance in terms of optimal solution as well as computation effort. Therefore DE based approaches can well be extended to other power system optimization and control problems.

References

Chiou, J.P., Chang, C.F., Su, C.T.: Ant direction hybrid differential evolution for solving large capacitor placement problems. IEEE Trans. on Power Syst. 19(4), 1794–1800 (2004)

Damoousis, I.G., Bakirtzis, A.G., Dokopoulos, P.S.: A solution to unit-commitment problem using integer-coded genetic algorithm. IEEE Trans. on Power Syst. 19, 1165–1172 (2004)

Gaing, Z.L.: Particle swarm optimization to solving the economic dispatch considering the generator constraints. IEEE Trans. on Power Syst. 18(3), 1187–1195 (2003)

Han, X.S., Gooi, H.B., Kirschen, D.S.: Dynamic economic dispatch: Feasible and optimal solution. IEEE Trans. on Power Syst. 16(1), 22–28 (2001)

Juste, K.A., Kita, H., Tanaka, E., Hasegawa, J.: An evolutionary programming to the unit commitment problem. IEEE Trans. on Power Syst. 14(4), 1452–1459 (1999)

Kannan, S., Slochanal, S.M.R., Padhy, N.P.: Application and comparison of metaheuristic techniques to generation expansion planning problem. IEEE Trans. Power Syst. 20(1), 466–475 (2005)

Kazarlis, S.A., Bakirtzis, A.G., Petridis, V.: A genetic algorithm solution to the unit commitment problem. IEEE Trans. on Power Syst. 11(1), 83–90 (1996)

Lakshminarasimman, L., Subramanian, S.: Short-term scheduling of hydrothermal power system with cascaded reservoirs by using modified differential evolution. IEE Proc. Gener. Transm. Distrib. 153(6), 693–700 (2006)

Lakshminarasimman, L., Subramanian, S.: Hydrothermal coordination using modified mixed integer hybrid differential evolution. Int. J. Energy Technology and Policy 5(4), 422–439 (2007)

Lakshminarasimman, L., Subramanian, S.: Hydrothermal optimal power flow using modified hybrid differential evolution. Caledonian J. Engg. 3(1), 8–14 (2007a)

Panigrahi, C.K., Chattopadhyay, P.K., Chakrabarti, R.N., Basu, M.: Simulated annealing technique for dynamic economic dispatch. Electric Power Components and Systems 34, 577–586 (2006)

Storn, R., Price, K.: Differential evolution – A simple and efficient heuristic for global optimization over continuous spaces. Journal of Global Optimization 11, 341–359 (1997)

Su, C.T., Lee, C.S.: Network reconfiguration of distribution systems using improved mixed integer hybrid differential evolution. IEEE Trans. on Power Delivery 18(3), 1022–1027 (2003)

Sun, L., Zhang, Y., Jiang, C.: A matrix real-coded genetic algorithm to the unit commitment problem. Electric Power Systems Research 76(9-10), 716–728 (2006)

Wood, A.J., Wollenberg, B.F.: Power Generation, operation and control. John Wiley & Sons, New York (1984)

Self-adaptive Differential Evolution Using Chaotic Local Search for Solving Power Economic Dispatch with Nonsmooth Fuel Cost Function

Leandro dos Santos Coelho and Viviana Cocco Mariani

[1] Production and Systems Engineering Graduate Program, PPGEPS
Pontifical Catholic University of Parana, PUCPR/CCET
Imaculada Conceicao, 1155, Zip code 80215-901, Curitiba, Parana, Brazil
leandro.coelho@pucpr.br
[2] Mechanical Engineering Graduate Program, PPGEM
Pontifical Catholic University of Parana, PUCPR/CCET
Imaculada Conceicao, 1155, Zip code 80215-901, Curitiba, Parana, Brazil
viviana.mariani@pucpr.br

Summary. The differential evolution (DE), proposed by Storn and Price, is a powerful population-based algorithm of evolutionary computation field designed for solving global optimization problems. The advantages of DE are its simple structure, easy use, convergence speed and robustness. However, the control parameters and learning strategies involved in DE are highly dependent on the problems under consideration. Choosing suitable parameter values requires also previous experience of the user. Despite its crucial importance, there is no consistent methodology for determining the control parameters of DE. In this chapter, different differential evolution approaches with self-adaptive mutation factor combined with a chaotic local search technique are proposed as alternative methods to solve the economic load dispatch problem of thermal units with valve-point effect. DE is used to produce good potential solutions, and the chaotic local search is used to fine-tune the DE run. DE and its variants with chaotic local search are validated for a test system consisting of 13 thermal units whose nonsmooth fuel cost function takes into account the valve-point loading effects. Numerical results indicate that performance of DE with chaotic local search presents best results when compared with previous optimization approaches in solving the load dispatch problem with the valve-point effect.

1 Introduction

The power economic dispatch problem (EDP) is one of the important problems for a power system. The objective of the EDP of electric power generation is to schedule the committed generating unit outputs so as to meet the required load demand at minimum operating cost while satisfying all unit and system equality and inequality constraints [1].

In traditional EDPs, the cost function of each generator is approximately represented by a simple quadratic function and the valve-points effects [2],[3] are ignored. These traditional EDPs are solved using mathematical programming based on several deterministic optimization techniques, such as lambda iteration, gradient method, dynamic programming, linear programming, nonlinear programming and quadratic programming [1]-[3].

However, the EDP problem with valve-point effects is represented as a nonsmooth optimization problem having complex and nonconvex features with heavy equality

U.K. Chakraborty (Ed.): Advances in Differential Evolution, SCI 143, pp. 275–286, 2008.
springerlink.com

and inequality constraints [2]. This kind of optimization problem is hard, if not impossible, to solve using traditional deterministic optimization algorithms. In other words, none of these mentioned methods may be able to provide an optimal solution, for they usually get stuck at a local optimum to the EDPs considering valve-point effects.

Recently, as an alternative to the conventional mathematical approaches, modern stochastic optimization techniques including genetic algorithms [3], evolutionary programming [4], evolution strategies [5], ant colony search algorithm [6], simulated annealing [7], and particle swarm optimization [1],[8] have been given much attention by many researchers due to their ability to find an almost global optimal solution.

In this chapter, an alternative hybrid method is proposed. The proposed hybrid method combines the differential evolution (DE) algorithm with self-adaptive mutation factor in the global search phase and a chaotic local search technique in the local search to solve the EDP associated with the valve-point effect.

DE as developed by Storn and Price [9] is one of the best evolutionary algorithms, and has proven to be a promising candidate to solve real-valued optimization problems [10]. The computational algorithm of DE is very simple and easy to implement, with only a few parameters required to be set by a user.

Chaos is a bounded unstable dynamic behavior, which exhibits sensitive dependence on initial conditions and includes infinite unstable periodic motions [11]. Optimization algorithms based on chaos theory are search methodologies that differ from all of the existing traditional stochastic optimization techniques. Due to the non-repetition of chaos, it can carry out overall searches at higher speeds than stochastic ergodic searches that depend on probabilities. The application of chaotic local search is a powerful strategy to prevent the premature convergence to local minima of DE approaches.

An EDP with 13 thermal units using nonsmooth fuel cost functions [4],[8] is employed in this chapter for demonstrate the performance of the proposed chaotic DE method. The results obtained with the DE approaches were analyzed and compared with those obtained in recent literature.

The remainder of this chapter is organized as follows. Section 2 describes the formulation of the EDP, while section 3 explains the concepts of validated optimization methods. Numerical simulation and comparisons are provided in section 5. Lastly, section 6 outlines the conclusion with a brief summary of results and future research.

2 Formulation of Economic Dispatch Problem

The objective of the economic dispatch problem is to minimize the total fuel cost at thermal power plants subjected to the operating constraints of a power system. Therefore, it can be formulated mathematically as an optimization problem (minimization) with an objective function and constraints. The equality and inequality constraints are represented by equations (1) and (2) given by:

$$\sum_{i=1}^{n} P_i - P_L - P_D = 0 \tag{1}$$

$$P_i^{min} \leq P_i \leq P_i^{max} \tag{2}$$

In the power balance criterion, an equality constraint must be satisfied, as shown in equation (1). The generated power should be the same as the total load demand plus total line losses. The generating power of each generator should lie between maximum and minimum limits represented by equation (2), where P_i is the power of generator i (in MW); n is the number of generators in the system; P_D is the system load demand (in MW); P_L represents the total line losses (in MW) and P_i^{min} and P_i^{max} are, respectively, the minimum and maximum power outputs of the i-th generating unit (in MW). The total fuel cost function is formulated as follows:

$$min\ f = \sum_{i=1}^{n} F_i(P_i) \tag{3}$$

where F_i is the total fuel cost for the generator unity i (in \$/h), which is defined by equation:

$$F_i(P_i) = a_i P_i^2 + b_i P_i + c_i \tag{4}$$

where a_i, b_i and c_i are cost coefficients of generator i.

Also in conventional methods the generating units cost functions are assumed to be convex and their incremental heat rate curves exhibit a monotonically increasing characteristics. But in reality large steam turbines have steam admission valves, which cause discontinuities in the incremental heat rate curves. Thus, the input–output characteristics of the generating units will become non-convex. Accurate modeling of the economic dispatch will be improved when the valve point loadings in the generating units are taken into account and furthermore they may generate multiple local optimum points in the cost function [1]. In this context, a more realistic cost function is obtained based on the ripple curve for more accurate modeling. This curve contains higher order nonlinearities and discontinuities due to the valve point effect, and should be refined by a sinusoidal function. Therefore, equation (4) can be modified [12], as:

$$\widetilde{F}_i(P_i) = F(P_i) + \left| e_i sin\left(f_i \left(P_i^{\min} - P_i\right)\right)\right| \quad \text{or} \tag{5}$$

$$\widetilde{F}_i(P_i) = a_i P_i^2 + b_i P_i + c_i + \left| e_i sin\left(f_i \left(P_i^{\min} - P_i\right)\right)\right| \tag{6}$$

where e_i and f_i are constants of the valve point effect of generators. Hence, the total fuel cost that must be minimized, according to equation (3), is modified to:

$$min\ f = \sum_{i=1}^{n} \widetilde{F}_i(P_i) \tag{7}$$

where $\widetilde{F}_i$ is the cost function of generator i (in \$/h) defined by equation (6). In the case study presented here, we disregarded the transmission losses, P_L; thus, $P_L = 0$.

3 Proposed Optimization Techniques

This section describes the proposed DE approaches. First, a brief overview of DE is provided, and then the DE with self-adaptive mutation factor and chaotic local search is detailed.

3.1 Differential Evolution

Evolutionary algorithms (EAs) are general-purpose stochastic search and optimization methods that find their inspiration in the biological world. EAs differ from other optimization methods, such as Newton method, conjugate gradient, simulated annealing, by the fact that EAs maintain a population of potential (or candidate) solutions rather than a single solution to a problem.

EAs in a general sense encompass a number of related paradigms, such as genetic algorithms, evolution strategies, evolutionary programming and recently the differential evolution, all of which are based on the natural selection paradigm.

In general, all EAs work as follows: a population of individuals is randomly initialized where each individual represents a potential solution to the problem. The quality of each solution is evaluated using a fitness function. A selection process is applied during each generation of an EA in order to form a new population. The selection process is biased toward the fitter individuals in order to increase their chances of being included in the new population. Individuals are altered using unary transformation (mutation) and higher-order transformation (crossover). This procedure is repeated until convergence is reached. The best solution found is expected to be a near-optimum solution [13].

DE is a population-based stochastic function minimizer (or maximizer) relating to EAs, whose simple yet powerful and straightforward features make it very attractive for numerical optimization.

DE combines simple arithmetical operators with the classical operators of recombination, mutation and selection to evolve from a randomly generated starting population to a final solution. DE uses mutation which is based on the distribution of solutions in the current population. In this way, search directions and possible step sizes depend on the location of the individuals selected to calculate the mutation values [14]. It evolutes generation by generation until the termination conditions have been met.

The different variants of DE are classified using the following notation: DE/α/β/δ, where α indicates the method for selecting the parent chromosome that will form the base of the mutated vector, β indicates the number of difference vectors used to perturb the base chromosome, and δ indicates the recombination mechanism used to create the offspring population. The *bin* acronym indicates that the recombination is controlled by a series of independent binomial experiments.

The fundamental idea behind DE is a scheme whereby it generates the trial parameter vectors. In each step, the DE mutates vectors by adding weighted, random vector differentials to them. If the cost of the trial vector is better than that of the target, the target vector is replaced by the trial vector in the next generation. The variant implemented here was DE/*rand*/1/*bin*, which involved the following steps and procedures:

Step 1: *Initialization of the parameter setup*: The user must choose the key parameters that control DE, i.e., population size, boundary constraints of optimization variables, mutation factor (f_m), crossover rate (CR), and the stopping criterion (t_{max}).
Step 2: *Initialize the initial population of individuals*: Initialize the generation's counter $t=0$ and also initialize a population of individuals (solution vectors) $x(t)$ with random values generated according to a uniform probability distribution in the n-dimensional problem space.
Step 3: *Evaluate the objective function value*: For each individual, evaluate its objective function (fitness) value.
Step 4: *Mutation operation (or differential operation)*: Mutate individuals according to the following equation:

$$z_i(t+1) = x_{i,r_1}(t) + f_m \cdot [x_{i,r_2}(t) - x_{i,r_3}(t)] \quad (8)$$

where $i = 1,2,...,N$ is the individual's index of population; t is the generation counter (time or iteration); $f_m > 0$ is a real parameter, called *mutation factor*, which controls the amplification of the difference between two individuals and it is usually taken form the range [0.1, 1]; $x_i(t) = \left[x_{i_1}(t), x_{i_2}(t), ..., x_{i_n}(t)\right]^T$ stands for the i-th individual of population of N real-valued n-dimensional vectors; $z_i(t) = \left[z_{i_1}(t), z_{i_2}(t), ..., z_{i_n}(t)\right]^T$ stands for the i-th individual of a *mutant vector*; r_1, r_2 and r_3 are mutually different integers and also different from the running index, i, randomly selected with uniform distribution from the set $\{1, 2, \cdots, i-1, i+1, \cdots, N\}$.

Step 5: *Crossover (recombination) operation*: Following the mutation operation, crossover is applied in the population. For each mutant vector, $z_i(t+1)$, an index $rnbr(i) \in \{1, 2, \cdots, n\}$ is randomly chosen using a uniform distribution, and a *trial vector*, $u_i(t+1) = \left[u_{i_1}(t+1), u_{i_2}(t+1), ..., u_{i_n}(t+1)\right]^T$, is generated via

$$u_{i_j}(t+1) = \begin{cases} z_{i_j}(t+1), & \text{if } randb(j) \le CR \text{ or } j = rnbr(i), \\ x_{i_j}(t), & \text{if } randb(j) > CR \text{ or } (j \ne rnbr(i). \end{cases} \quad (9)$$

where $j=1,2,..., n$ is the parameter index; $x_{ij}(t)$ stands for the i-th individual of *j-th* real-valued vector; $z_{ij}(t)$ stands for the i-th individual of *j-th* real-valued vector of a *mutant vector*; $u_{ij}(t)$ stands for the i-th individual of *j-th* real-valued vector after crossover operation; *randb(j)* is the j-th evaluation of a uniform random number generation with [0, 1]; CR is a *crossover rate* in the range [0, 1].

To decide whether or not the vector $u_i(t+1)$ should be a member of the population comprising the next generation, it is compared to the corresponding vector $x_i(t)$. Thus, if f denotes the objective function under minimization, then

$$x_i(t+1) = \begin{cases} u_i(t+1), & \text{if } f(u(t+1)) < f(x_i(t)), \\ x_i(t), & \text{otherwise} \end{cases} \quad (10)$$

Step 6: *Update the generation's counter*: $t = t + 1$;

Step 7: *Verification of the stopping criterion*: Loop to **Step 2** until a stopping criterion is met, usually a maximum number of iterations (generations), t_{max}.

3.2 Self-adaptive Differential Evolution Approaches

The parameters *CR* and *fm* of DE are generally the key factors affecting the DE's convergence [13],[15],[16]. In this chapter, we use a self-adaptive control mechanism to change the mutation factor *fm* during the run. The control parameters *M* and *CR* are not changed during the run. In this context, the DE/*rand*/1/*bin* algorithm based on self-adaptive mutation factor is proposed in this work. Several DE-variants are used in this work for comparison purposes:

- DE(1): classical DE using a constant mutation factor of $f_m = 0.50$;
- DE(2): classical DE using a constant mutation factor of $f_m = 0.75$;
- DE(3): classical DE using a constant mutation factor of $f_m = 1.00$;
- ADE(1): adaptive DE using a linear increase of f_m with initial and final values of 0.5 and 1.0, respectively;
- ADE(2): adaptive DE using a linear reduction of f_m with initial and final values of 1.0 and 0.5, respectively;
- ADE(3): adaptive DE using a mutation factor f_m generated by random number with uniform distribution in the range [0.5, 1];
- ADE(4): adaptive DE using a mutation factor f_m generated by random number with Gaussian distribution and normalized in the range [0.5, 1].

3.3 Chaotic Local Search

Chaos theory is recognized as very useful in many optimization applications. An essential feature of chaotic systems is that small changes in the parameters or the starting values for the data lead to vastly different future behaviors, such as stable fixed points, periodic oscillations, bifurcations, and ergodicity.

This sensitive dependence on initial conditions of chaotic systems is generally exhibited by systems containing multiple elements with nonlinear interactions, particularly when the system is forced and dissipative. Sensitive dependence on initial conditions is not only observed in complex systems, but even in the simplest logistic equation [17].

The application of chaotic sequences in DE approaches can be a good alternative to maintain the search diversity in an optimization procedure. Due to the non-repetition of chaos, it can carry out overall searches at higher speeds than stochastic ergodic searches that depend on probabilities [18]-[20].

Different types of equations of chaotic systems have been considered in the literature for applications in optimization methods. The logistic equation and other equations, such as sinusoidal iterator, Chua's oscillator, Lorenz system, Ikeda map, and others, have been adopted instead of generation of random numbers using a uniform distribution and very interesting results have emerged [18]-[20]. The design of approaches to improve the convergence of chaotic optimization is a challenging issue. A chaotic local search approach is proposed here based on Lozi map [19].

The Lozi's piecewise liner model is a simplification of the Hénon map [21] and it admits strange attractors. The Lozi map is given by

$$y_1(k)=1-a\cdot|y_1(k-1)|+y(k-1) \quad (11)$$

$$y(k)=b\cdot y_1(k-1) \quad (12)$$

where k is the iteration number. In this work, the values of y are normalized in the range [0,1] to each i-th decision variable. This transformation is given by

$$w_i(k)=\frac{y(k)-\alpha}{\beta-\alpha} \quad (13)$$

where $y\in[-0.6418,0.6716]$ and (α,β) = (-0.6418,0.6716). The parameters used in this work are a=1.7 and b=0.5, as these values have been suggested by [19].

The chaotic search procedure based on the Lozi map can be illustrated as follows:

Notation:
$X=[x_1,x_2,...,x_n]$: solution vector consisting of n variables x_i , i = 1,…,n bounded by lower (L_i) and upper limits (U_i).

Input:
M_L: maximum number of iterations of chaotic *L*ocal search;
λ: step size in chaotic local search.

Output:
X_i*: best solution of *j-th* variable from current run of chaotic search;
f*: best objective function (minimization problem).

Chaotic optimization algorithm:
Step 1: *Initialization of variables*: Set k = 0, where k represents the iteration number. Set the initial conditions $y_1(0)$,$y(0)$, a=1.7 and b=0.5 of Lozi map. Set the initial best objective function f*. In this work, the best objective function is the best individual of differential evolution in current generation t;

Step 2: *Exploitation phase of chaotic search*:

Begin
While $k \le M_L$ do
  For i =1 to n
    If $r<0.5$ then
    (where r is a uniformly distributed random variable in [0, 1])
    $$x_i(k)=X_i^*+\lambda\cdot w_i(k)\cdot\left|U_i-X_i^*\right|$$
    Else If
    $$x_i(k)=X_i^*-\lambda\cdot w_i(k)\cdot\left|X_i^*-L_i\right|$$
    End If
  End
  If $f(X(k))<f^*$ then
    $X^*=X(k)$
    $f^*=f(X(k))$

```
        End If
          k = k + 1;
      End
    End
```

During the exploitation phase of chaotic search, the step size λ is an important parameter for the convergence behavior of the optimization method, which adjusts small ranges around X^*. A suitable value for the step size usually provides a balance between global and local search abilities and consequently a reduction on the number of iterations required to locate the optimum solution. In this work, the step size λ = 0.0001 is adopted in chaotic local search (CLS).

3.4 Differential Evolution with Chaotic Local Search

The approaches configuration composite by DE hybridized with stochastic techniques is a promising alternative in optimization and must be evaluated. DE and the proposed chaotic local method have supplementary potentialities. In this work, the following way of hybridizing of DE combined with CLS was tested: after having solved the EDP use the best solution from DE as a starting point and solve the EDP using CLS method.

4 Simulation Results

In this section, we judge the performance of the DE and DE-CLS algorithms using a case study of power economic dispatch using 13 thermal units.

This case study consisted of 13 thermal units of generation with the effects of valve-point loading, as given in Table 1. The data shown in Table 1 is also available in [4] and [22]. In this case, the load demand expected to be determined was $P_D = 1800$ MW. This EDP has many local minima, and the global minimum is difficult to determine.

Table 1. Data for the 13 thermal units

Thermal unit	P_i^{min}	P_i^{max}	a	b	c	e	f
1	0	680	0.00028	8.10	550	300	0.035
2	0	360	0.00056	8.10	309	200	0.042
3	0	360	0.00056	8.10	307	150	0.042
4	60	180	0.00324	7.74	240	150	0.063
5	60	180	0.00324	7.74	240	150	0.063
6	60	180	0.00324	7.74	240	150	0.063
7	60	180	0.00324	7.74	240	150	0.063
8	60	180	0.00324	7.74	240	150	0.063
9	60	180	0.00324	7.74	240	150	0.063
10	40	120	0.00284	8.60	126	100	0.084
11	40	120	0.00284	8.60	126	100	0.084
12	55	120	0.00284	8.60	126	100	0.084
13	55	120	0.00284	8.60	126	100	0.084

Each optimization method was implemented in Matlab (MathWorks). All the programs were run on a 3.2 GHz Pentium IV processor with 2 GB of random access memory. In each case study, 50 independent runs were made for each of the optimization methods involving 50 different initial trial solutions for each optimization method.

A key factor in the application of DE approaches is how the algorithm handles the constraints relating to the problem. In this work, a penalty-based method proposed in [23] was used for the equality constraints.

The population size *N* was 20 and the stopping criterion t_{max} was 800 generations (16000 evaluations of the objective function) for classical DE.

In the DE-CLS, the population size of DE was 12 and the stopping criterion t_{max} was 500 generations. CLS procedure is adopted using 12 cost function evaluations ($M_L = 12$) in each generation of DE. In this case, the DE-CLS routine is adopted using 16000 cost function evaluations in each run. The crossover rate of $CR = 0.8$ was adopted for both the classical DE and DE-CLS approaches.

The results obtained for this case study are given in Table 2, which shows that the DE(3)-CLS succeeded in finding the best solution for the tested methods. The best result obtained for solution vector P_i, i=1,..,13 with DE(3)-CLS is the minimum cost of 17963.9571 which is given in Table 3. However, the ADE(1)-CLS approach shows a performance which is clearly better than that of DE(3)-CLS in terms of mean cost.

It also observed that the classical DE approaches outperformed the other tested DE-CLS methods in terms of solution time.

Table 4 compares the results obtained in this chapter with those of other studies reported in the literature. Note that in the case studied here, the best result reported using DE(3)-CLS is comparatively lower than recent studies presented in the literature.

Table 2. Convergence results (50 runs) of DE and DE-CLS approaches

Optimization Method	Mean Time (s)	Minimum Cost ($/h)	Mean Cost ($/h)	Maximum Cost ($/h)
DE(1)	1.78	18095.7270	18323.9653	**18637.0927**
DE(2)	1.77	18091.1464	18315.6026	18682.1625
DE(3)	1.82	18377.7128	18752.4246	19116.6163
ADE(1)	1.78	18052.7891	18294.6310	18645.2262
ADE(2)	1.77	18069.1528	18419.8325	18903.2219
ADE(3)	1.79	18097.9214	18302.1210	18646.4057
ADE(4)	1.79	18070.2032	18337.7369	18782.8841
DE(1)-CLS	5.39	18085.5078	18427.0199	18815.4248
DE(2)-CLS	5.38	18089.7461	18327.8504	18623.3178
DE(3)-CLS	5.37	**17963.9571**	18431.1479	18892.7540
ADE(1)-CLS	5.37	18001.7035	**18274.9005**	18524.9235
ADE(2)-CLS	5.37	18093.4723	18424.7626	18782.6906
ADE(3)-CLS	5.36	18101.2664	17320.5504	17683.0652
ADE(4)-CLS	5.36	18057.9074	18371.7782	18786.6667

Table 3. Best result (50 runs) obtained for the case study using DE(3)-CLS

Power	Generation (MW)	Power	Generation (MW)
P_1	628.3180	P_8	60.0000
P_2	149.1094	P_9	109.8664
P_3	223.3226	P_{10}	40.0000
P_4	109.8650	P_{11}	40.0000
P_5	109.8618	P_{12}	55.0000
P_6	109.8656	P_{13}	55.0000
P_7	109.7912	$\sum_{i=1}^{13} P_i$	1800.0000

Table 4. Comparison of best results for fuel costs presented in the literature

Optimization Technique	Best Objective Function
Evolutionary programming [4]	17994.07
Particle swarm optimization [1]	18030.72
Hybrid evolutionary programming with SQP [1]	17991.03
Hybrid particle swarm with SQP [1]	17969.93
Genetic algorithms [24]	17975.3437
Improved genetic algorithm with multiplier updating [24]	17963.9848
Best result of this chapter using DE(3)-CLS	**17963.9571**

5 Conclusion and Future Research

In this chapter, DE and DE-CLS methods have been successfully introduced to solve a case study of EDP considering 13 thermal units with valve-point effect. In this case study, DE, DE-CLS and ADE-CLS can provide accurate dispatch solutions in reasonable time.

In relation to procedure of solution of the economic dispatch problem of electric energy with effect of valve point, the results with the DE(3)-CLS for optimization of the equations (1) and (2) were best that the results presented in [1], [4] and [24].

Future research will investigate theoretically the effect of chaos incorporation into DE further and apply the DE-CLS methods for solving the multiobjective economic dispatch problems in power systems.

Acknowledgments

This work was supported by the National Council of Scientific and Technologic Development of Brazil — CNPq — under Grant 309646/2006-5/PQ.

References

1. Victoire, T.A.A., Jeyakumar, A.E.: Hybrid PSO-SQP for economic dispatch with valve-point effect. Electric Power Systems Research 71(1), 51–59 (2004)
2. Al-Othman, A.K., El-Naggar, K.M.: Application of pattern search method to power security constrained economic dispatch with non-smooth cost function. Electric Power Systems Research 78(4), 667–675 (2008)
3. Walters, D.C., Sheble, G.B.: Genetic algorithm solution of economic dispatch with valve point loading. IEEE Transactions on Power Systems 8(3), 1325–1332 (1993)
4. Sinha, N., Chakrabarti, R., Chattopadhyay, P.K.: Evolutionary programming techniques for economic load dispatch. IEEE Transactions on Evolutionary Computation 7(1), 83–94 (2003)
5. Gomes, J.R., Saavedra, O.R.: A Cauchy-based evolution strategy for solving the reactive power dispatch problem. Electrical Power and Energy Systems 24(4), 277–283 (2002)
6. Sum-im, T.: Economic dispatch by ant colony search algorithm. In: Proceedings of the IEEE Conference on Cybernetics and Intelligent Systems, Singapore, pp. 416–421 (2004)
7. Wong, K.P., Wong, Y.W.: Thermal generator scheduling using hybrid genetic/simulated-annealing approach. IEE Proc.-Generation, Transmission and Distribution 142(4), 372–380 (1995)
8. Park, J.-B., Lee, K.-S., Shin, J.-R., Lee, K.Y.: A particle swarm optimization for economic dispatch with nonsmooth cost function. IEEE Transactions on Power Systems 20(1), 34–42 (2005)
9. Storn, R., Price, K.: Differential evolution: a simple and efficient adaptive scheme for global optimization over continuous spaces. Technical Report TR-95-012, International Computer Science Institute, Berkeley, USA (1995)
10. Storn, R.: Differential evolution – a simple and efficient heuristic for global optimization over continuous spaces. Journal of Global Optimization 11(4), 341–359 (1997)
11. Ho, S.J., Shu, L.S., Ho, S.Y.: Optimizing fuzzy neural networks for tuning PID controllers using an orthogonal simulated annealing algorithm OSA. IEEE Transactions on Fuzzy Systems 14(3), 421–434 (2006)
12. Wood, A.J., Wollenberg, B.F.: Power generation, operation and control. John Wiley & Sons, New York (1994)
13. Anzi, F.S., Allahverdi, A.: A self-adaptive differential evolution heuristic for two-stage assembly scheduling problem to minimize maximum lateness with setup times. European Journal of Operation Research (accepted for future publication, 2007)
14. Montes, E.M., Reyes, J.V., Coello, C.A.C.: A comparative study of differential evolution variants for global optimization. In: Proceedings of Genetic and Evolutionary Computation Conference, Seattle, Washington, USA (2006)
15. Liu, J., Lampinen, J.: On setting the control parameter of the differential evolution method. In: Proceeding of 8th International Conference on Soft Computing (MENDEL 2002), Brno, Czech Republic, pp. 11–18 (2002)
16. Brest, J., Saso, G., Mernik, M., Zumer, V.: Self-adapting control parameters in differential evolution: a comparative study on numerical benchmark problems. IEEE Transactions on Evolutionary Computation (accepted for future publication, 2007)
17. Yan, X.F., Chen, D.Z., Hu, S.X.: Chaos-genetic algorithms for optimizing the operating conditions based on RBF-PLS model. Computers and Chemical Engineering 27(10), 1393–1404 (2003)

18. Pan, H., Wang, L., Liu, B.: Chaotic annealing with hypothesis test for function optimization in noisy environments. Chaos, Solitons & Fractals (accepted for future publication, 2007)
19. Caponetto, R., Fortuna, L., Fazzino, S., Xibilia, M.G.: Chaotic sequences to improve the performance of evolutionary algorithms. IEEE Transactions on Evolutionary Computation 7(3), 289–304 (2003)
20. Li, L., Yang, Y., Peng, H., Wang, X.: Parameters identification of chaotic systems via chaotic ant swam. Chaos, Solitons & Fractals 28(5), 1204–1211 (2006)
21. Hénon, M.: A two dimensional mapping with a strange attractor. Communications in Mathematical Physics 50, 69–77 (1976)
22. Wong, K.P., Wong, Y.W.: Genetic and genetic/simulated-annealing approaches to economic dispatch. IEE Proc. Control, Generation, Transmission and Distribution 141(5), 507–513 (1994)
23. Coelho, L.S., Mariani, V.C.: Combining of chaotic differential evolution and quadratic programming for economic dispatch optimization with valve-point effect. IEEE Transactions on Power Systems 21(2), 989–996 (2006)
24. Chiang, C.L.: Improved genetic algorithm for power economic dispatch of units with valve-point effects and multiple fuels. IEEE Transactions on Power Systems 20(4), 1690–1699 (2005)

An Adaptive Differential Evolution Algorithm with Opposition-Based Mechanisms, Applied to the Tuning of a Chess Program

Borko Bošković, Sašo Greiner, Janez Brest, Aleš Zamuda, and Viljem Žumer

University of Maribor
Faculty of Electrical Engineering and Computer Science
Smetanova 17, 2000 Maribor, Slovenia
borko.boskovic@uni-mb.si

Summary. This chapter describes an algorithm for the tuning of a chess program which is based on Differential Evolution using adaptation and opposition based optimization mechanisms. The mutation control parameter F is adapted according to the deviation of search parameters in each generation. Opposition-based optimization is included in the initialization, and in the evolutionary process itself. In order to demonstrate the behaviour of our algorithm we tuned our BBChess chess program with a combination of adaptive and opposition-based optimization. Tuning results show that adaptive optimization with an opposition-based mechanism increases the robustness of the algorithm and has a comparable convergence to the algorithm which uses only adaptation optimization.

Keywords: Differential Evolution, Adaptation, Tuning of a Chess Program, Opposition-Based mechanisms.

1 Introduction

Computer chess games have a long history of research in the field of artificial intelligence. Computer chess has advanced to a remarkable degree where computers now play against other computers and humans. With ever growing computer strength, we are witnessing more and more matches between computers and humans where computers usually win.

The reasons why the computer is beating humans are mainly hardware improvements and chess algorithm optimizations. The first computer that won against a human world champion chess player was Deep Blue which defeated the world champion chess player Garry Kasparov in 1996. In 2006 the Deep Fritz 10 computer program which ran on a PC, defeated world champion Vladimir Kramnik. So why are chess program developers trying to improve already very strong chess programs, even further? Many professional human chess players use chess programs to improve their own playing skills. Chess programs are also very useful in correspondence and freestyle chess. Matches between programs are also gaining popularity. As far as artificial intelligence is concerned, chess is regarded as a very useful environment for testing different approaches.

In this chapter we describe a Differential Evolution (DE) based algorithm for tuning chess programs. Using evolutionary concepts, this algorithm tunes chess programs and

U.K. Chakraborty (Ed.): Advances in Differential Evolution, SCI 143, pp. 287–298, 2008.
springerlink.com

makes them stronger without any interaction with humans and without humans' expert knowledge. In order to improve the tuning process our algorithm includes the adaptation of DE control parameters and opposition-based optimization mechanisms. Because our DE uses adaptation and opposition-based optimization it is called 'AODE'.

The chapter is structured as follows. Section 2 gives an overview on tuning chess programs and briefly describes the basic DE and ODE (Opposition-Based Differential Evolution Algorithm). Section 3 describes the structure of those chess programs and parameters that may be tuned. Section 4 describes the details of our evolutionary algorithm AODE. Section 5 presents three experiments which tune the chess program by the use of AODE optimizations. We then show how these optimizations influence the tuning process. Section 6 concludes the chapter with final remarks.

2 Related Work

One of the possible improvements of a chess program is achieved by parameter tuning, but with conventional approaches this becomes a very difficult task. Developers have to change program parameters and then choose the best values through out the testing phase. The nature of such a task is very time consuming.

Another method is automated tuning or "learning". When we talk about automated tuning in computer chess we focus on algorithms such as hill climbing, simulated annealing, temporal difference learning [1, 2], and evolutionary algorithms [7, 8]. All approaches enable tuning on the basis of the program's own experiences, i.e. final result of a chess games competition: win, lose, or draw.

The pioneer of computer chess was Shannon (1949). He advocated the idea that computer chess programs would require an evaluation function and search algorithm to successfully play a game against human players [14]. In the beginning computer chess programs were designed "by hand" by the developers. The most important part of every chess program is its evaluation function. Evaluation functions contains a lot of parameters in the form of expressions and weights. In order to obtain a good evaluation function the developers had first to test it by playing numerous games and then modify it according to the produced results. Finding a proper evaluation function was a difficult and very time consuming task, because this was a recurring cycle. This is the main reason why current research has become involved in finding a method for automatically improving the evaluation function's parameters. Additionally, developers can tune the parameters of the search algorithm alone or together with those of the evaluation function.

Samuel [13] shows that a computer can be programmed so that it will learn to play better game of checkers than can be played by the person who wrote the program. The NeuroChess [17] is a program which learns to play chess from the final outcome of games. It learns its evaluation function, represented by artificial neural networks. This learning approach included inductive neural network learning, temporal differencing, and a variation of explanation-based learning. Another important work on learning is KnightCap [1] chess program. It learns parameters of its evaluation function using combination of Temporal Differences learning and on-line play on FICS and ICC chess servers. The program started with blitz rating 1650 and after 3 days of learning and 308 games played the program obtained blitz rating of 2150. The principles of evolution have also been used in the tuning of a chess evaluation function. Kendall and Whitwell [8] presented one such approach by using population dynamics. Fogel et al. [7] presented an evolutionary algorithm which has managed to improve a chess program by almost 400 rating points. Last two approaches used a population of individuals which consist of evaluation function parameters and new individuals are generated using mutation, crossover, and selection operators.

The DE [15, 9, 16] algorithm was proposed by Storn and Price, and since then it has been used in many practical cases. The original DE was modified and many new versions have been proposed [5]. Rahnamayan, Tizhoosh and Salama proposed an opposition-based DE (ODE) algorithm [11, 12]. ODE includes opposition based optimizations in order to improve the efficiency of classical DE algorithm. DE has also been used for chess program tuning [4, 6] because it converges quickly and improves playing ability during the evolutionary process.

3 Chess Program

The basic components of all modern chess programs are the search algorithm, evaluation function, move generator, transposition table, representation of game, opening book, and the end-game database [3]. These components enable a chess program to play equally well against the strongest human players, or even better. To improve an existing chess program with automated tuning, we can tune its parameters. The most tunable components are the evaluation function and the search algorithm.

The evaluation function contains a lot of expressions and parameters as weights of expressions. Expressions and parameters together represent all the chess knowledge of a chess program. Like the evaluation function, the search algorithm also contains parameters. However the number of parameters of a search algorithm is lower than in an evaluation function. The parameters are responsible for pruning the search tree and for selective searching.

Search algorithms only have a few parameters and their values have been tuned by the conventional approach (by hand and expert knowledge). On the other hand, an evaluation function has many more parameters which depend on each other and have been set by the developer according to experience and expert human instructions. Because an evaluation function contains complex expressions, the values of the parameters are approximated. Therefore, using automated tuning we can obtain better parameter values and improve the evaluation function and, consequently, the efficiency of chess programs.

4 AODE Algorithm for Tuning a Chess Program

Our tuning algorithm is based on Differential Evolution which uses adaptation and opposition-based optimization techniques. DE is a floating-point encoding evolutionary algorithm for global optimization over continuous spaces [9, 10]. Each generation of our AODE contains a current population P_g (g is a number of current generation), which further contains NP D-dimensional vectors (individuals) $\overrightarrow{X}_{g,i}$ with parameter values that represent the weights of a chess program.

$$\overrightarrow{X}_{g,i} = \{X_{g,i,1}, X_{g,i,2}, ..., X_{g,i,D}\},$$

$$i = 1, 2, ..., NP \ , \quad g = 1, 2, 3, ...$$

DE employs mutation, crossover, and selection operations during the evolutionary process, in each generation. Our algorithm uses the idea of adaptation and opposition-based optimization as shown in the algorithm below. P_0 represents the initial population, $P_{U,0}$ is an opposition population of P_0, P_1 is the first population, P_g is the current population, $P_{V,g}$ is the mutant population, $P_{U,g}$ is the trial population, and P_{g+1} is the population of the next generation. CR and JR are control parameters defined by the user.

Algorithm 1. AODE Algorithm

1: Initialization(P_0);
2: $P_{U,0}$ = Opposition(P_0);
3: Evaluation($P_0, P_{U,0}, depth$);
4: P_1 = Selection($P_0, P_{U,0}$);
5: **while** continue tuning **do**
6:    **if** rand(0,1) < JR **then**
7:       $P_{U,g}$ = DynamicOpposition(P_g);
8:    **else**
9:       $P_{V,g}$ = AdaptiveMutation(P_g);
10:       $P_{U,g}$ = Crossover($P_g, P_{V,g}, CR$);
11:    **end if**
12:    Evaluation($P_g, P_{U,g}$);
13:    P_{g+1} = Selection($P_g, P_{U,g}$);
14: **end while**

4.1 Initialization

At the beginning, the population P_0 is initialized with parameter values that are distributed uniform-randomly between parameter bounds ($X_{j,low}$, $X_{j,high}$; $j = 1, 2, ..., D$). The bound values are problem-specific. In chess programs the parameters are set to approximate values by the developers. Developers can also intuitively determine those intervals which effectively define the bounds for parameters tuning. Accurately-defined bounds enable the algorithm to search through much smaller space and, consequently, find better parameters more quickly. If the search space is too limited, our algorithm can not find solution because it is out of bounds.

4.2 Opposition

The efficiency of the tuning process depends on the distance between the solution and the individuals in the initial population. After initialization, an opposite population $P_{U,0}$

is generated from the initial population. This mechanism together with evaluation (Section 4.3) and selection (Section 4.4), increases the probability of first generation containing individuals closer to the solution and, thus, accelerates convergence [11, 12].

The opposition population contains opposite individuals of the initial population and is defined by the following equations:

$$\vec{U}_{0,i} = \{U_{0,i,1}, U_{0,i,2}, ..., U_{0,i,D}\}$$

$$U_{0,i,j} = X_{j,low} + X_{j,high} - X_{0,i,j}$$

$$i = 1, 2, ..., NP, \quad j = 1, 2, ..., D,$$

where $\vec{U}_{0,i}$ represent the opposition individuals of the corresponding initial individuals $\vec{X}_{0,i}$.

4.3 Evaluation

Using trial $P_{U,g}(P_{U,0})$ and current $P_g(P_0)$ populations we have to evaluate their individuals. To do this we calculate the relative efficiencies of individuals according to both populations. Relative efficiency is measured according to the collected points and number of played games, as shown with the following equation:

$$\text{efficiency} = \frac{\text{collected points}}{2 \cdot \text{number of played games}}.$$

We can use more strategies to play games. Firstly, each individual of a trial population can play a specific number of games (N) against randomly chosen individuals of the current population. Secondly each individual of a trial population can play two games (one as white and one as black) against a corresponding individual of the current population. Other strategies are also possible.

An individual plays each game with a specific search depth and gets 2 points for winning, 1 for a draw, and 0 for losing. An individual wins when opponent's King is mate. The game is a draw if the position is a known draw position or the same position is obtained three times in one game, or because of the 50-moves rule. Games are limited to 150 moves for both players. Therefore, if the game has 150 moves, the result is a draw. An individual loses if its opponent wins.

4.4 Selection

The selection operation selects according to the relative efficiency of those individuals among the i-th current population and their corresponding individuals in the trial population. Selection dictates which individuals will survive into the next generation. In our case we used the following selection rule for a maximization problem:

$$\vec{X}_{g+1,i} = \begin{cases} \vec{U}_{g,i}, & \text{efficiency}(\vec{U}_{g,i}) > \text{efficiency}(\vec{X}_{g,i}), \\ \vec{X}_{g,i}, & \text{otherwise.} \end{cases}$$

$$i = 1, 2, ..., NP,$$

where $\vec{U}_{g,i}$ is an i-th individual from the trial population, $\vec{X}_{g,i}$ is an i-th individual from the current population, and $\vec{X}_{g+1,i}$ is an i-th individual from the population of the next generation.

4.5 Dynamic Opposition

As proposed in [11, 12], we can also use opposition-based optimization during the evolutionary process. This optimization is applied using a jump rate JR, as shown in Algorithm 1, and dynamic interval bounds ($X^g_{j,low}$, $X^g_{j,high}$; $j = 1, 2, ..., D$), as shown by the following equation:

$$\vec{U}_{g,i} = \{U_{g,i,1}, U_{g,i,2}, ..., U_{g,i,D}\}$$

$$U_{g,i,j} = X^g_{j,low} + X^g_{j,high} - X_{g,i,j}$$

$$i = 1, 2, ..., NP\,, \quad j = 1, 2, ..., D,$$

where $\vec{U}_{g,i}$ represents an opposition individual of a corresponding current individual $\vec{X}_{g,i}$ and $X^g_{j,low}$, $X^g_{j,high}$ are bound values for each parameter in the current population.

4.6 Adaptive Mutation

Adaptive mutation generates a mutant population $P_{V,g}$ from the current population P_g, using mutant strategy and adaptive mutation scale factor F. For each vector from the current population, mutation (using one of the mutation strategies) creates a mutant vector $\vec{V}_{g,i}$, which is an individual of mutant population.

$$\vec{V}_{g,i} = \{V_{g,i,1}, V_{g,i,2}, ..., V_{g,i,D}\}, \quad i = 1, 2, ..., NP.$$

DE includes various mutation strategies for global optimization. In our algorithm we used the $rand/2$ mutation strategy, which is given by the equation:

$$\vec{V}_{g,i} = \vec{X}_{g,r1} + F_g \cdot (\vec{X}_{g,r2} - \vec{X}_{g,r3}) + F_g \cdot (\vec{X}_{g,r4} - \vec{X}_{g,r5})$$

The indexes $r1$, $r2$, $r3$, $r4$, $r5$ are random and mutually different integers generated within the range $[1, NP]$ and also different from index i. F_g is a mutation scale factor in the g-th generation within the range $[0, 2]$ but usually less than 1.0. Because F_g scales the distance between the new and old individuals, it is responsible for exploration and exploitation balance in the evolutionary process. Therefore, we used adaptive F_g defined as the ratio of the standard deviations between parameters of the initial and current populations, as shown in the following equations:

$$F_g = \frac{\sum_{i=1}^{D} \sigma_{g,i}}{\sum_{i=1}^{D} \sigma_{0,i}}$$

$$\sigma_{g,i} = \sqrt{\frac{\sum_{j=1}^{NP} (X_{g,i,j} - \overline{X}_{g,i})^2}{NP - 1}}.$$

where $\sigma_{g,i}$ is a standard deviation of the i-th parameter in the current population.

4.7 Crossover

After mutation, a "binary" crossover forms a trial population $P_{U,g}$. According to the i-th population vector and its corresponding mutant vector, crossover creates trial vectors $\overrightarrow{U}_{g,i}$ using the following rule:

$$\overrightarrow{U}_{g,i} = \{U_{g,i,1}, U_{g,i,2}, ..., U_{g,i,D}\}$$

$$U_{g,i,j} = \begin{cases} V_{g,i,j}, & rand_j(0,1) \leq CR \, or \, j = j_{rand}, \\ X_{g,i,j}, & otherwise. \end{cases}$$
$$i = 1, 2, ..., NP \,, \quad j = 1, 2, ..., D.$$

CR is a crossover factor within the range [0,1) and determines the probability of creating parameters of the trial vector from the mutant vector. Index j_{rand} is a randomly chosen integer within the range $[1, NP]$ and is responsible for the trial vector containing at least one parameter from the mutant vector. After crossover, the parameters of trial vector may be out of bounds ($X_{j,low}$, $X_{j,high}$). In this case the parameters can be mapped inside an interval, set to bounds or used as they are – out of bounds.

5 Experiments

Our algorithm was tested for tuning a simplified chess evaluation function of the chess program, BBChess. The evaluation function contains only material (values of pieces) and mobility (number of available moves for pieces) information, as shown in the following equation:

$$chess\_evaluation = X_m(M_w - M_b) + \sum_{i=0}^{5} X_i(N_{i,w} - N_{i,b}).$$

In this equation X_i represents material weights for all piece types without king and X_m the mobility weight. M_w represents mobility for white and M_b for black pieces. $N_{i,w}$ is the number of specific white pieces (i.e. the number of white pawns) and $N_{i,b}$ for specific black pieces. The principal reason for using such a simple and straightforward evaluation function was to demonstrate how the weight parameters of the function can be tuned by applying our tuning algorithm. In addition the behavior and features of the AODE algorithm were also presented. To do this, three experiments were performed. In the first experiment only adaptation optimization was used, in the second opposition-based optimization was included, and the third included all optimizations including opposition-based optimization during the evolutionary process.

In all experiments pawn material weight was fixed to 100 and the search depth set to 5 ply (half move). Experiments were run 30 times and tuning performed throughout 50 generations. The size of the population NP was 20 because larger NP would substantially increase the number of required games in one generation. The control parameter CR was set to 0.9 and the parameter bounds for all parameters were set to $X_{j,low} = 0$ and $X_{j,high} = 1000$ for all experiments. If the parameters were out of

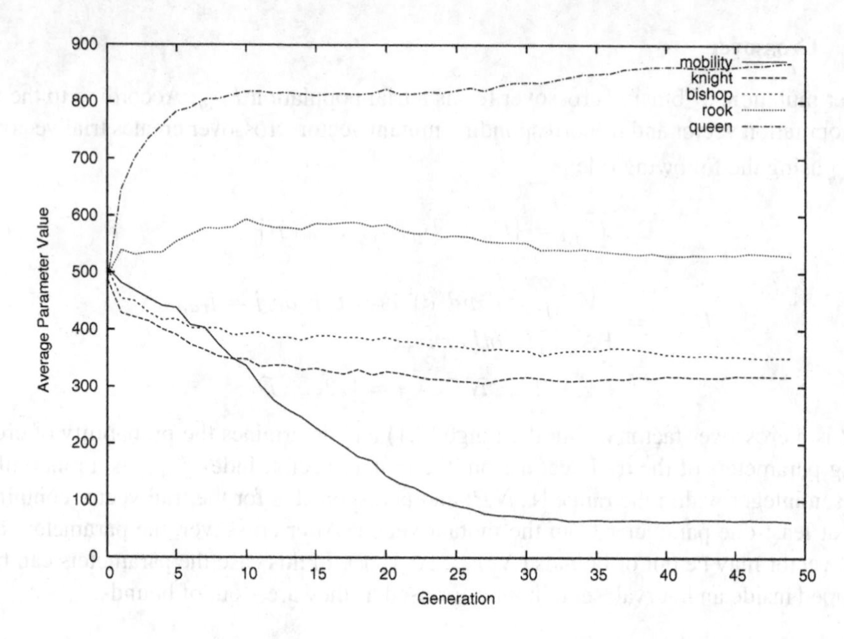

Fig. 1. Average parameter values along generations for AODE without opposition based optimization mechanisms

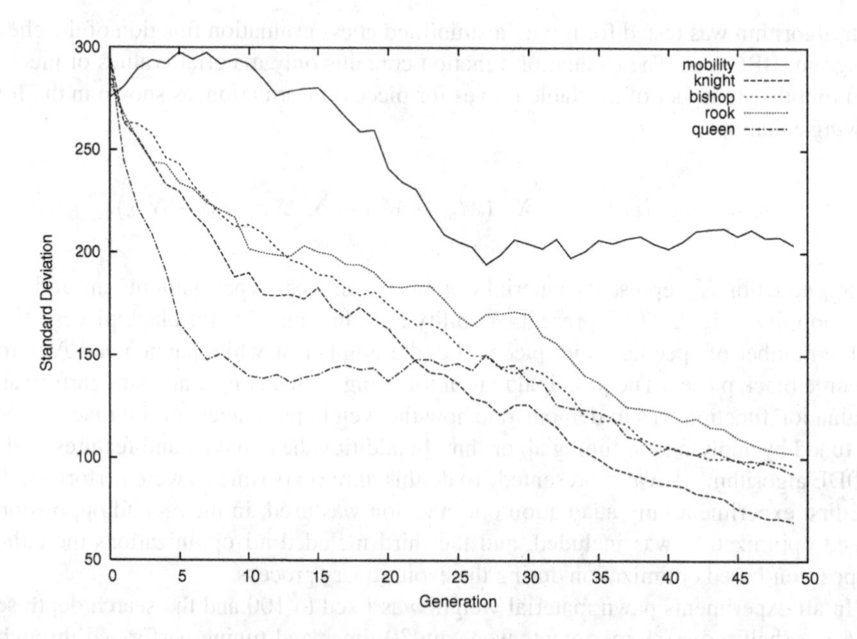

Fig. 2. Standard deviation of parameters along generations for AODE without opposition based optimization mechanisms

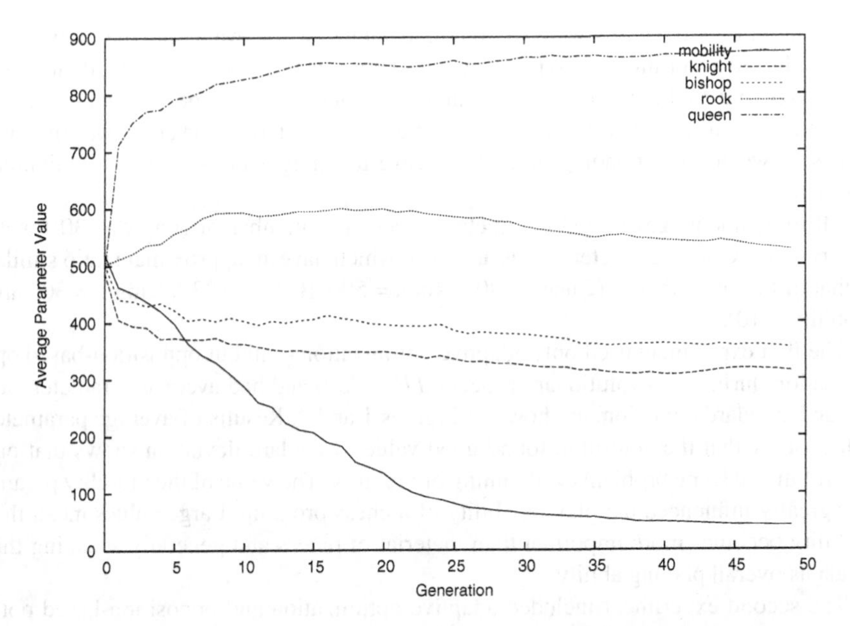

Fig. 3. Average parameter values along generations for AODE with initial population opposition and JR = 0.0

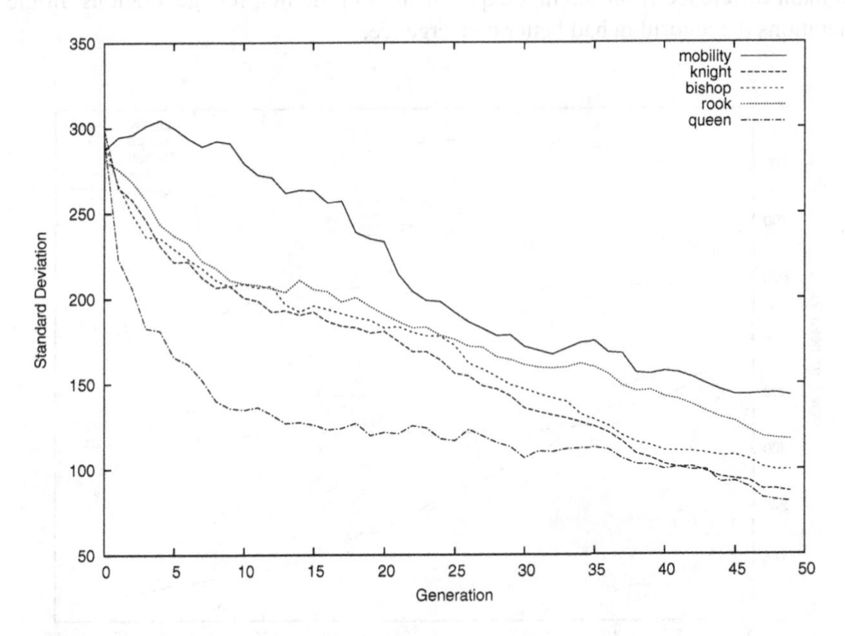

Fig. 4. Standard deviation of parameters along generations for AODE with initial population opposition and JR = 0.0

bounds after crossover they were set to bound values. When evaluating the individuals they evaluated according to two played games between corresponding individuals from the current and trial populations. Each individual played one game as a white player. This strategy was used because of the simplified evaluation function and it was that two games between corresponding individuals gave fair judgment as to which individual was better.

All experiments gave good parameter values. The number of runs was 30 for all experiments. Good parameter values are those which have an approximate ratio similar to that of the chess theory (Queen = 900, Rook = 500, Bishop = 330, Knight = 300 and mobility = 10).

The first experiment used only adaptive optimization (without opposition-based optimization during the evolutionary process $JR = 0.0$) and had average parameter values and standard deviation, as shown in Figures 1 and 2. Results of average parameter values show that the algorithm found good values. Standard deviation shows that our algorithm had some problems with tuning of mobility. The value of the mobility parameter greatly influenced the playing ability of a chess program. Large values mean that mobility becomes more important than material of pieces and generally speaking this weakens overall playing ability.

The second experiment included adaptive optimization and opposition-based optimization during initialization and had average parameter values and their standard deviation, as shown in Figures 3 and 4. This experiment also found good parameters values. The main difference from the first experiment is in the first few generations. In these generations the algorithm had better convergence.

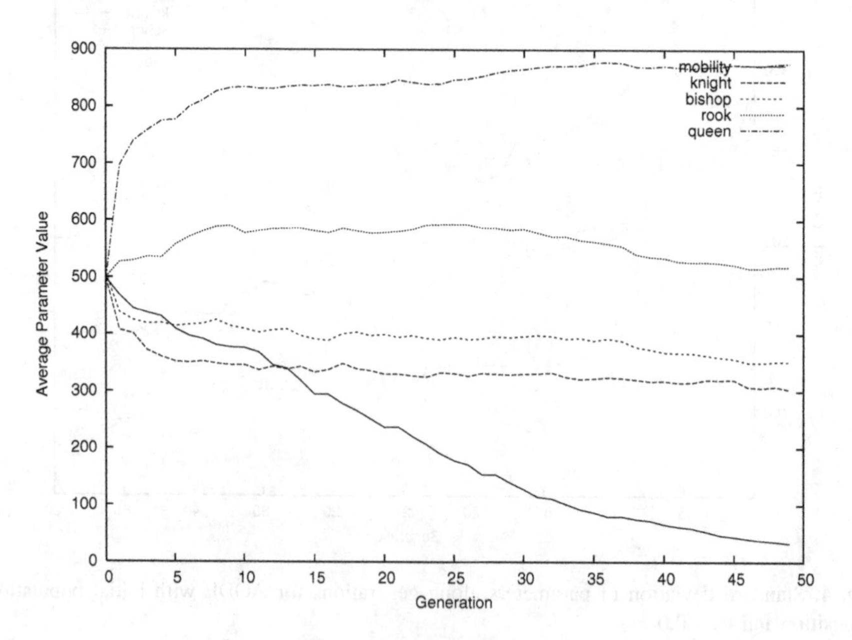

Fig. 5. Average parameter values along generations for AODE with initial population opposition and JR = 0.1

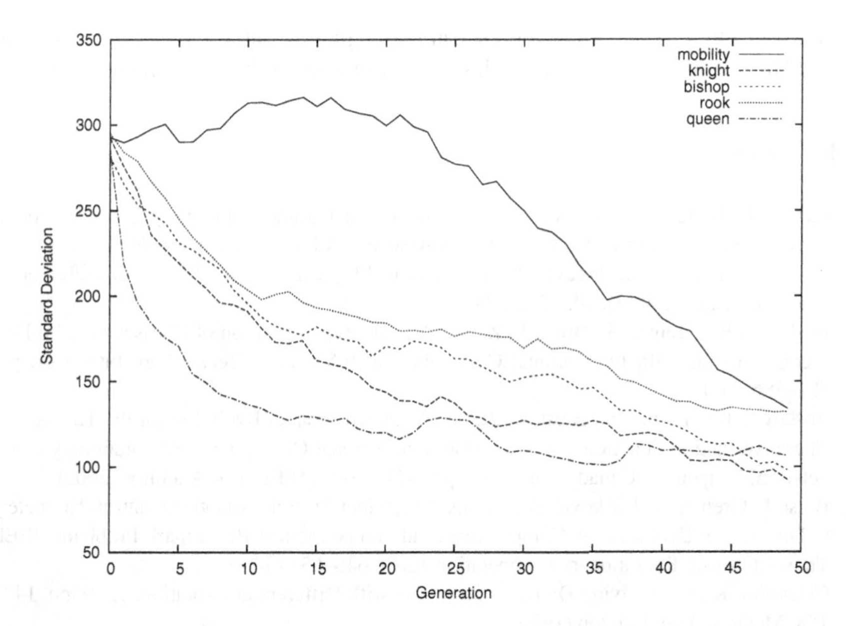

Fig. 6. Standard deviation of parameters along generations for AODE with initial population opposition and JR = 0.1

The third experiment included all optimizations ($JR = 0.1$) and achieved average parameter values and standard deviation are shown in Figures 5 and 6. In comparison with first two experiments this algorithm had poor convergence in the beginning but at the end obtained equally good parameters.

As shown on Figures 1, 3, and 5, the most critical parameter is mobility. Using additional analysis, it was discovered that the AODE in the first two experiments converged to local optima over two runs and in the third experiment only over one run. In all these runs, the algorithm found mobility parameter values that are considered inadequate in chess theory. In the third experiment we also observed that two populations had equal individuals sequentially because of repositioning in the dynamic opposition. Although JR was 0.1, the $rand(0, 1)$ was smaller than JR sequentially and, therefore, the algorithm generated a lot of equal individuals.

6 Conclusions

We have proposed an algorithm for the tuning of a chess program based on Differential Evolution. In the chess program we tuned only the parameters of its evaluation function. The algorithm included adaptation and opposition-based optimization mechanisms. Using different combinations of these mechanisms inside DE, which already includes adaptation, we have demonstrated the behavior of our algorithm. With opposition based-optimization only in initialization the algorithm has better convergence at the beginning. With opposition-based optimization during the entire evolutionary process the algorithm has poor convergence at the beginning but at the end obtains equal results. The results also show that such settings of the algorithm make it more robust.

References

1. Baxter, J., Tridgell, A., Weaver, L.: Experiments in Parameter Learning Using Temporal Differences. International Computer Chess Association Journal 21(2), 84–99 (1998)
2. Baxter, J., Tridgell, A., Weaver, L.: Learning to Play Chess Using Temporal Differences. Machine Learning 40(3), 243–263 (2000)
3. Bošković, B., Greiner, S., Brest, J., Žumer, V.: The Representation of Chess Game. In: Proceedings of the 27th International Conference on Information Technology Interfaces, pp. 381–386 (2005)
4. Bošković, B., Greiner, S., Brest, J., Žumer, V.: A Differential Evolution for the Tuning of a Chess Evaluation Function. In: CEC 2006: International Conference on Evolutionary Computation, Vancouver, Canada, July 2006, pp. 6742–6747. IEEE, Los Alamitos (2006)
5. Brest, J., Greiner, S., Bošković, B., Mernik, M., Žumer, V.: Self-Adapting Control Parameters in Differential Evolution: A Comparative Study on Numerical Benchmark Problems. IEEE Transactions on Evolutionary Computation 10(6), 646–657 (2006)
6. Chisholm, K.: Co-evolving Draughts Strategies with Differential Evolution, ch. 9, pp. 147–158. McGraw-Hill, London (1999)
7. Fogel, D.B., Hays, T.J., Hahn, S.L., Quon, J.: A Self-Learning Evolutionary Chess Program. Proceedings of the IEEE 92(12), 1947–1954 (2004)
8. Kendall, G., Whitwell, G.: An Evolutionary Approach for the Tuning of a Chess Evaluation Function Using Population Dynamics. In: Proceedings of the 2001 Congress on Evolutionary Computation CEC 2001, COEX, World Trade Center, 159 Samseong-dong, Gangnam-gu, Seoul, Korea, pp. 995–1002. IEEE Press, Los Alamitos (2001)
9. Price, K., Storn, R.: Differential Evolution: A Simple Evolution Strategy for Fast Optimization. Dr. Dobb's Journal of Software Tools 22(4), 18–24 (1997)
10. Price, K.V., Storn, R.M., Lampinen, J.A.: Differential Evolution, A Practical Approach to Global Optimization. Springer, Heidelberg (2005)
11. Rahnamayan, S., Tizhoosh, H.R., Salama, M.M.: Opposition-Based Differential Evolution Algorithms. In: CEC 2006: International Conference on Evolutionary Computation, Vancouver, Canada, July 2006, pp. 7363–7370. IEEE, Los Alamitos (2006)
12. Rahnamayan, S., Tizhoosh, H.R., Salama, M.M.A.: Opposition-based differential evolution. IEEE Transactions on Evolutionary Computation 12(1) (2008)
13. Samuel, A.L.: Some studies in machine learning using the game of checkers. IBM Journal of Research and Development (3), 211–229 (1995)
14. Shannon, C.: Programming a computer for playing chess. Philosophical Magazine 41(4), 256 (1950)
15. Storn, R., Price, K.: Differential Evolution - a simple and efficient adaptive scheme for global optimization over continuous spaces. Technical Report TR-95-012, Berkeley, CA (1995)
16. Storn, R., Price, K.: Differential Evolution – A Simple and Efficient Heuristic for Global Optimization over Continuous Spaces. Journal of Global Optimization 11, 341–359 (1997)
17. Thrun, S.: Learning To Play the Game of Chess. In: Tesauro, G., Touretzky, D., Leen, T. (eds.) Advances in Neural Information Processing Systems 7, pp. 1069–1076. The MIT Press, Cambridge (1995)

Differential Evolution for the Offline and Online Optimization of Fed-Batch Fermentation Processes

Rui Mendes[1], Isabel Rocha[2], José P. Pinto[1], Eugénio C. Ferreira[2], and Miguel Rocha[1]

[1] Departament of Informatics / CCTC - University of Minho
Campus de Gualtar, 4710-057 Braga - Portugal
azuki@di.uminho.pt, mrocha@di.uminho.pt

[2] IBB - Institute for Biotechnology and Bioengineering
Centre of Biological Engineering - University of Minho
Campus de Gualtar, 4710-057 Braga - Portugal
irocha@deb.uminho.pt, ecferreira@deb.uminho.pt

Summary. The optimization of input variables (typically feeding trajectories over time) in fed-batch fermentations has gained special attention, given the economic impact and the complexity of the problem. Evolutionary Computation (EC) has been a source of algorithms that have shown good performance in this task. In this chapter, Differential Evolution (DE) is proposed to tackle this problem and quite promising results are shown. DE is tested in several real world case studies and compared with other EC algorihtms, such as Evolutionary Algorithms and Particle Swarms. Furthermore, DE is also proposed as an alternative to perform online optimization, where the input variables are adjusted while the real fermentation process is ongoing. In this case, a changing landscape is optimized, therefore making the task of the algorithms more difficult. However, that fact does not impair the performance of the DE and confirms its good behaviour.

1 Introduction

In recent years, many efforts have been devoted to the optimization of processes in biotechnology and bioengineering, since a number of valuable products such as recombinant proteins, antibiotics and amino-acids are produced using fermentation techniques.

A problem that has received major attention is the dynamic optimization of fed-batch bioreactors, which has traditionally been done on the substrate feed rates as key manipulated variables. The optimization problem is usually solved before the beginning of the fermentation process (open-loop optimal control) and may consist on finding an expression or a sequence of values for the feeding rate, that maximize a given objective function. This function will typically be a performance index that measures the process productivity, subject to the constraints represented by a dynamical model.

U.K. Chakraborty (Ed.): Advances in Differential Evolution, SCI 143, pp. 299–317, 2008.
springerlink.com

Several optimization methods have been applied to solve this class of problems. It has been shown that for relatively simple bioreactor systems, which are expressed as differential equation models, the optimization problem can be solved analytically from the Hamiltonian function, by applying the Minimum Principle of Pontryagin [3]. However, in the majority of the cases reported, determination of the optimal feed rate profile has a problem of singular control, because the control variable (feed rate) often appears linearly in the system of differential equations. Thus, this approach fails to provide a complete solution. Also, those methodologies become too complex when the number of state and control variables increase.

Numerical methods make a distinct approach to dynamic optimization. The gradient algorithms are used to adjust the control trajectories in order to iteratively improve the objective function [4]. In contrast, dynamic programming methods discretize both time and control variables to a predefined number of values. A systematic backward search method in combination with the simulation of the system model equations is then used to find the optimal path through the defined grid. However, in order to achieve a global minimum, the computational burden is very high [22].

An alternative approach comes from the use of algorithms from the Evolutionary Computation (EC) field, which have been used in the past to optimize nonlinear problems with a large number of variables. These techniques have been applied with success to the optimization of feeding or temperature trajectories [14][1], and, when compared with traditional methods, usually perform better [19][6].

In this chapter, the application of Differential Evolution (DE) to the optimization of input variables in fed-batch fermentation processes is proposed. The DE is implemented and tested in several distinct variants and compared to other algorithms from the EC field, such as Evolutionary Algorithms (EAs) and Particle Swarm Optimization (PSO) .

Three case studies were used to illustrate and validate the approach and to compare the performance of the different algorithms. Each algorithm was allowed to run for a given number of function evaluations and the comparison among the methods was based on their final result and on the convergence speed.

This work also tackled the complementary issue of online optimization . In fact, in a real environment, even when the mathematical models used for open-loop optimization are reliable and validated by experimentation, several sources of noise can contribute to changes in the observed values of the state variables. These issues are of particular importance when dealing with recombinant high-cell density fermentations, as the process, besides the nonlinearities exhibited, tends to change dramatically upon some events, like induction. Also, it is likely that there exists a time-variance of both yield and kinetic parameters not contemplated in most process models. These scenarios have an important impact on the experimental results, that end up being worse than the ones predicted after running the offline optimization.

An alternative to cope with these model inaccuracies is the use of online optimization algorithms that periodically generate new solutions as the process is running, making use of the measurement of relevant state variables for update of the internal model. In this case, the optimization is performed simultaneously, taking into account values of the state variables measured by sensors within the fermentation process.

The performance of DE in this online optimization task was evaluated and compared to the results of a real-valued EA. The same case studies used in offline optimization, are now used in order to test the performance of both algorithms. These are firstly used to perform an offline optimization and then a simulation of a real-world fermentation is conducted. The relevant state variables are, in each case, disturbed by adding noise, at regular periods of time. The behavior of both algorithms is compared, as well as the degradation in performance when the initial offline solution is subjected to perturbations.

This chapter is organized as follows: firstly, the fed-batch fermentation case studies are presented; next, the optimization algorithms are described; the results of the application of the different algorithms to the case studies are presented, followed by a discussion of the results; online optimization is described, followed by the description of the experiments conducted and the discussion of the results; finally, the conclusions and further work are presented.

2 Case Studies: Fed-Batch Fermentation Processes

In fed-batch fermentations there is an addition of certain nutrients to the bioreactor along the time, in order to prevent the accumulation of toxic products, allowing the achievement of higher product concentrations.

During this process the system states change considerably, from a low initial to a very high biomass and product concentrations. This dynamic behavior motivates the development of optimization methods to find the optimal input feeding trajectories in order to improve the process. The typical inputs in this process are the substrate inflow rates time profiles.

For the proper optimization of the process, a white box mathematical model is typically developed, based on differential equations that represent the mass balances of the relevant state variables.

2.1 Case Study I

In previous work by the authors, a fed-batch recombinant *Escherichia coli* fermentation process was optimized by *EAs* [16][17]. This was considered as the first case study in this work.

During the aerobic growth of the bacterium, with glucose as the only added substrate, the microorganism can follow three main different metabolic pathways:

- Oxidative growth on glucose:

$$k_1 S + k_5 O \xrightarrow{\mu_1} X + k_8 C \tag{1}$$

- Fermentative growth on glucose:

$$k_2S + k_6O \xrightarrow{\mu_2} X + k_9C + k_3A \tag{2}$$

- Oxidative growth on acetic acid:

$$k_4A + k_7O \xrightarrow{\mu_3} X + k_{10}C \tag{3}$$

where S, O, X, C and A represent glucose, dissolved oxygen, biomass, dissolved carbon dioxide and acetate components, respectively. In the sequel, the same symbols are used to represent the state variables' concentrations (in g/kg); μ_1 to μ_3 are time variant specific growth rates that nonlinearly depend on the state variables, and k_i are constant yield coefficients.

The associated dynamical model can be described by the following equations:

$$\frac{dX}{dt} = (\mu_1 + \mu_2 + \mu_3)X - DX \tag{4}$$

$$\frac{dS}{dt} = (-k_1\mu_1 - k_2\mu_2)X + \frac{F_{in,S}S_{in}}{W} - DS \tag{5}$$

$$\frac{dA}{dt} = (k_3\mu_2 - k_4\mu_3)X - DA \tag{6}$$

$$\frac{dO}{dt} = (-k_5\mu_1 - k_6\mu_2 - k_7\mu_3)X + OTR - DO \tag{7}$$

$$\frac{dC}{dt} = (k_8\mu_1 + k_9\mu_2 + k_{10}\mu_3)X - CTR - DC \tag{8}$$

$$\frac{dW}{dt} \simeq F_{in,S} \tag{9}$$

being D the dilution rate, $F_{in,S}$ the substrate feeding rate (in kg/h), W the fermentation weight (in kg), OTR the oxygen transfer rate and CTR the carbon dioxide transfer rate.

The kinetic behavior, expressed in the rates μ_1 to μ_3, was given by specific functions of the state variables, whose description is out of the scope of the present work but can be found in [15].

The purpose of the optimization is to determine the feeding rate profile ($F_{in,S}(t)$) that maximizes the productivity of the process, defined as the units of product (recombinant protein) formed per unit of time. In this case, this is related with the final biomass obtained, when the duration of the process is pre-defined. Thus, a *performance index (PI)* is defined by the following expression:

$$PI = \frac{X(T_f)W(T_f) - X(0)W(0)}{T_f} \tag{10}$$

The relevant state variables are initialized with the following values: $X(0) = 5$, $S(0) = 0$, $A(0) = 0$, $W(0) = 3$. Due to limitations in the feeding pump capacity, the value of $F_{in,S}(t)$ must be in the range $[0.0; 0.4]$. Furthermore, the following constraint is defined over the value of W: $W(t) \leq 5$. The final time (T_f) is set to 25 hours.

2.2 Case Study II

This system is a fed-batch bioreactor for the production of ethanol by *Saccharomyces cerevisiae*, firstly studied by Chen and Huang [5]. The aim is to find the substrate feed rate profile that maximizes the final amount of ethanol.

The model equations are the following:

$$\frac{dx_1}{dt} = g_1 x_1 - u\frac{x_1}{x_4} \tag{11}$$

$$\frac{dx_2}{dt} = -10 g_1 x_1 + u\frac{150 - x_2}{x_4} \tag{12}$$

$$\frac{dx_3}{dt} = g_2 x_1 - u\frac{x_3}{x_4} \tag{13}$$

$$\frac{dx_4}{dt} = u \tag{14}$$

where x_1, x_2 and x_3 are the cell mass, substrate and ethanol concentrations (g/L), x_4 the volume of the reactor (L) and u) the feeding rate (L/h).

On the other hand, the kinetic variables g_1 and g_2 are given by:

$$g_1 = \frac{0.408}{1 + \frac{x_3}{16}} \frac{x_2}{0.22 + x_2} \tag{15}$$

$$g_2 = \frac{1}{1 + \frac{x_3}{71.5}} \frac{x_2}{0.44 + x_2} \tag{16}$$

The *performance index (PI)* is given by: $PI = x_3(T_f)x_4(T_f)$.

The final time is set to $T_f = 54$ hours, and the initial values for the state variables are the following: $x_1(0) = 1$, $x_2(0) = 150$, $x_3(0) = 0$ and $x_4(0) = 10$. Additionally, there are physical constraints over the variables, namely: $0 \leq x_4(t) \leq 200$ for all time points and the feeding rate $0 \leq u(t) \leq 12$.

2.3 Case Study III

This case study handles a hybridoma reactor described by the equations [19]:

$$\frac{dX_v}{dt} = (\mu - k_d)X_v - \frac{F_1 + F_2}{V}X_v \tag{17}$$

$$\frac{dGlc}{dt} = \frac{F_1}{V}Glc_{in} - \frac{F_1 + F_2}{V}Glc - q_{Glc}X_v \tag{18}$$

$$\frac{dGln}{dt} = \frac{F_2}{V}Gln_{in} - \frac{F_1 + F_2}{V}Gln - q_{Gln}X_v \tag{19}$$

$$\frac{dLac}{dt} = q_{Lac}X_v - \frac{F_1 + F_2}{V}Lac \tag{20}$$

$$\frac{dAmm}{dt} = q_{Amm}X_v - \frac{F_1 + F_2}{V}Amm \tag{21}$$

$$\frac{dMab}{dt} = q_{Mab}X_v - \frac{F_1 + F_2}{V}Mab \tag{22}$$

$$\frac{dV}{dt} = (F_1 + F_2) \tag{23}$$

where the state variables X_v, Glc, Gln, Lac, Amm, Mab, V are the concentrations of viable cells, glucose, glutamine, lactate, ammonia, monoclonal antibodies and culture volume, respectively. The control variables F_1 and F_2 are the volumetric feed rates. The complete kinetic expressions for μ, k_d, q_{Glc}, q_{Gln}, q_{Lac}, q_{Amm} and q_{Mab} are given in [19].

The target of the optimization process, in this case, is to increase the total amount of monoclonal antibodies produced. So, the PI is given by:

$$PI = \int_0^{T_f} -q_{Mab}X_v(t)V(t) \tag{24}$$

Initialization values for the state variables are the following: $X_v = 2.0 \times 10^8 cells/L$, $Glc = 25g/L$, $Gln = 4g/L$, $Lac = 0g/L$, $Amm = 0g/L$, $Mab = 0g/L$, $V = 0.8L$. T_f is 10 days and the value of $V(t)$ is constrained by $V(t) \leq V_{max}$.

3 Algorithms

3.1 Solution Representation and Evaluation

The optimization task addressed in this chapter is to find the best trajectory of some input variables (e.g. substrate feed), that yield the maximum performance index, defined in each specific case. A solution to the problem will consist in a real-valued vector, that encodes a temporal sequence of values, one per each time unit.

As mentioned above, the typical input variable in fed-batch fermentation processes is the feeding trajectory or trajectories, i.e. the amount of a given substrate to be introduced into the bioreactor, in a given time unit. Case studies I and II have only one input variable, given by the substrate feeding rate ; case study III has two feeding rates F_1 and F_2.

The size of the solution will be determined by the final time of the process (T_f), the discretization step (d) considered in the numerical simulation of the model and the number of input variables NV, given by the expression: $NV\frac{T_f}{d}$.

However, as the resulting genome would be very large (e.g. 5000 genes, for case study I), feeding values were defined only at certain equally spaced points, and the remaining values are linearly interpolated . The size of the genome (G) becomes:

$$G = NV(\frac{T_f}{dI} + 1) \tag{25}$$

where I stands for the number of points within each interpolation interval. The value of d used in the experiments was $d = 0.005$, for all case studies.

The evaluation of each solution is performed by running a numerical simulation of the defined model, given as input the feeding values. The numerical simulation is performed using *ODEToJava*, a package of ordinary differential equation solvers, using a linearly implicit implicit/explicit (IMEX) Runge-Kutta scheme used for stiff problems [2]. The fitness value is then calculated from the values of the state variables according to the *PI* defined for each case.

3.2 Differential Evolution

Differential Evolution (DE) is a population-based approach to function optimization that generates trial individuals by calculating vector differences between other randomly selected members of the population.

Given a function $f : \mathbb{R}^n \rightarrow \mathbb{R}$ to be minimised, a DE begins by randomly generating p n-dimensional vectors. These vectors (called individuals) form a population that will evolve over the course of the algorithm's run. The algorithm then proceeds to manipulate the population until a termination criterion is met. The termination condition can be that a fixed number of function evaluations have elapsed or no sufficient improvement is achieved.

The following is an outline of DE that uses a binomial crossover [21]. For clarity, the computation of the new trial vector has been shown separately from the crossover operation that selects only some of the dimensions of the trial vector.

1. Initialize the population;
2. Evaluate the population;
3. Generate a new population where each candidate individual i is generated in parallel according to:
   (i) Randomly select 3 distinct individuals r_1, r_2, r_3 from the population that are different from i;
   (ii) Generate a trial vector based on the scheme
   (iii) Perform crossover between this vector and the vector of the current individual, with probability CR, using at least one dimension of the trial vector.
   (iv) If the candidate is not valid, change its invalid coordinates by resetting them to the closest bound;
   (v) Evaluate the candidate;
   (vi) Use the candidate in the new generation if it is at least as good as the current individual;
   (vii) Replace the current individual by the candidate if the candidate is at least as good.
4. Loop to 3 unless the termination criterion is met.

Various schemes are currently in use for DEs [20]. Each scheme varies according to the number of difference vectors used and to whether or not the current individual or the global best individual will be used as part of that computation. Four schemes are considered in this paper. These are shown below along with the corresponding trial vector generation formula. The variables x_{r_j}, $2 \leq j \leq 5$ represent distinct randomly selected individuals that are different from the current individual x_i and x_{best} is the best individual. The parameter $F \in \mathbb{R}$ is the scale of the difference vectors and is usually set between 0 and 2 and CR is the crossover probability.

DE/rand/1 $\mathbf{t} = \mathbf{x}_{r_1} + F(\mathbf{x}_{r_2} - \mathbf{x}_{r_3})$
DE/rand/2 $\mathbf{t} = \mathbf{x}_{r_1} + F(\mathbf{x}_{r_2} + \mathbf{x}_{r_3} - \mathbf{x}_{r_4} - \mathbf{x}_{r_5})$
DE/best/1 $\mathbf{t} = \mathbf{x}_{best} + F(\mathbf{x}_{r_2} - \mathbf{x}_{r_3})$
DE/best/2 $\mathbf{t} = \mathbf{x}_{best} + F(\mathbf{x}_{r_2} + \mathbf{x}_{r_3} - \mathbf{x}_{r_4} - \mathbf{x}_{r_5})$

3.3 Real-Valued EA

A real-valued Evolutionary Algorithm (EA) was also considered, since it provided good results in previous work [17][18]. The overall structure of the *EA* is given by:

1. Initialize time ($t = 0$), generate and evaluate the initial population (P_0).
2. While the termination criteria is not met:
   (i) Select from P_t a subset of individuals for reproduction.
   (ii) Apply the genetic operators to the individuals in order to breed the offspring and evaluate them.
   (iii) Insert the offspring into the next population (P_{t+1}).
   (iv) Select the survivors from P_t to be kept in P_{t+1}.
   (v) Increase current time ($t = t + 1$).

Regarding the reproduction step, this EA uses the following mutation and crossover operators:

- *Random Mutation*, which replaces one gene by a new randomly generated value, within the range $[min_i, max_i]$ [13]; and
- *Gaussian Mutation*, which adds to a given gene a value taken from a Gaussian distribution, with a zero mean and a standard deviation given by $\frac{max_i - min_i}{4}$ (i.e., small perturbations will be preferred over larger ones).
- *Two-Point crossover*, a standard *Genetic Algorithm* operator [13], applied in the traditional way;
- *Arithmetical crossover*, where each gene in the offspring will be a linear combination of the values in the ancestors' chromosomes [13];

where $[min_i; max_i]$ is the range of values allowed for gene i.

Both mutation operators are applied to a variable number of genes (a value that is randomly set between 1 and 10 in each application). In previous work, the best result was obtained using an alternative that contemplates the use of all genetic operators described above [17]. All operators are used with equal probabilities to breed the offspring.

The selection procedure is done by converting the fitness value into a linear ranking in the population, and then applying a roulette wheel scheme. In each generation, 50% of the individuals are kept from the previous generation, and 50% are bred by the application of the genetic operators.

3.4 Fully Informed Particle Swarm

A particle swarm optimizer (PSO) uses a population of particles that evolve over time by flying through the search space. Particles imitate their neighbors by approaching their best positions. In the canonical particle swarm, the two sources of imitation are the individual's previous best position and the best position found by the most successful neighbor.

Due to the fact that in previous studies [11] the Fully Informed Particle Swarm (FIPS) [12] clearly outperformed the canonical particle swarm in this class of problems, this method will be used in this study. In this case, each particle is defined by:

$$P_t^{(i)} = \langle x_t, v_t, p_t, e_t \rangle$$

where $x_t \in \mathbb{R}^d$ is the current position in the search space; $p_t \in \mathbb{R}^d$ is the position visited by the particle in the past that had the best function evaluation; $v_t \in \mathbb{R}^d$ is a vector that represents the direction in which the particle is moving, it is called the 'velocity'; e_t is the evaluation of p_t under the function $f : \mathbb{R}^d \to \mathbb{R}$ being optimized, i.e. $e_t = f(p_t)$.

Particles are connected to others in the population via a predefined topology. This can be represented by the adjacency matrix of a directed graph $M = (m_{ij})$, where $m_{ij} = 1$ if there is an edge from particle i to particle j and $m_{ij} = 0$ otherwise.

In FIPS, each particle moves in the direction of the stochastic barycenter of the previous best position of all the neighboring particles (excluding the particle itself). As in the canonical particle swarm, the neighbors of a particle are the ones that share a vertex in the graph that represents the topology.

The following is an outline of a generic PSO:

1. Set the iteration counter, $t = 0$.
2. Initialize each $x_0^{(i)}$ and $v_0^{(i)}$ randomly. Set $p_0^{(i)} = x_0^{(i)}$.
3. Evaluate each particle and set $e_0^{(i)} = f(p_0^{(i)})$.
4. Let $t = t + 1$ and generate a new population, where each particle i is moved to a new position in the search space according to:
   (i) $v_t^{(i)} = velocity\_update(v_{t-1}^{(i)})$.
   (ii) $x_t^{(i)} = x_{t-1}^{(i)} + v_t^{(i)}$.
   (iii) Evaluate the new position, $e = f(x_t^{(i)})$.
   (iv) If the new position is better than the previous best, update the particle's previous best position. i.e if $e < e_{t-1}^{(i)}$ then let $p_t^{(i)} = x_t^{(i)}$ and $e_t^{(i)} = e$ else let $p_t^{(i)} = p_{t-1}^{(i)}$ and $e_t^{(i)} = e_{t-1}^{(i)}$.
   (v) Loop to 4 until the termination criterion is met.

Clerc and Kennedy [7] introduced the use of a factor called the 'constriction factor', symbolized by χ, into the velocity update equation. The velocity update equation for FIPS is given by:

$$velocity\_update(v_{t-1}^{(i)}) = \chi(v_t^{(i)} + \sum_{j \in N(i)} U(0,1) \cdot \frac{\varphi}{|N(i)|} \cdot (p_{t-1}^{(j)} - x_{t-1}^{(i)}))$$

where U is the generator of pseudo-random numbers following the uniform distribution, $\varphi = 4.1$, $\chi = 0.729$, $N(i)$ is the neighborhood (the set of the particles) of particle i. The values of φ and χ are given by Clerc's formula. In this study, the population is organized according to the *von Neumann* topology [10], where each particle is connected to four others, in a torus configuration.

4 Offline Optimization

4.1 Methodology

The results reported in this text are the means of 30 runs and are presented with 95% confidence intervals. Additionally, the use of t-tests [8] for two-sample comparisons was adopted. In order to improve the readibility of the analysis, a symbolic encoding of the p-values resulting from the t-tests was used. To enhance readability of the tables and allow a straighforward comparison between the approaches tested, different symbols are used to report whether the mean of approach $A1$ is greater than the mean of $A2$ or vice-versa. The encoding used is presented in Table 1.

Table 1. Encoding used in the presentation of p-values of the pairwise t-tests comparing approaches $A1$ and $A2$

p-value	condition	symbol
$p \leq 0.001$	$mean(A1) > mean(A2)$	+++
$p \leq 0.001$	$mean(A1) < mean(A2)$	- - -
$0.001 < p \leq 0.01$	$mean(A1) > mean(A2)$	++
$0.001 < p \leq 0.01$	$mean(A1) < mean(A2)$	- -
$0.01 < p \leq 0.05$	$mean(A1) > mean(A2)$	+
$0.01 < p \leq 0.05$	$mean(A1) < mean(A2)$	-
$p \geq 0.05$		O

Given that multiple pairwise comparisons were performed, the authors used the *Holm* correction for the p-values [9]. Sometimes statistical tests cannot find a significant difference between two algorithms (e.g., because the confidence interval of one of them is too wide). Nonetheless, we are interested in a *reliable* method: one that consistently yields good results. Thus, an algorithm with a good average and a narrow confidence interval is preferred in these cases.

4.2 Parameter Settings and Test Conditions

When solving a real world problem, the main concern is to have a tool that may be applied to the problem with as few fine-tuning as possible. The main focus of this work will be in the results and not in a thorough study about the parameterization of the algorithms involved. It was not an aim of this work to go through the cumbersome task of testing the valuation of all the parameters of these algorithms until a suitable setting for the problem at hand could be found. Furthermore, these experiments take a long time (typically a few hours per run) and there are usually time constraints. Thus, it was decided to use standard configurations for each algorithm that were either validated by experimental results or suggested by previous studies.

Due to the previous experience of the authors with the real-valued EA, each run was stopped after 200,000 function evaluations. In the case of FIPS the population size was 20 and the other parameters have the usual values given in the literature. The neighborhood topology selected was the *von Neumann* [10].

For all *DE* algorithms, the population size was set to 20, F was set to 0.5, CR to 0.6 and the schemes used were *DE/rand/1*, *DE/rand/2*, *DE/best/1* and *DE/best/2*. In terms of the real-valued EA, the population size was set to 200.

For each case study, 30 runs were performed with each algorithm. The value of I was determined, for each case study, based on preliminary results, and set in the following way: $I = 200$ in case studies 1 and 2, and $I = 100$ for the case study III.

Thus, the solution sizes are equal to 26, 55 and 21 for the three case studies.

4.3 Results

For all case studies, the results will be shown in two distinct tables. The first will present the results obtained by each of the algorithms showing the mean and the 95% confidence intervals for the *PI*. These will be shown for three distinct steps of the optimization process: when 50,000, 100,000 and 200,000 function evaluations were performed by each algorithm. It was decided to probe PI at these time-steps to estimate the possibility of terminating the runs earlier whilst still maintaining good quality solutions.

The second set of tables will help to further validate the results, showing the pairwise t-test results, when 200 k FEs have elapsed, using the methodoloy aforementioned. This will show the statistically significant differences among the algorithms. In these tables, the algorithm that appears on each row will correspond to *A1* on Table 1 and the algorithm given by the column to *A2*.

Tables 2 and 3, 4 and 5 and finally 6 and 7, present the results obtained by each of the algorithms on the case studies I, II and III, respectively.

Table 2. Results for case study I: mean and confidence intervals of the PI

Algorithm	PI 50,000 FEs	PI 100,000 FEs	PI 200,000 FEs
DE/rand/1	9.4726 ± 0.0005	9.4727 ± 0.0005	9.4727 ± 0.0003
DE/rand/2	9.0669 ± 0.0390	9.4074 ± 0.0102	9.4728 ± 0.0001
DE/best/1	5.1580 ± 0.4795	5.2274 ± 0.4470	5.2315 ± 0.4443
DE/best/2	9.4423 ± 0.0626	9.4729 ± 0.0000	9.4729 ± 0.0000
EA	8.4762 ± 0.0731	8.7891 ± 0.0613	9.0037 ± 0.0497
FIPS	9.4716 ± 0.0014	9.4729 ± 0.0000	9.4729 ± 0.0000

Table 3. Pairwise t-test with the Holm p-value adjustment for the algorithms of case study I

	DE/rand/1	DE/rand/2	DE/best/1	DE/best/2	EA
DE/rand/2	O				
DE/best/1	- - -	- - -			
DE/best/2	O	+++	+++		
EA	- - -	- - -	+++	- - -	
FIPS	O	+++	+++	O	+++

Table 4. Results for case study II: mean and confidence intervals of the PI

Algorithm	PI 50,000 FEs	PI 100,000 FEs	PI 200,000 FEs
DE/rand/1	20386 ± 7	20400 ± 7	20409 ± 6
DE/rand/2	20348 ± 8	20366 ± 6	20382 ± 6
DE/best/1	19702 ± 128	19723 ± 128	19751 ± 134
DE/best/2	20229 ± 86	20263 ± 80	20281 ± 84
EA	20119 ± 48	20280 ± 35	20373 ± 17
FIPS	19821 ± 120	19822 ± 120	19822 ± 120

Table 5. Pairwise t-test with the Holm p-value adjustment for the algorithms of case study II

	DE/rand/1	DE/rand/2	DE/best/1	DE/best/2	EA
DE/rand/2	O				
DE/best/1	- - -	- - -			
DE/best/2	O	O	+++		
EA	- -	O	+++	O	
FIPS	- - -	- - -	O	- - -	- - -

Table 6. Results for case study III: mean and confidence intervals of the PI

Algorithm	PI 50,000 FEs	PI 100,000 FEs	PI 200,000 FEs
DE/rand/1	392.81 ± 3.81	393.93 ± 3.20	394.99 ± 3.13
DE/rand/2	391.66 ± 0.48	394.18 ± 0.33	395.73 ± 0.20
DE/best/1	276.40 ± 10.74	283.50 ± 12.25	289.37 ± 12.82
DE/best/2	372.90 ± 12.44	375.08 ± 12.60	378.67 ± 11.86
EA	374.83 ± 1.67	382.49 ± 0.86	387.62 ± 0.52
FIPS	362.45 ± 15.10	370.66 ± 13.68	375.69 ± 10.79

The first conclusion to be drawn from the results is a superiority of some of the DE schemes (specially DE/rand/1 and DE/rand/2) over the EA and FIPS as soon as 50,000 FEs. FIPS converges fast, but the quality of the solutions in cases II and III is not at the level of the results of DE and even the EA. The EA

Table 7. Pairwise t-test with the Holm p-value adjustment for the algorithms of case study III

	DE/rand/1	DE/rand/2	DE/best/1	DE/best/2	EA
DE/rand/2	O				
DE/best/1	- - -	- - -			
DE/best/2	O	O	+++		
EA	- - -	- - -	+++	O	
FIPS	-	- -	+++	O	O

is usually the algorithm with the slowest convergence, although it steadly improves over the entire run. The worst algorithm in all problems is the DE/best/1 scheme. DE/best/2 is much better showing that in some problems having a noisier setup with a greedier scheme can pay off. However, it is still a step behind the DE/rand alternatives.

In case study I, DE/rand/1 has already obtained good solutions with only 50,000 FEs, closely followed by FIPS and DE/best/2. When 100,000 FEs have elapsed, the quality of the solutions of the three approaches is very similar, with the EA still trailing far behind and DE/best/1 out of the competition. Finally, with 200,000 FEs the three approaches (DE/rand/1, FIPS and DE/best/2) maintain similar performance and the EA is still trailing behind, although steadily improving.

In case study II, there are some changes. FIPS is not as good as the EA and presents results similar to the ones of DE/best/1. DE/rand/1 and DE/rand/2 have similar results with 50,000 FEs, slowly improving with a larger number of FEs. The EA is the algorithm that shows the steadier improvement but still exhibits a somewhat lower performance when compared to DE/rand at 200,000 FE.

Case study III shows a similar performance for DE/rand/1 and DE/rand/2, with DE/rand/2 having somewhat better performance for 100,000 and 200,000 FEs. In this case, the EA is again the third best alternative and the worst performer is still DE/best/1.

5 Online Optimization

5.1 Description

During the fermentation process, some of the state variables can be measured, but its values are scarcely used for closed-loop optimization purposes, and are rather employed to evaluate qualitatively the performance of the process. However, it is possible to develop dynamic optimization algorithms capable of timely reacting to this new knowledge generated by updating the corresponding internal model and generating new solutions.

EC is a promising approach to this real-time optimization task, since the algorithms keep a population of solutions that can be easily adapted to perform re-optimization. Indeed, a population of solutions previously obtained can be evaluated under the new scenario and better adapted solutions can be created through the use of evolution. The fact that a set of solutions is kept, and not only the best solution, makes a faster adaptation to new conditions possible, while taking advantage of previous optimisation efforts.

In this work, two online optimization strategies based on *EAs* and *DE* are proposed, working in two stages: before the fermentation process starts, an offline optimization is conducted as it is described in the previous sections. After this preliminary step, online optimization algorithms use information gathered by measuring the value of relevant state variables in certain points in time during the real fermentation. These algorithms react by updating their internal model and reaching an improved solution, that is available to be sent back to the fermentation monitoring software.

The version of the *EA/DE* used to perform online optimization is similar to the ones described for the offline problem. The DE scheme selected was DE/rand/1 due to the fact that it is the simplest to implement and the one that usually gave the best results. When new information regarding the state variables is received, the following steps are followed by both *EA* and *DE*:

1. a starting point (in time) is determined for the re-optimized solution, by adding the time label of the received data with the predicted time necessary to compute a new solution (since it is impossible to reach and therefore apply a solution before that time);
2. the last available population is adapted by removing from the genome of each individual the genes that encode feeding values for elapsed time periods;
3. half of the individuals in the population are replaced by new randomly generated solutions (these individuals are chosen randomly, although the best individual is always kept). This helps in maintaining genetic diversity, a useful feature for the optimization in changing landscapes;
4. the internal model of the fermentation used by the *EA/DE* is updated with the new information available from the real process and each of the individuals is re-evaluated taking this new knowledge into consideration;
5. the normal process of the *DE* or *EA* proceeds for a given number of iterations;
6. the best solution obtained is sent to the fermentation software and can be used in the real process.

5.2 Experimental Setup

In this study, and given time and physical constraints, real fermentations were not conducted and instead these were replaced by simulating the fermentation process and adding noise to the value of the state variables. This process is implemented by considering two interacting components: an *optimizer*, that implements the optimization algorithm (*DE* or *EA*), and a *noise simulator (NS)*, that simulates the real fermentation process adding noise to the state variables.

This is performed by considering that there is a deviation between the model prediction and the behaviour of the process due to several reasons (e.g. model innacuracies or parameters changing over time). Therefore, for each sampling time, the state variables that represent the real process are obtained from the simulated variables by adding noise. These new values of the state variables would originate a deviation of the process from its optimal behavior, which had been defined during offline optimization. To compensate for this deviation, the new values of the state variables will be used by the optimization algorithm to reach a new feeding profile.

The following sequence of events takes place:

1. an offline optimization is performed by the *optimizer* and its results are passed on to the *NS*, used to compute the predicted values of the state variables. The *optimizer* stops and waits for new information.
2. the variable t, which stores the simulated time in the *NS* is set to $t = 0$.
3. while $t < T_f$ (where T_f denotes the final time of the fermentation process) the following steps are executed:
   (i) the values of all state variables at time t are disturbed by the *NS* by adding/ subtracting noise, given by the original value multiplied by a value taken from an uniform distribution with range $[0, U]$. The new values of the state variables are sent to the *optimizer*.
   (ii) the *optimizer* receives this information and runs the steps for online optimization listed in the previous section. The best solution reached is sent to the *NS* that updates its model accordingly.
   (iii) the *NS* updates $t = t + \Delta t$.

Each run for the initial optimization is stopped after 200,000 function evaluations and the re-optimization process is allowed 20,000 function evaluations. The parameters of the *DE* and *EA* keep the values of the offline optimization given in the previous sections The value of Δt was set to 1 (h.) in case study I, 2 (h.) in II and 0.5 (d.) in III.

5.3 Results

The results will be presented in terms of the mean of the *PI* values obtained in 30 runs, as well as 95% confidence intervals. The Tables 8, 9 and 10 show the results of the algorithms obtained on case studies I, II and III, respectively. In every case, the first column represents the parameter U used to generate noise (an increase in this parameter implies noisier setups). The next two columns show the results for the *DE* and the *EA* during offline optimization; columns 4 and 5 show the results obtained for the same algorithms, but applying the noise disturbances without changing the solutions of offline optimization (simulating the case where there are discrepancies between model predictions and real processes but without intervention of online optimizers) and, finally, the last two columns show the results obtained by the online optimization.

The first conclusion to draw from the results is that, in every case study, even a low level of noise is enough to clearly disturb the results, although that effect is clearly more visible in case study I.

The levels of noise studied are certainly within the range of the differences observed between model predictions and experimental results in biotechnological processes. However, the consequences in terms of process performance when an

Table 8. Results obtained by the *DE* and *EAs* in case study I

U	Initial optim.		Initial+noise		Online opt.	
	DE	EA	DE	EA	DE	EA
0.01	9.47 ± 0.00	8.85 ± 0.04	4.67 ± 0.70	4.79 ± 0.73	9.11 ± 0.14	8.72 ± 0.14
0.02	9.47 ± 0.00	8.83 ± 0.05	4.41 ± 0.75	4.69 ± 0.78	8.80 ± 0.24	8.53 ± 0.25
0.03	9.47 ± 0.00	8.81 ± 0.05	4.20 ± 0.76	4.35 ± 0.81	8.47 ± 0.34	8.17 ± 0.35

Table 9. Results obtained by the *DE* and *EAs* in case study II

U	Initial optim.		Initial+noise		Online opt.	
	DE	EA	DE	EA	DE	EA
0.01	20405 ± 4	20374 ± 9	20097 ± 133	20236 ± 108	20421 ± 115	20408 ± 119
0.02	20407 ± 3	20379 ± 7	19832 ± 305	19986 ± 244	20404 ± 243	20392 ± 242
0.03	20405 ± 5	20376 ± 9	19711 ± 357	19938 ± 393	20282 ± 317	20236 ± 335

Table 10. Results obtained by the *DE* and *EAs* in case study III

U	Initial optim.		Initial+noise		Online opt.	
	DE	EA	DE	EA	DE	EA
0.01	394.7 ± 0.2	386.3 ± 0.8	371.7 ± 8.5	367.9 ± 7.1	386.2 ± 4.8	379.8 ± 3.8
0.02	394.7 ± 0.2	385.2 ± 0.7	353.9 ± 14.9	351.2 ± 12.3	374.1 ± 9.2	371.8 ± 8.3
0.03	394.7 ± 0.2	386.1 ± 0.9	330.0 ± 23.5	343.0 ± 15.4	364.5 ± 13.0	367.6 ± 11.0

open-loop fermentation (without online optimization) is performed are quite extreme, implying that in many cases the utility of even relatively good models for process optimization with current state-of-the-art optimization techniques (mostly offline approaches) is quite low.

Therefore, the results obtained with online optimization strategies indicate that the reward obtained in terms of process productivity is probably more than enough to justify its implementation and the corresponding costs. In fact, he results obtained for all 3 case studies are quite close to the ones predicted by offline optimization without added noise, thus implying that the optimization scheme is robust to the levels of noise studied in this work. Furthermore, the degradation of the results that is caused by the increase of U is quite graceful, as an increase in U does not cause dramatic effects in the PI.

A comparison of the results obtained by both optimization algorithms show that *DE* seems to be more effective than the *EAs*. The difference is very clear when offline optimization is performed, but decreases when the level of noise increases. In fact, the differences for $U = 0.02$ and 0.03 are not significant from a statistical perspective and in case study III, the *EA* displays a better mean than *DE* for $U = 0.03$. Nevertheless, if an alternative has to be chosen the *DE* still has an advantage, since it shows the best results (mean) in almost all scenarios.

6 Conclusions and Further Work

This chapter compares FIPS, a real-valued EA (*EA*) and four distinct schemes of *Differential Evolution (DE)* in three case studies of optimizing the feeding trajectory in fed-batch fermentation processes. The best overall algorithms in these tasks were the *DE/rand/1* and *DE/rand/2*, that consistently obtained good results in all the case studies and furthermore had a good convergence speed. If a single configuration was to be chosen, the DE/rand/1 scheme would be the selected one, since it represents the simpler alternative to implement and obtains good results.

Fips was a good contender in one of the cases where it found good results and was as fast as DE. However, it got stuck on local optima on the other ones. EA was slower to converge but reliable. If one can afford the computational time needed, it always finds good solutions. However, in some problems (specially case study I) it requires a large number of function evaluations to achieve a good result. Given that the computational time needed for these problems is quite large, it is a good reason to choose *DE* instead.

In this work, the task of optimizing feed profiles for fed-batch fermentation problems was also approached by proposing optimization algorithms, such as *EAs* and *DE*, that are able to implement online optimization strategies, i.e., to perform the optimization simultaneously with the real process. The proposed approach was validated by conducting a number of experiments that used a noise simulator to emulate the differences between the values predicted by the mathematical model and the real values in the fermentation process. The results of the experiments show that even small differences lead to important disruptions in the behavior that was predicted by offline optimization.

The proposed approach to online optimization deals well with the noise and exhibits properties of graceful degradation. When comparing the optimization algorithms, the *DE* seems the best alternative, but its superiority seems to decrease when noisier settings are considered.

In future work, the priority is to validate these results by implementing the approach to online optimization with a real fed-batch fermentation process. Furthermore, other case studies will be tested and distinct optimization algorithms will be taken into account.

Previous work by the authors [18] developed a new representation in EAs in order to allow the optimization of a time trajectory with automatic interpolation. It would be interesting to develop a similar approach within DE.

Acknowledgments

This work was supported in part by the Portuguese Foundation for Science and Technology under project POSC/EIA/59899/2004. The authors wish to thank Project SeARCH (Services and Advanced Research Computing with HTC/HPC clusters), funded by FCT under contract CONC-REEQ/443/2001, for the computational resources made available.

References

1. Angelov, P., Guthke, R.: A Genetic-Algorithm-based Approach to Optimization of Bioprocesses Described by Fuzzy Rules. Bioprocess Engin. 16, 299–303 (1997)
2. Ascher, Ruuth, Spiteri: Implicit-explicit runge-kutta methods for time-dependent partial differential equations. Applied Numerical Mathematics 25, 151–167 (1997)
3. Banga, J.R., Moles, C., Alonso, A.: Global Optimization of Bioprocesses using Stochastic and Hybrid Methods. In: Floudas, C.A., Pardalos, P.M. (eds.) Frontiers in Global Optimization - Nonconvex Optimization and its Applications, vol. 74, pp. 45–70. Kluwer Academic Publishers, Dordrecht (2003)
4. Bryson, A.E., Ho, Y.C.: Applied Optimal Control - Optimization, Estimation and Control. Hemisphere Publication Company, New York (1975)
5. Chen, C.T., Hwang, C.: Optimal Control Computation for Differential-algebraic Process Systems with General Constraints. Chemical Engineering Communications 97, 9–26 (1990)
6. Chiou, J.P., Wang, F.S.: Hybrid Method of Evolutionary Algorithms for Static and Dynamic Optimization Problems with Application to a Fed-batch Fermentation Process. Computers & Chemical Engineering 23, 1277–1291 (1999)
7. Clerc, M., Kennedy, J.: The particle swarm - explosion, stability, and convergence in a multidimensional complex space. IEEE Transactions on Evolutionary Computation 6(1), 58–73 (2002)
8. Goulden, C.H.: Methods of Statistical Analysis, 2nd edn. John Wiley & Sons Ltd., Chichester (1956)
9. Holm, S.: A simple sequentially rejective multiple test procedure. Scandinavian Journal of Statistics 6, 65–70 (1979)
10. Kennedy, J., Mendes, R.: Topological structure and particle swarm performance. In: Fogel, D.B., Yao, X., Greenwood, G., Iba, H., Marrow, P., Shackleton, M. (eds.) Proceedings of the Fourth Congress on Evolutionary Computation (CEC 2002), Honolulu, Hawaii, May 2002. IEEE Computer Society, Los Alamitos (2002)
11. Mendes, R., Rocha, I., Ferreira, E., Rocha, M.: A comparison of algorithms for the optimization of fermentation processes. In: 2006 IEEE Congress on Evolutionary Computation, Vancouver, BC, Canada, July 2006, pp. 7371–7378 (2006)
12. Mendes, R., Kennedy, J., Neves, J.: The fully informed particle swarm: Simple, maybe better. IEEE Transactions on Evolutionary Computation 8(3), 204–210 (2004)
13. Michalewicz, Z.: Genetic Algorithms + Data Structures = Evolution Programs, 3rd edn. Springer, USA (1996)
14. Moriyama, H., Shimizu, K.: On-line Optimization of Culture Temperature for Ethanol Fermentation Using a Genetic Algorithm. Journal Chemical Technology Biotechnology 66, 217–222 (1996)
15. Rocha, I.: Model-based strategies for computer-aided operation of recombinant E. coli fermentation. PhD thesis, Universidade do Minho (2003)
16. Rocha, I., Ferreira, E.C.: On-line Simultaneous Monitoring of Glucose and Acetate with FIA During High Cell Density Fermentation of Recombinant E. coli. Analytica Chimica Acta 462(2), 293–304 (2002)
17. Rocha, M., Neves, J., Rocha, I., Ferreira, E.: Evolutionary algorithms for optimal control in fed-batch fermentation processes. In: Raidl, G.R., Cagnoni, S., Branke, J., Corne, D.W., Drechsler, R., Jin, Y., Johnson, C.G., Machado, P., Marchiori, E., Rothlauf, F., Smith, G.D., Squillero, G. (eds.) EvoWorkshops 2004. LNCS, vol. 3005, pp. 84–93. Springer, Heidelberg (2004)

18. Rocha, M., Rocha, I., Ferreira, E.: A new representation in evolutionary algorithms for the optimization of bioprocesses. In: Proceedings of the IEEE Congress on Evolutionary Computation, pp. 484–490. IEEE Press, Los Alamitos (2005)
19. Roubos, J.A., van Straten, G., van Boxtel, A.J.: An Evolutionary Strategy for Fed-batch Bioreactor Optimization: Concepts and Performance. Journal of Biotechnology 67, 173–187 (1999)
20. Storn, R.: On the usage of differential evolution for function optimization. In: 1996 Biennial Conference of the North American Fuzzy Information Processing Society (NAFIPS 1996), pp. 519–523. IEEE, Los Alamitos (1996)
21. Storn, R., Price, K.: Minimizing the real functions of the icec'96 contest by differential evolution. In: IEEE International Conference on Evolutionary Computation, pp. 842–844. IEEE, Los Alamitos (1996)
22. Tholudur, A., Ramirez, W.F.: Optimization of Fed-batch Bioreactors Using Neural Network Parameters. Biotechnology Progress 12, 302–309 (1996)

Worst Case Analysis of Control Law for Re-entry Vehicles Using Hybrid Differential Evolution

P.P. Menon[1], D.G. Bates[1], I. Postlethwaite[1], A. Marcos[2], V. Fernandez[2], and S. Bennani[3]

[1] Control and Instrumentation Research Group, Department of Engineering, University of Leicester, Leicester, LE1 7RH, UK
[2] Advanced Projects Division, DEIMOS-SPACE S.L., Madrid, Spain
[3] Guidance Navigation and Control Group, ESA/ESTEC, Noordwijk, The Netherlands
ppm6@le.ac.uk

Summary. The development and application of the differential evolution (DE) optimisation algorithm to the problem of worst-case analysis of nonlinear control laws for hypersonic re-entry vehicles is described. The algorithm is applied to the problem of evaluating a proposed nonlinear handling qualities clearance criterion for a detailed simulation model of a hypersonic re-entry vehicle (also known as a reusable launch vehicle (RLV)) having a full-authority nonlinear dynamic inversion (NDI) flight control law. A hybrid version of the differential evolution algorithm, incorporating local gradient-based optimisation, is also developed and evaluated. Comparisons of computational complexity and global convergence properties reveal the significant benefits which may be obtained through hybridisation of the standard differential evolution algorithm. The proposed optimisation-based approach to worst-case analysis is shown to have significant potential for improving both the reliability and efficiency of the flight clearance process for next generation RLV's.

1 Introduction

Atmospheric re-entry is an important and safety-critical part of the reusable launch vehicle mission. During the re-entry flight phase, the space vehicle follows a predefined trajectory towards the designated landing point, travelling from space to the dense atmosphere of earth. As a result, the vehicle is subjected to high levels of uncertainty and variations in key flight parameters during the course of its mission. A primary requirement for re-entry guidance and flight control laws is that they exhibit sufficient levels of robustness to allow close tracking of the pre-defined trajectory in spite of high levels of uncertainty and disturbances. In order to demonstrate that this requirement is satisfied, maximum deviations from the prescribed trajectory due to uncertainty in flight parameters such as mass, centre-of-gravity locations, inertias and aerothermodynamic parameters, as well as actuator and sensor uncertainties need to be precisely evaluated in simulation , prior to any test flight. This process of "flight clearance must be carried out in all normal and various failure conditions, and in the presence of all possible parameter variations.

U.K. Chakraborty (Ed.): Advances in Differential Evolution, SCI 143, pp. 319–333, 2008.
springerlink.com

The task of analysing and quantifying the robustness properties of the RLV flight control algorithms is a very lengthy and expensive one, where different combinations of large numbers of uncertain parameters must be investigated such that an estimate about the worst case stability and performance of the control laws can be made. For nonlinear flight clearance problems, the current industrial standard is to use a gridding approach, where either the clearance criteria are evaluated for all combinations of the extreme points of the vehicle's uncertain parameters or Monte-Carlo simulation is employed to randomly sample the parameter space, [1]. Unfortunately, the computational effort involved in the resulting clearance assessment increases exponentially with the number of uncertain parameters that are to be considered (combinations of extreme points) or with the desired confidence levels for the clearance results (Monte-Carlo simulation). Another difficulty with these approaches is the fact that there is no guarantee that the worst case uncertainty combination has in fact been found, since it is possible that the worst-case combination of uncertain parameters does not lie on the extreme points, or in the sampled set used by Monte-Carlo approaches. A promising approach to address the above difficulties is to use advanced optimisation algorithms to search the parameter space for worst-cases that violate the particular clearance criterion under investigation. Clearly, given that the parameter space for this type of problem will in general be highly nonlinear and non-convex, [6], global optimisation methods will be required to avoid getting trapped in locally optimal solutions. Previous work by the authors has explored the applicability of various evolutionary optimisation methods to the flight clearance problem for high-performance aircraft, and has shown that, when hybridised with appropriate gradient-based algorithms, they have the potential to improve significantly both the reliability and efficiency of the flight clearance process, [24, 25].

In this chapter, the flight clearance problem for a highly detailed simulation model of a generic RLV over a lower atmospheric phase of its re-entry trajectory is considered. The flight control law included in the model has been designed using nonlinear dynamic inversion (NDI) methods to provide robust trajectory tracking over the specified flight phase. The clearance problem is solved using differential evolution and a hybrid version of differential evolution. Differential evolution is a relatively new global optimisation method, introduced by Storn and Price in [11]. This method belongs to the same class of evolutionary global optimisation techniques as GA, but unlike GA it does not require either a selection operator or a particular encoding scheme. To reduce the computational complexity of the approach, the DE algorithm is hybridised with a local gradient-based optimisation method 'fmincon'. The contributions of this chapter are as follows. We demonstrate conclusively, for a realistic, industry-standard re-entry vehicle simulation model with an NDI control law, the ability of the DE global optimisation algorithm to avoid getting trapped in local solutions to the flight clearance problem. We also show, however, that incorporation of local optimisation methods into global algorithms can drastically reduce computation times *and* improve convergence to the global solution. To the authors knowledge, this is the first time that advanced optimisation methods of this type have been applied to the problem of worst-case analysis for space applications.

2 RLV Model, Control Law and Clearance Criterion

The generic RLV high-fidelity simulation model is based on the HL-20 aerodynamic database and X38-type geometric and aerodynamic surface configuration, and has a dry mass of 19,100-lb. This simulation model has been developed by DEIMOS Space S.L. for the European Space Agency (ESA) to act as a research platform for the investigation of re-entry and autoland guidance, navigation and control systems, [23].

The model consists of a reference trajectory generator, a nonlinear dynamic inversion (NDI)-based flight control system, nonlinear actuator models, the RLV dynamics, sensors such as gyros and accelerometers, and detailed environment models (US standard 1969 and Earth gravity and geoid models). Figure 1 shows a block diagram schematic of the RLV simulation model, which is implemented in the Matlab Simulink environment.

The reference trajectory is defined in terms of Angle of Attack (AoA or α), Angle of Side Slip (AoSS or β), and bank angle ϕ. The NDI controller provides the elevator, aileron, rudder and brake control inputs according to the desired dynamics. The controller also includes actuator allocation functions depending on the commanded moments, altitude and velocity of the RLV. More details of the model and its associated flight control system are available in [23]. The parameters in the model, and associated uncertainty values, are accessible through a database consisting of a collection of XML files accessible by the user.

Table 1. RLV Model Uncertain Parameters

Parameter	Bound	Description
Δ_{mass}	[-2313.3, 2313.3]	variation in dry mass from nominal (11566.55 kg)
Δ_{Ixx}	[-1627, +1627]	variation in M.I about X (8135.0 $4kgm^2$)
Δ_{Iyy}	[-15185, +15185]	variation in M.I about Y (75926.0 kgm^2)
Δ_{Izz}	[-15863, +15863]	variation in M.I about Z (79315.0 kgm^2)
Δ_{Ixz}	[-628.8, +628.8]	variation in Product of inertia XZ (3144.0 kgm^2
$\Delta_{x_{cog}}$	[-0.4912, +0.4912]	variation in X c.g from nominal (4.9213 m)
$\Delta_{y_{cog}}$	[-0.01, +0.01]	variation in Y c.g from nominal (0.0 m)
$\Delta_{z_{cog}}$	[-0.1009, +0.1009]	variation in Z c.g from nominal (1.0094976 m)

The complete re-entry trajectory for the vehicle takes 1680 seconds of simulation time and is divided in flight phases based on dynamic pressure and atmospheric layer. The present analysis focuses on a lower atmosphere phase starting at 1588 seconds and ending at 1675 seconds that covers the 32 to 20 km

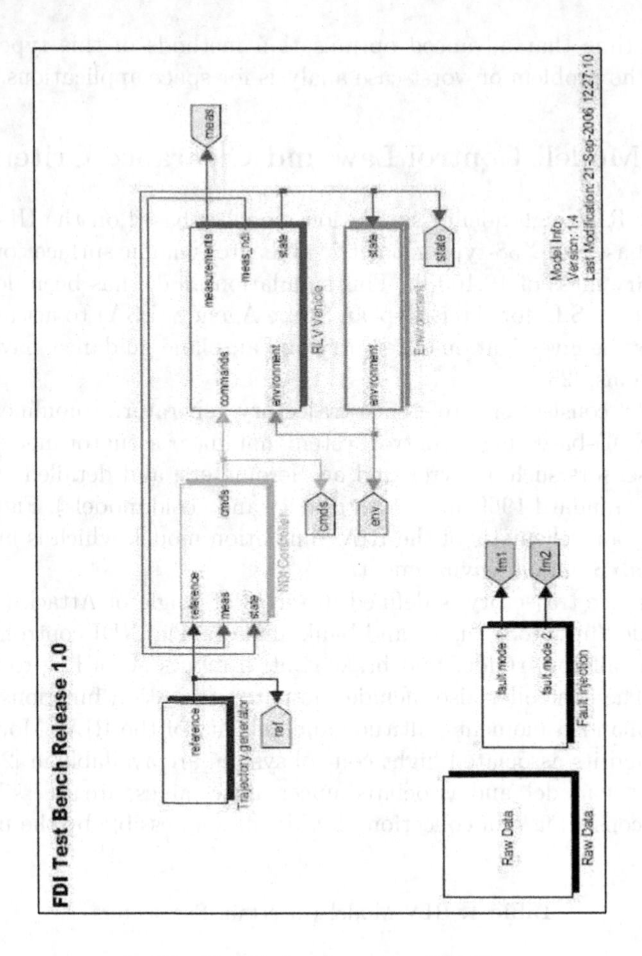

Fig. 1. Block schematic of RLV

altitude range. The reference trajectory in this segment includes a reduction of AoA from 30 degrees to nearly 20 degrees, while keeping a zero AoSS and with a defined bank angle variation. The description and allowed ranges of the uncertain parameters considered for the present analysis are given in Table 1. As can be seen from the table, in the present analysis we focus mainly on uncertainty in the parameters representing the vehicle's mass, inertias and centre-of-gravity.

2.1 Clearance Criterion

To analyse the robustness of the NDI control law in tracking AoA trajectories over the considered flight phase, a cost J is defined by Equation (1),

$$J = \|\alpha_{ref} - \alpha_{\Delta}\|_{\infty} \tag{1}$$

where α_{ref} represents the reference AoA trajectory and α_{Δ} represents the actual AoA trajectory followed by the vehicle in simulation in the presence of any uncertainty Δ. This particular clearance criterion was chosen for this study as criteria of this type are widely used throughout the European aerospace industry for the clearance of flight control laws for high performance aircraft [1, 6]. The uncertain parameter vector Δ consists of the parameters defined in Table 1, and its dimension is hence fixed at 8. The worst-case analysis problem is posed as identifying the Δ^* vector such that the following maximisation problem is solved.

$$max J = \|\alpha_{ref} - \alpha_{\Delta}\|_{\infty} \tag{2}$$

$$sub.to \ \underline{\Delta} \leq \Delta \leq \overline{\Delta} \tag{3}$$

where $\underline{\Delta}$ and $\overline{\Delta}$ define the lower and upper bounds on the uncertain parameters. The maximum cost value J^* corresponds to the uncertain parameters Δ^* that give the maximum deviation from the reference trajectory α_{ref}. The resulting optimisation problem is obviously nonlinear and nonconvex in general. Note that in this chapter we focus on a clearance criterion involving AoA only. However, the optimisation framework proposed is generic, and thus many other types of clearance could be assessed in a similar way.

3 Optimisation Based Worst Case Analysis

In this chapter the robustness analysis of an NDI flight control law for a RLV is formulated as an optimisation problem and solved using a global optimisation algorithm, DE, and its hybrid version. The optimisation problem itself is to find the combination of real parametric uncertainties that gives the worst value of the criterion defined in Eq. 1. Since this and many other clearance criteria must be checked over a huge number of conditions and re-entry vehicle configurations, it is imperative to find the most computationally efficient approach to the problem. Previous efforts to apply optimisation methods to similar problem, [1] Chapter 7, have revealed that the nonlinear optimisation problems arising in flight clearance, while having relatively low order, often have multiple local optima and expensive function evaluations. Therefore, the issue of whether to use local or global optimisation, and the associated impact on computation times is a key consideration for this problem.

In [1] Chapter 21, local optimisation methods such as SQP (Sequential Quadratic Programming), and L-BFGS-B (Limited memory Broyden-Fletcher-Goldfarb-Shanno method with Bounded constraints) were used to evaluate a range of linear clearance criteria for the HIRM+ (High Incidence Research Model) aircraft model. In [1] Chapter 22, global optimisation schemes such as Genetic Algorithms (GA), Adaptive Simulated Annealing (ASA) and Multi Coordinate Search (MCS) were also applied to evaluate nonlinear clearance criteria for the same aircraft model. In [5, 6] global optimisation methods such as GA and ASA were applied to the ADMIRE model with a different flight clearance criterion. In [25], a number of optimisation schemes were employed and compared, evaluating a nonlinear clearance criterion for ADMIRE aircraft.

4 Differential Evolution

The global optimisation method considered in this study is differential evolution , a relatively new global optimisation method, introduced by Storn and Price in [11]. This method belongs to the same class of evolutionary global optimisation techniques as Genetic Algorithm (GA) [15], but unlike GA it does not require either a selection operator or a particular encoding scheme. Essentially a subtype of GA, despite its apparent simplicity, the quality of the solutions computed using this approach has been claimed to be often better than that achieved using other evolutionary algorithms, both in terms of accuracy and computational overhead [11].

The DE method has recently been applied to several problems in different fields of engineering design, with promising results. In [17], for example, it was applied to find the optimal solution for a mechanical design example formulated as a mixed integer discrete continuous optimisation problem. In [18], DE was successfully applied in system design application, in particular handling the nonlinear design specification constraints. In [10], the DE method was applied and compared with other local and global optimisation schemes in an aerodynamic shape optimisation problem for an aerofoil. The application of differential evolution and its hybridised versions with neural networks and local search methods for aerodynamic shape optimisation has been reported in [27]. In [25], a nonlinear flight clearance criterion for a modern high performance aircraft was posed and solved using both standard and hybrid GA and DE optimisation methods. In that study, it was demonstrated that a hybrid version of the DE algorithm significantly outperforms the corresponding GA method.

The DE method consists of the following four main steps 1) Random initialisation, 2) Mutation 3) Crossover 4) Evaluation and Selection. There are different schemes of DE available based on the operators. The one used in the present studies is referred as "$DE/rand/1/bin$". The steps of this scheme will be described in detail in the sequel.

4.1 Random Initialisation

Like other evolutionary algorithms, DE works with a fixed number, N_p, of potential solution vectors, initially generated at random according to

$$\mathbf{x}_i = \mathbf{x}^L + \rho_i(\mathbf{x}^U - \mathbf{x}^L), \; i = 1, 2, ..., N_p \tag{4}$$

where $\mathbf{x}^U$ and $\mathbf{x}^L$ are the upper and lower bounds of the parameters of the solution vector and ρ_i is a vector of random numbers in the range [0 1]. N_p is fixed at 30 in the current study. Each $\mathbf{x}_i$ consists of elements $(x_{1i}, x_{2i}, ..., x_{di})$, which are the uncertain parameters defined in Table 1. The dimension d of the optimisation problem considered is, therefore, 8. The fitness of each of these N_p solution vectors is evaluated using the cost function given in Eq. 1.

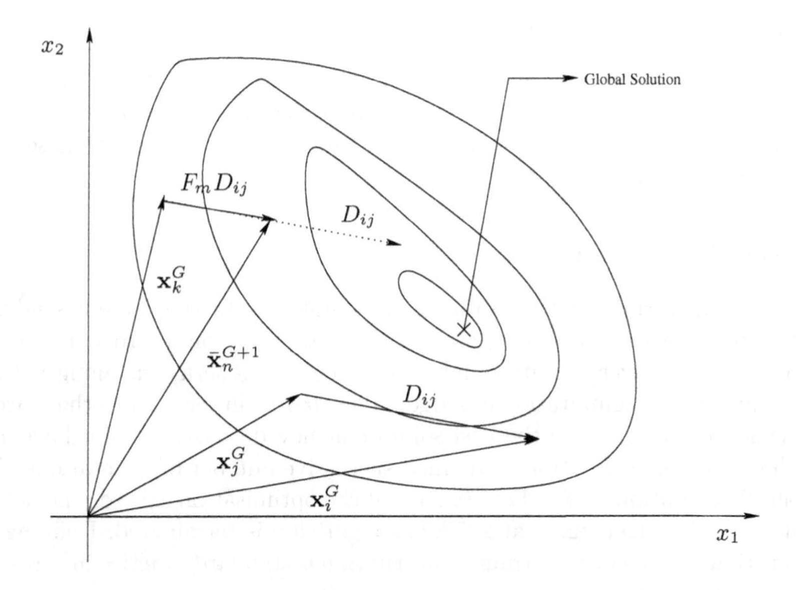

Fig. 2. DE mutation strategy

4.2 Mutation

The scaled difference vector $F_m D_{ij}$ between two random solution vectors $\mathbf{x}_i$ and $\mathbf{x}_j$ is added to another randomly selected solution vector $\mathbf{x}_k$ to generate the new mutated solution vector $\bar{\mathbf{x}}_n^{G+1}$, i.e.,

$$\bar{\mathbf{x}}_n^{G+1} = \mathbf{x}_k^G + F_m D_{ij}, \quad D_{ij} = \mathbf{x}_i^G - \mathbf{x}_j^G \tag{5}$$

where F_m is the mutation scale factor, a real valued number in the range [0, 1], (fixed at 0.8 in this study), and G represents the iteration number. Fig. 2 shows a simple two dimensional example of the mutation operation used in the DE scheme. The difference vector D_{ij} determines the search direction and F_m determines the step size in that direction from the point $\mathbf{x}_k^G$.

4.3 Crossover

During crossover, each element of the nth solution vector of the new iteration, $\mathbf{x}^{G+1}$, is reproduced from the mutant vector $\bar{\mathbf{x}}_n^{G+1}$ and a chosen parent individual $\mathbf{x}_n^G$ as given in Eq. 6,

$$x_{ji}^{G+1} = \begin{cases} x_{ji}^G, & if a generated random number > \rho_c \\ \bar{x}_{ji}^{G+1}, & otherwise; \end{cases} \tag{6}$$

where $j = 1, 2, \ldots, d$ and $i = 1, 2, \ldots, N_p$. Note that $\bar{\mathbf{x}}_n^{G+1}$ has elements $(\bar{x}_{1n}^{G+1}, \bar{x}_{2n}^{G+1}, ..., \bar{x}_{dn}^{G+1})$ and $\mathbf{x}_n^G$ has elements $(x_{1n}^G, x_{2n}^G, ..., x_{dn}^G)$. $\rho_c \in [0, 1]$ is the crossover factor, which is fixed at 0.8 in the present study.

4.4 Evaluation and Selection

After crossover, the fitness of the new candidate $\mathbf{x}_n^{G+1}$ is evaluated and if the new candidate $\mathbf{x}_n^{G+1}$ has a better fitness than the parent candidate $\mathbf{x}_n^G$, then $\mathbf{x}_n^{G+1}$ is selected to become part of the next iteration. Otherwise $\mathbf{x}_n^G$ is selected and subsequently identified as $\mathbf{x}_n^{G+1}$.

4.5 Termination Criterion

Many different termination criteria can be employed. In the present study, an adaptive termination criterion is used that is dependent on improvement in the solution accuracy over a finite number of successive generations along with an upper limit on the computational budget. The algorithm terminates the search if there is no improvement on the best solution achieved (above a defined accuracy level, here chosen as 10^{-6}) for a defined successive number of generations. This number of generations is fixed at 20. Also, if the optimisation exceeds the defined computational budget, fixed at 2250, the algorithm is terminated. Defining the computational budget as a termination criterion is standard practice in aerospace industry applications.

5 Hybrid Optimisation

Global optimisation methods based on evolutionary principles are generally accepted as having a high probability of converging to the global or near global solution, if allowed to run for a long enough time with sufficient initial candidates and reasonably appropriate probabilities for the evolutionary optimisation parameters. As shown by the preceding results, however, the rate of convergence can be very slow, and moreover, there is still no guarantee of convergence to the true global solution. Local optimisation methods, on the other hand, can very rapidly find optimal solutions, but the quality of those solutions entirely depends on the starting point chosen for the optimisation routine. In order to try to extract the best from both schemes, several researchers have proposed combining the two approaches [16], [19], [20]. In such hybrid schemes there is the possibility of incorporating domain knowledge, which gives them an advantage over a pure blind search based on evolutionary principles. In [25], a Hybrid GA (HGA) scheme was developed using a switching strategy originally proposed in [20], and applied to a nonlinear flight clearance problem. The performance of the HGA scheme was compared to that of a novel Hybrid DE (HDE) scheme. For a recent comprehensive overview of other approaches to hybrid optimisation (also known as memetic algorithms), the reader is referred to [21].

5.1 Hybrid DE

In [22], the conventional DE methodology was augmented by combining it with a downhill simplex local optimisation scheme. This hybrid scheme was applied to an aerofoil shape optimisation problem and was found to significantly improve the convergence properties of the method. At each iteration, local optimisation was applied to the best individual in a current random set. The hybrid DE scheme employed in this study applies gradient-based local optimisation, again using *"fmincon"*, to a solution vector *randomly* selected from the current set - for our problem, this was seen to give better results than using the *best* solution vector, as proposed in [22]. Use of the local optimisation method based on gradient estimation, specifically the function *"fmincon"* provided in [7], is considered in present study to hybridise the DE algorithm. Local optimisation methods can, of course, get locked into a local optimum in the case of nonconvex and/or multimodal surfaces, however, they are also much more computationally efficient than global optimisation approaches. Whether a local method converges to a local or global optimum completely depends on the initial starting point in the search space, and the convexity of the search space. In the present context, the aim is to obtain local improvements in the search space and thereby accelerate the search to global solution. Crucially however, in typical flight clearance problems very little information is available as to where to start the optimisation - the number of uncertain parameters and strong nonlinearity of the system mean that even advanced knowledge of flight mechanics provides little insight into how to choose initial values for the uncertain parameters. In such case, a hybrid version of DE will be very beneficial in finding the true global solution, through ensuring an adequate coverage of search space. The function *"fmincon"* finds the constrained minimum of a scalar function of several variables starting

Table 2. Hybrid Differential Evolution Algorithm - Pseudo-Code:

1. Initialize random candidate solutions in search space
2. Evaluate fitness of each solution and choose best fitness
3. Apply DE for a fixed number of initial iterations(say 10); Update best fitness value in each iteration
4. While termination criteria not satisfied
   a) calculate the improvement in best fitness
      i. If *Improvement* in best fitness
      ii. Continue DE
      iii. else
      iv. Choose a random solution from current set, say X_0
      v. Apply local optimisation with X_0 as initial point (termination occurs when exceeding the defined maximum number of function evaluations)
      vi. If *Improvement* in best fitness
      vii. Replace X_0 with the new solution
      viii. else
      ix. Keep X_0 in the set
      x. end
      xi. end
5. end of While

at an initial estimate. In the present analysis, constraints are due only to the upper and lower bounds of the uncertainty in the variables. A medium scale optimisation scheme is chosen where the gradients are estimated by the function itself using the finite difference method. The function uses the sequential quadratic programming (SQP) method - for further details of the *"fmincon"* optimisation strategy, the reader is referred to [7]. When the local scheme is chosen, the optimisation starts from the given initial condition and continues until it either converges or reaches a defined maximum number of cost function evaluations. The algorithm is simple, and tries to search for the global optimum in a "greedy" way, demanding improvement in the achieved optimum value in every iteration. A pseudo-code for the hybrid DE algorithm is given in Table 2.

6 Worst-Case Analysis Results

The optimisation-based worst-case analysis procedure is implemented in the Matlab 2006A and Simulink 6.1 environments. The various uncertain parameters listed in Table 1 are considered as the optimisation parameters and these variables are normalised by multiplication with an appropriate scaling factor. Prior to simulation, the respective entries of the uncertain variables in the XML database are accessed and updated with the new set of values provided by the optimisation algorithm. The cost function as given in Eqn.(1) is evaluated at the end of every simulation. The optimisation algorithm iterates, identifies the

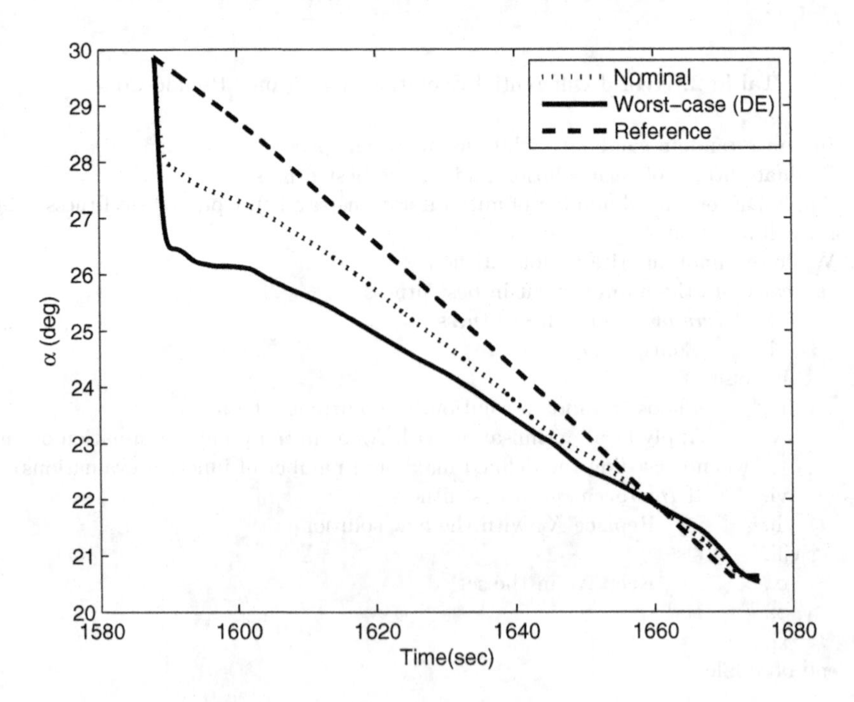

Fig. 3. Re-entry Angle of Attack trajectory

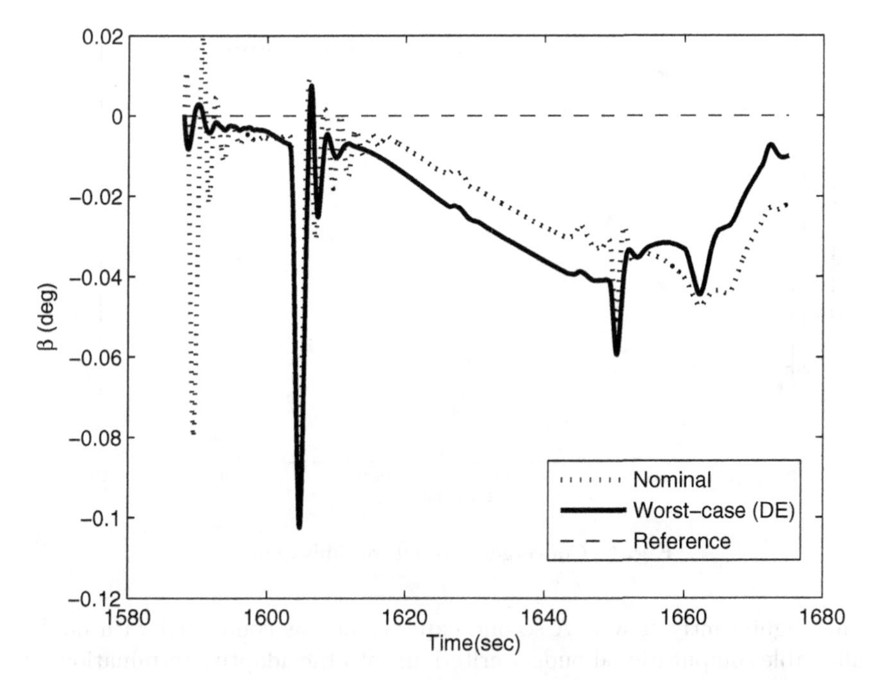

Fig. 4. Re-entry Angle of Sideslip trajectory

potential solutions and eventually converges to the global solution. However, to control the computational complexity we defined an adaptive termination criterion for the worst case analysis problem. In addition, an upper bound on the computational budget is also provided, fixed at 2250.

For the problem considered here, the DE algorithm took a total number of 2250 simulations, which is the computational budget termination criterion. The normalised worst-case obtained is $[\Delta_{mass}, \Delta_{Ixx}, \Delta_{Iyy}, \Delta_{Izz}, \Delta_{Ixz}, \Delta_{x_{cog}}, \Delta_{y_{cog}}, \Delta_{z_{cog}}] = [-0.9995, 0.9897, 0.9937, 0.3428, -0.9914, -0.9986, -0.1967, 0.9949]$.

Figure 3 shows the corresponding reference trajectory, worst-case and nominal angle-of-attack responses for the RLV model. Figure 4 shows the corresponding nominal and worst-case deviations from the desired zero value of $\beta(t)$. Interestingly, although the present cost function depends only on the value of $\alpha(t)$, the significant amount of cross-coupling between longitudinal and lateral dynamics at high AoA results in the worst-case β trajectory also being significantly different from the nominal response. To explore this issue further, a multi objective clearance criteria can be considered in this same framework, to identify the set of worst-case uncertain parameters for all of the controlled variables that define the reference trajectory.

Figure 5 shows the convergence of the DE algorithm. The x-axis indicates the function evaluations and the y-axis represents the maximum best cost value achieved over iterations. It can be seen from this figure that the DE algorithm shows good performance in the initial runs but subsequently the convergence rate

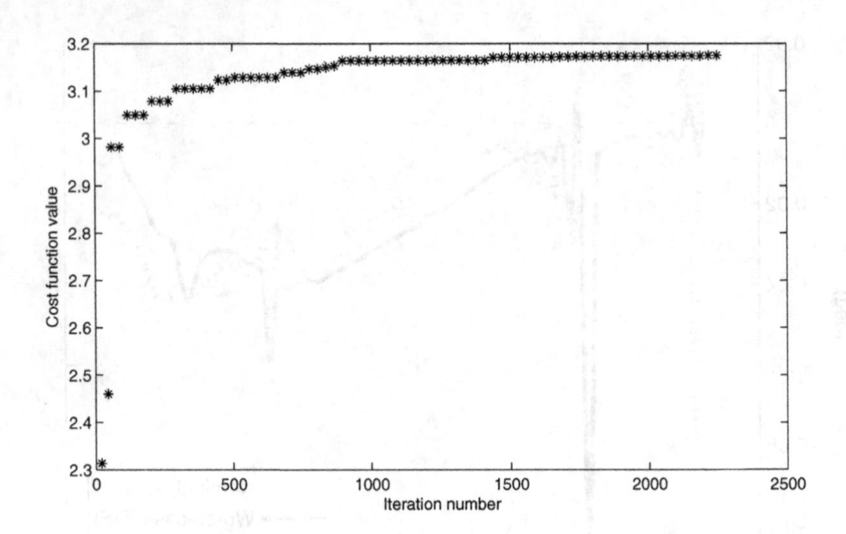

Fig. 5. Convergence of DE optimisation

becomes significantly slower. A second experiment was conducted with double the allowable computational budget criterion, with the adaptive termination criterion part left untouched. In this case, the maximum allowable computational budget was 5000. The repeated optimisation took a total number of 3250 simulations and the normalised worst-case obtained was $[\Delta_{mass}, \Delta_{Ixx}, \Delta_{Iyy}, \Delta_{Izz}, \Delta_{Ixz}, \Delta_{x_{cog}}, \Delta_{y_{cog}}, \Delta_{z_{cog}}]=[-1, 1, 1, 1, -1, -1, 1, 1]$, producing a worst-case cost value 3.178. The difference in the worst-case solution obtained from the two experiments can be explained with the help of a sensitivity analysis about the solution obtained from the optimisation. Figure 6 shows the results of a sensitivity analysis conducted about the global solution, by varying one parameter at a time and fixing all the other parameter values to their worst-case values. It can be noticed that the parameter $\Delta_{x_{cog}}$ has the greatest influence on the dynamics of the model, while the dynamics are relatively insensitive to the parameters Δ_{Ixx} and Δ_{Izz}. The presence of such insensitive parameters can make the optimisation convergence very slow. A possible way to avoid such a situation is by providing a termination condition of variation for insensitivity in parameter space.

The fact that the worst-case value of the uncertainties describing mass, centre-of-gravity and inertia variations are all on their maximum or minimum bounds is not surprising, and agrees well both with flight mechanics intuition and with the results of previous studies. The situation will becomes much more complex however when stability derivatives, sensor errors, etc are included, since in this case the corresponding worst-cases will not necessarily lie on the uncertain parameter bounds.

When compared with the standard DE algorithm, the HDE algorithm took a total number of 1775 simulations to converge. The normalised worst-case

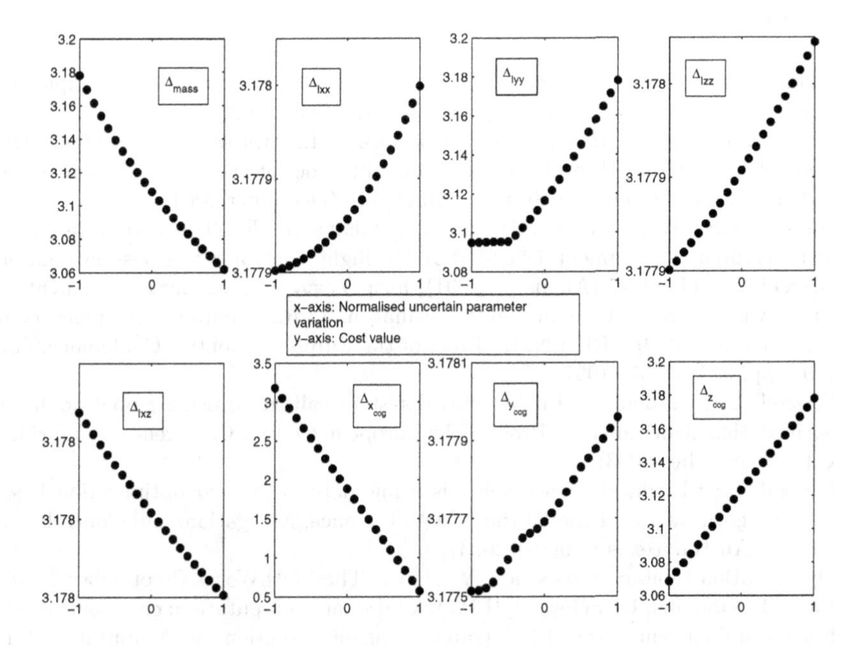

Fig. 6. Sensitivity analysis about the global solution

obtained is $[\Delta_{mass}, \Delta_{Ixx}, \Delta_{Iyy}, \Delta_{Izz}, \Delta_{Ixz}, \Delta_{x_{cog}}, \Delta_{y_{cog}}, \Delta_{z_{cog}}]=[-1,\ 1,\ 1,\ 1,\ -1,\ -1,\ 1,\ 1]$. Thus, to obtain a solution of the same quality, HDE has taken 45% less simulations than those required for the standard DE algorithm. This clearly demonstrates the significant computational savings which may be made by hybridising global optimisation algorithms with local gradient-based methods. Such savings are particularly crucial in the context of flight clearance, where computational cost is one of the key drivers for industrial applications.

7 Conclusions

In this chapter, differential evolution and hybrid differential evolution algorithms were applied to perform a worst-case analysis of a nonlinear-dynamic inversion (NDI) flight control law for a realistic simulation model of a re-entry vehicle over a particular phase of the trajectory for re-entry flight. A clearance criterion was defined based on the maximisation of the infinity norm of the error vector between the reference trajectory in Angle-of-Attack and the actual trajectory obtained by simulation of the model. The results of the study suggest that the proposed optimisation-based approach has the potential to improve significantly both the reliability and efficiency of the flight clearance process for future Reusable Launch Vehicles.

References

1. Fielding, C., Varga, A., Bennani, S., Selier, M. (eds.): Advanced techniques for clearance of flight control laws. Springer, Heidelberg (2002)
2. Menon, P.P., Kim, J., Bates, D.G., Postlethwaite, I.: Improved Clearance of Flight Control Laws Using Hybrid Optimisation. In: Proc. of the IEEE Conference on Cybernetics and Intelligent Systems, Singapore (December 2004)
3. Forssell, L.S., Hovmark, G., Hyden, Å., Johansson, F.: The aero-data model in a research environment (ADMIRE) for flight control robustness evaluation, GARTUER/TP-119-7 (August 1, 2001), http://www.foi.se/admire/main.html
4. Rundqwist, L., Stahl-Gunnarsson, K., Enhagen, J.: Rate limiters with phase compensation in JAS39 GRIPEN. In: Proc. of the European Control Conference, July 1997, pp. 2451–2457 (1997)
5. Forssell, L.S., Hyden, Å.: Flight control system validation using global nonlinear optimisation algorithms. In: Proc. of the European Control Conference, Cambridge, U.K. (September 2003)
6. Forssell, L.S.: Flight clearance analysis using global nonlinear optimisation based search algorithms. In: Proc. of the AIAA Guidance, Navigation, and Control Conference, Austin, Texas (August 2003)
7. Optimization toolbox users guide, Version 2, The MathWorks (September 2000)
8. Back, T., Hammel, U., Schwefel, H.P.: Evolutionary computation: comments on the history and current state. IEEE Transactions on Evolutionary Computation 1(1), 3–17 (1997)
9. Fleming, P.J., Purshouse, R.C.: Evolutionary algorithms in control systems engineering: a survey. Control Engineering Practice 10, 1223–1241 (2002)
10. Rogalsky, T., Derksen, R.W., Kocabiyik, S.: Differential evolution in aerodynamic optimization. Canadian Aeronautics and Space Institute Journal 46, 183–190 (2000)
11. Storn, R., Price, K.: Differential evolution: a simple and efficient heuristic for global optimization over continuous space. Journal of Global Optimization 11, 341–369 (1997)
12. Clerc, M., Kennedy, J.: The particle swarm - explosion, stability, and convergence in a multidimensional complex space. IEEE Transactions on Evolutionary Computation 6(1), 58–73 (2002)
13. Dorigo, M., Gambardella, L.M.: Ant colony system: a cooperative learning approach to the traveling salesman problem. IEEE Transactions on Evolutionary Computation 1(1), 53–66 (1997)
14. Zelinka, I., Lampinen, J.: SOMA - self-organizing migrating algorithm. In: 6th International Conference on Soft Computing, Brno, Czech Republic (2000) ISBN 80-214-1609-2
15. Goldberg, D.E.: Genetic algorithms in search, optimization and machine learning. Addison-Wesley, Reading (1989)
16. Davis, L. (ed.): Handbook of genetic algorithms. Van Nostrand Reinhold, New York (1991)
17. Lampinen, J., Zelinka, I.: Mechanical engineering design by differential evolution. In: Corne, D., Dorigo, M., Glover, F. (eds.) New Ideas in Optimisation, pp. 127–146. McGraw-Hill, London (1999)
18. Storn, R.: System design by constraint adaptation and differential evolution. IEEE Transactions on Evolutionary Computation 3(1), 22–34 (1999)

19. Yen, J., Liao, J.C., Randolph, D., Lee, B.: A hybrid approach to modeling metabolic systems using genetic algorithm and simplex method. In: Proc. of the 11th IEEE Conference on Artificial Intelligence for Applications, Los Angeles, CA, Feburary 1995, pp. 277–283 (1995)
20. Lobo, F.G., Goldberg, D.E.: Decision making in a hybrid genetic algorithm, IlliGAL Report No. 96009 (September 1996)
21. Krasnogor, N., Smith, J.: A tutorial for competent memetic algorithms: model, taxonomy, and design issues. IEEE Transactions on Evolutionary Computation 9(5), 474–488 (2005)
22. Rogalsky, T., Derksen, R.W.: Hybridization of differential evolution for aerodynamic design. In: Proc. of the 8th Annual Conference of the Computational Fluid Dynamics Society of Canada, pp. 729–736 (2000)
23. FDI Test Bench Software User Manual, FDITB-DME-SUM, Version 1.1 (September 18, 2006)
24. Menon, P.P., Bates, D.G., Postlethwaite, I.: Robustness Analysis of Nonlinear Flight Control Laws over Continuous Regions of the Flight Envelope. In: Proceedings of the IFAC Symposium on Robust Control Design (July 2006)
25. Menon, P.P., Kim, J., Bates, D.G., Postelthwaite, I.: Clearance of nonlinear flight control laws using hybrid evolutionary optimisation. IEEE Transactions on Evolutionary Computation 10(6), 689–699 (2006)
26. Menon, P.P., Bates, D.G., Postlethwaite, I.: A Deterministic Hybrid Optimisation Algorithm for Nonlinear Flight Control Systems Analysis. In: Proceedings of the American Control Conference (June 2006)
27. Madavan, N.: Aerodynamic shape optimisation using hybrid differential evolution, AIAA-2003-3792, 21st AIAA Applied aerodynamic conference, Orlando, Florida, USA (2003)

Index

1st De Jong 163
α constrained method 141
ε constrained differential evolution 140, 141, 144
ε constrained method 140
ε-level 141
ε-level comparison 141
ϵ-MyDE 185
εDE, *see* ε constrained differential evolution 140

acceleration operation 265
accuracy 113
Ackley's Problem 163
adaptation 288, 290
adaptive 166
Alpine Function 163
antitheses 155
antithetic variables 155
array factor 242
array synthesis 240
automated tuning 288, 289
Axis Parallel Hyper-Ellipsoid 163

B coefficients 259
benchmark functions 155
best solution 166
boundary constraints 115
box plots 122
buried structures 251

capacitor placement problem 258
chess program 288
classical DE 155
combinatorial problem 18
ComCrit 119
constrained optimization 139
constraint-handling 115
constraint violation 141
constriction factor in PSO algorithms 307
control parameter 90
convergence rate 155
convergence speed 159
crossover 2, 159, 261, 290, 293
crossover probability constant 164
crowding distance 136, 180
cultured differential evolution 189
current population 160, 290

Data Mining 219
DEMORS, *see* differential evolution for multiobjective optimization with rough sets 188
DE operators 198, 232
deterministic 166
dialectic 155
Diff 119
Diff_MaxDistQuick 120
difference pattern 241
differential amplification factor 164
differential equation models 300
differential evolution 89, 140, 143, 155, 240, 258, 287, 289, 300, 305, 320, 324, 331
  variants, 175
differential evolution for multi-objective optimization 182, 187

differential evolution for multiobjective optimization with rough sets 188
digital filter design 24
distribution-based criteria 116
dither 11
diversity 179
DNA microarrays 227
drift-free DE 56
drift bias 34, 40
dualism 155
dynamic economic dispatch 263
dynamic interval 165
dynamic optimization of bioreactors 299
dynamic programming methods 300

economic dispatch 259
effectiveness 162
elitism 180
elitist 115
equality constraints 139
evaluation and selection 261
evaluation function 288, 289
evolutionary algorithms 173, 288, 300
evolutionary computation 300
evolutionary computation techniques 258
evolutionary optimization 157
evolutionary programming 258
Evolution Strategies 92
evolution strategy 143
exploitation 170
exploration 170
exponential problem 163

fed-batch bioreactors 299
fed-batch fermentation processes 301
fermentation techniques 299
fitness function 158, 243, 248
fitness sharing 180
fitness value 91
flight clearance 319, 320, 323, 324, 327, 331
Fully Informed Particle Swarm algorithm 307
function optimization 201

GD3 183
generality 162
generalized differential evolution 184
generation expansion planning 259
generation jumping 155
genetic algorithm 117, 141, 258
genetic programming 230
gradient algorithms for dynamic optimization 300
greedy 115
Green's function 247
Griewangk's function 163

half space 251
Hamiltonian function 300
hill climbing 117
hybrid differential evolution 259, 327, 331
hybrid optimisation 326
hydrothermal coordination 259
hydrothermal optimal power flow 259
hydrothermal scheduling 259
hypervolume 136

ImpAv 117
ImpBest 116
improvement-based criteria 116
inequality constraints 139
initialization 260
initial population 160
integer coded genetic algorithm 272
inverse scattering 246, 248

jitter 12
jumping rate constant 164

Levy Function 163
LimFuncEval 112
linear interpolation of input variables 304
linearly implicit implicit/explicit (IMEX) Runge-Kutta method 305
Lippmann-Schwinger integral equation 247
local optimization 117

matrix real coded genetic algorithm 272
MaxDist 118
MaxDistQuick 118
maximum number of function calls 164
memetic Pareto artificial neural networks 184

Michalewicz function 163
migration operation 266
minimum down time 270
Minimum Principle of Pontryagin 300
minimum uptime 270
mixed integer hybrid differential evolution 259
MODE/D, *see* multi-objective differential algorithm based on decomposition 182
monopulse array antennas 241
movement-based criteria 116
MovObj 118
MovPar 117
multi-objective differential algorithm based on decomposition 182
multi-objective differential evolution 179, 186
  convergence properties, 189
  research trends, 191
  taxonomy, 182
multi-objective evolutionary algorithm 181
multi-objective optimization 115, 178, 205
multimodal 163
mutant population 290
mutation 2, 159, 199, 261, 290
mutation differential 38
mutation strategy 164

network reconfiguration problem 259
neural network training 216–218
NoAcc 117
No Free Lunch (NFL) Theorem 5
noisy objective function 23
nondominated sorting differential evolution 183, 186
nondominated sorting genetic algorithm 181
nonlinear optimization 139
NSDE, *see* nondominated sorting differential evolution 183
NSGA, *see* nondominated sorting genetic algorithm 181
NSGA-II 182, 186
number of function calls 163
numerical methods for dynamic optimization 300

objective space 117
online optimization 300
online optimization of fed-batch bioreactors 311
open-loop optimal control 299
opposite numbers 155
opposite of current population 160
opposite of initial population 160
opposite population 290
opposite solution 157
opposition 155
Opposition-Based Differential Evolution 288
opposition-based optimization (OBO) 155
opposition day 155

parallel DE 200
parameter space 117
Pareto-based differential evolution 186
Pareto-optimal front 135
Pareto differential evolution 183
Pareto front 206
Pareto optimal set 183
Pareto optimal solutions 206
Pareto ranking 181
Particle Swarm Optimization 117, 141, 174, 262, 300
penalty factor 261
population 1
population-based optimization 159
population initialization 155
population size 164
problem independency 162
prohibited operating zones 260

ramp rate limits 260
random guess 157
Rastrigin's function 163
re-entry vehicle 320, 323, 331
real-time optimization 311
real-valued evolutionary algorithms 306
real-world problems 112
recobmination differential 38
recombination operator 199
representation 119
reproduction operators for evolutionary algorithms 306
robustness 162

robustness analysis 323
rotationally invariant 37

Salomon problem 163
scale factor 35
Schwefel's Problem 1.2 163
Schwefel's Problem 2.22 163
search algorithm 288
selection 2, 159, 290, 291
selection differential 50
selection operator 200
self-adaptive 166
self-adaptive mechanism 89
self-adaptive Pareto differential evolution 184
side lobe level 243
simplicity 162
simulated annealing 269
simulation 319, 321, 328
simulation of real fermentations for online optimization 312
single objective optimization 201
singular control problem 300
solution quality 162
start-up cost 270
StdDev 118
StdDevQuick 119
step function 163
stopping 112
subarray configuration 241
substrate feeding rate 304
substrate feeding rates 299
success performance 163
success rate 163, 263
sum of different power 163
sum pattern 241
switch-over point 117

Taijitu 157
test suite 155
theses 155
time varying jumping rate (TVJR) 166
traveling salesman problem (TSP) 18
trial population 290, 291

uncertainty 319, 322, 323, 328
unimodal 163
unit commitment 269
unsupervised k–windows clustering 212

value-to-reach 163
valve point loading 264
vector evaluated differential evolution 188
vector evaluated genetic algorithm 188
VEDE, *see* vector evaluated differential evolution 188
VEGA, *see* vector evaluated genetic algorithm 188

white-box models 301
worst-case analysis 321, 323, 328, 331

Yin-Yang 155

Zakharov function 163

Author Index

Bates, D.G. 319
Bennani, S. 319
Bošković, Borko 89, 287
Brest, Janez 89, 287

Coelho, Leandro dos Santos 275
Coello, Carlos A. Coello 173

Fernandez, V. 319
Ferreira, Eugénio C. 299

Greiner, Sašo 89, 287

Lakshminarasimman, L. 257
Laur, Rainer 111

Marcos, A. 319
Mariani, Viviana Cocco 275
Massa, A. 239
Mendes, Rui 299
Menon, P.P. 319
Mezura-Montes, Efrén 173

Pastorino, M. 239
Pinto, José P. 299
Plagianakos, V.P. 197
Postlethwaite, I. 319
Price, Kenneth V. 33

Rahnamayan, Shahryar 155
Randazzo, A. 239
Reyes-Sierra, Margarita 173
Rocha, Isabel 299
Rocha, Miguel 299

Sakai, Setsuko 139
Salama, Magdy M.A. 155
Storn, Rainer 1
Subramanian, S. 257

Takahama, Tetsuyuki 139
Tasoulis, D.K. 197
Tizhoosh, Hamid R. 155

Vrahatis, M.N. 197

Zamuda, Aleš 89, 287
Zielinski, Karin 111
Žumer, Viljem 89, 287